Mathematical Research

Systems Analysis and Simulation 1988 II

edited by
A. Sydow · S. G. Tzafestas
R. Vichnevetsky

Band 47

AKADEMIE-VERLAG BERLIN

In this series original contributions of mathematical research in all fields are contained, such as

— research monographs
— collections of papers to a single topic
— reports on congresses of exceptional interest for mathematical research.

This series is aimed at promoting quick information and communication between mathematicians of the various special branches.

In diese Reihe werden Originalbeiträge zu allen Gebieten der mathematischen Forschung aufgenommen wie

— Forschungsmonographien
— Sammlungen von Arbeiten zu einem speziellen Thema
— Berichte von Tagungen, die für die mathematische Forschung besonders aktuell sind.

Die Reihe soll die schnelle Information und gute Kommunikation zwischen den Mathematikern der verschiedenen Fachgebiete fördern.

Manuscripts
in English and German comprising at least 100 pages and not more than 500 pages can be admitted to this series. With respect to a quick publication the manuscripts are reproduced photomechanically. Authors who are interested in this series please turn directly to the 'Akademie-Verlag'. Here you will get more detailed information about the form of the manuscripts and the modalities of publication.

Manuskripte
in englischer und deutscher Sprache, die mindestens 100 Seiten und nicht mehr als 500 Seiten umfassen, können in diese Reihe aufgenommen werden. Im Interesse einer schnellen Publikation werden die Manuskripte auf fotomechanischem Weg reproduziert. Autoren, die an der Veröffentlichung entsprechender Arbeiten in dieser Reihe interessiert sind, wenden sich bitte direkt an den Akademie-Verlag. Sie erhalten dort genauere Informationen über die Gestaltung der Manuskripte und die Modalitäten der Veröffentlichung.

Systems Analysis and Simulation 1988

II: Applications

Proceedings of the International Symposium held in Berlin (GDR), September 12–16, 1988

edited by

Achim Sydow
Spyros G. Tzafestas
Robert Vichnevetsky

Akademie-Verlag Berlin 1988

Herausgeber:

Prof. Dr. Achim Sydow, Zentralinstitut für Kybernetik und Informationsprozesse der Akademie der Wissenschaften der DDR, Berlin

Prof. Dr. Spyros G. Tzafestas, Computer Engineering Division, National Technical University of Athens, Athens

Prof. Dr. Robert Vichnevetsky, Dept. of Computer Science, Rutgers University, New Brunswick

---

Die Titel dieser Schriftenreihe werden vom Originalmanuskript der Autoren reproduziert.

ISBN 3-05-500655-0
ISSN 0138-3019

Erschienen im Akademie-Verlag Berlin, DDR-1086 Berlin,Leipziger Str.3-4

Lizenznummer: 202·100/501/88
Printed in the German Democratic Republic
Gesamtherstellung: VEB Kongreß- und Werbedruck, 9273 Oberlungwitz
LSV 1095
Bestellnummer: 763 960 2 (2182/47)

05600

PREFACE

The present volume contains the papers which were accepted for presentation at the 3rd International Symposium for Systems Analysis and Simulation held in Berlin (GDR), September 12-16, 1988.

It is already a tradition to meet a broad international community of experts in systems analysis, modelling and simulation at this symposium. This fact shows the requirements for a forum of presentation and discussion of new developments and applications of modelling and simulation in systems analysis.

To realize the great interest in this field one has to take into consideration the developed role of computer simulation as a powerful tool of problem solving. More and more areas in sciences and production have been investigated by mathematical models and computer simulation. Biological sciences and social sciences are even by now influenced by this trend.

The model use on the computer has been very much improved in decision support systems. Parallel simulation will provide drastic shortening of computing time. Parallel simulation and model based decision support systems are brought in the focus of international activities.

Numerical mathematics, systems theory and control sciences provide with algorithms supporting the modelling process itself based on simulation or analytic methods. Such simulation systems equipped with tools for modelling and graphics for representing results are real model support systems.

A new important impact comes from artificial intelligence by knowledge processing. Expert systems may help decision making in case of missing mathematical models. Expert systems may also support teaching and using simulation systems.

New application areas are investigated. Complex systems with multicriteria control problems are in the scope of the symposium as well as problems of qualitative analysis of small scale nonlinear systems. Applications in engineering sciences, economy and management, natural sciences and social sciences are examined but also mixed problems from different areas.

The state of computer technique and programming environment set up efficient conditions for simulations. Personal computer are even used for simulation more and more.

The symposium reflects the state of the art and trends in systems analysis, modelling and simulation.

The 3rd International Symposium is organized by the Central Institute of Cybernetics and Information Processes of the Academy of Sciences of the GDR (ZKI) with cosponsorship of the

- International Association for Mathematics and Computers in Simulation (IMACS),
- International Federation of Automatic Control (IFAC),
- International Institute for Applied Systems Analysis Laxenburg (IIASA),
- Scientific Society of Measurement and Automation (WGMA) in the Chamber of Technology (KdT) of the GDR,
- Mathematical Society (MG) of the GDR.

The papers included in these proceedings were not formally refereed. The authors themselves are fully responsible.

The international Program Committee consisted of:
W. Ameling (FRG), P. Borne (France), L. Dekker (The Netherlands), S. Deng (PRC), A. A. Dorodnicyn (USSR), K. H. Fasol (FRG), W. Findeisen (Poland), O. I. Franksen (Denmark), V. Hamata (Czechoslovakia), C. Hu (PRC), A. Javor (Hungary), K. Kabes (Czechoslovakia), V. V. Kalashnikov (USSR), V. Kempe (GDR), E. J. H. Kerckhoffs (The Netherlands), R. Klötzler (GDR), R. Kulikovski (Poland), A. Kurzhansky (Austria, USSR), N. Levan (USA), A. H. Levis (USA), T. I. Ören (Canada), M. Peschel (GDR), P. D. Dieu (Vietnam), F. Pichler (Austria), K. Reinisch (GDR), W. Schirmer (GDR), B. Schmidt (FRG), V. V. Solodovnikov (USSR), F. Stanciulescu (Romania), J. M. Svirezhev (USSR), M. Thoma (FRG), I. Troch (Austria), S. G. Tzafestas (Greece), G. C. Vansteenkiste (Belgium), R. Vichnevetsky (USA), A. Sydow (GDR).

Many thanks should be given to the members of this committee for the very helpful cooperation.

Special thanks are said to Prof. Dr. V. Kempe, Director of the ZKI, for his great support in preparing and performing the symposium. Furthermore, great gratitude is to express to
Prof. Dr. R. Vichnevetsky (USA), IMACS-President,
Prof. Dr. B. Tamm (USSR), IFAC-President,
Prof. Dr. R. H. Pry (USA), IIASA-Director,
Prof. Dr. W. Richter (GDR), Chairman of WGMA,
Prof. Dr. R. Klötzler (GDR), Chairman of MG
for help and encouragement.

A lot of the hard preparation work was done by the Department for Systems Analysis and Simulation of the Central Institute of Cybernetics and Information Processes. The editor expresses his thanks to all col-

Mathematical Research · Mathematische Forschung

Wissenschaftliche Beiträge
herausgegeben von der
Akademie der Wissenschaften der DDR
Karl-Weierstraß-Institut für Mathematik

Band 47

Systems Analysis and Simulation 1988

II

Systems Analysis and Simulation 1988

II

leagues and friends who were very much engaged in the research work and in the preparation. First of all I would like to name Dr. P. Rudolph and Dr. A. Wittmüß who helped to prepare the proceedings. Furthermore I thank these colleagues and Dr. K. Bellmann, Dr. W. Jansen, Dr. E. Matthäus, Dr. R. Straubel and all the other colleagues for engaged cooperation for years in developing this research area. Mrs. Ch. Fröhlich and Mrs. J. Obretenov should be named for speedy service in preparing the manuscript. Mrs. S. Böttcher made an excellent job as organizer.
Last not least I would give my thanks to the publishers, especially Mrs. R. Helle and Mrs. G. Reiher, for their assistance and cooperation.
Finally I would like to express my expectation also on behalf of the co-editors Prof. Dr. S. G. Tzafestas and Prof. Dr. R. Vichnevetsky that also the third symposium will be a contribution to the further development in systems analysis, modelling and simulation as well as a place for cooperation and communication like the first both.

Achim Sydow
On behalf of the editors

April 1988

TABLE OF CONTENTS

1. Complex Systems Analysis for Environmental Problems
1.1. Basic Problems

Page,B. /INVITED PAPER/: Review of Applied Informatics in Environmental Protection 17

Dorrer,G.A.; Ovezgeldyev,O.G.; Dmitriev,M.G.; Mansimov,M.R.: Application of Optimal Control Methods to Simulation of Nature Processes in Arid Zone 21

Puta,H.; Gerecke,D.: A Repetitive Working Irrigation Algorithm Using Actual Climate Data 24

Korbicz,J.; Parmaste,I.L.; Gawłowicz,P.: Simulation of Urban Air Pollutant Processes and Monitoring Stations Location Problem 29

Carl,P.; Stenchikov,G.L.: Structural Analysis of the Climatic Response to a Nuclear War 33

1.2. Acid Rain

Sydow,A.; Bellmann,K.; Straubel,R.; Imming,I.; Kaschenz,H.; Damrath,U.; Hofmann,G.; Anders,S.: PEMU - An Impact Model Based Environment Protection Decision Support System 37

Lasch,P.; Model,N.; Bellmann,K.: The PEMU/Air Pollutant transport Model Based on Emittent-Receptor-Point-Transmission Calculation 42

1.3. Forestry

Bellmann,K.; Lasch,P.; Hofmann,G.; Anders,S.; Schulz,H.: The PEMU Forest-Impact-Model FORSTK. A Pine Stand Decline and Wood Supply Model 45

1.4. Water Systems

Richter,J. /INVITED PAPER/: Water and Matter Dynamics in Agro-Ecosystems 51

Graeber,P.-W.: Automatical Control Strategy of Non-Technical Systems in the Groundwater Zone 61

Popescu,Th.D.; Demetriu,S.; Marinescu,V.: Modelling and Forecasting of River Flows Using Stochastic Models 65

Bereziński,M.; Petryczek,G.: Application of Trajectory Decomposition Method of the Water Quality Control Problem 69

Braun,P.; Rudolph,P.; Albrecht,K.-F.: Computer Aided Decision for Water Quality Management 75

Gnauck,A.; Rathke,P.; Straubel,R.; Wittmuess,A.: Pareto-Optimal Cost Division for the Design of Sewage Water Treatment Plants by means of the DSS REH 79

2. Complex Systems Analysis for Industrial Automation
2.1. Basic Problems

Gutenbaum,J. /INVITED PAPER/: Two-Level Heuristic Algorithm for Parallel Machines Scheduling 84

Zhou Shichang; Cao Xinming; Yao Guiyou: The State Variable Diagram of a Hydraulic Control System 90

Valdés Pardo, V.G.: Decision-Making Simulations with the Micro-CED Programming System 94

Geschke, S.; Kempe, N.; Schlawatzky, H.: Application of a Hierarchical Computer System for Picosecond Technology 98

Wrycza, St.: Some Comments on Constituents of Contemporary Information Systems Development Methodologies 102

Singer, D.: Model Based Default Data Generation for Net Systems 107

Winnicki, A.; Płoszajski, G.: Optimisation Problem for By-Pass Routes Creation in the Case of a Teletransmission Line Break-Down 121

Wydro, K.: Substitutional Routing for Trunk Line Break-Down in the Teletransmission Network 125

Holeňa, M.: Object System Structure and Behaviour Analysis in Database Design 131

Schiele, K.; Koch, M.; Nachbar, M.: Computer-Aided Elaboration of Research Strategies - The Expert System PHARMEX 136

2.2. Manufacturing Systems and CIM-Structures

Bobeanu, C.; Neagu, G.; Filip, F.: A Simulation Based Optimization Algorithm for Operations Control in FMS 140

Lorenz, P.; Tolujev, J.: Simulation of Manufacturing Systems. Goals, Methods, Tools and Problems 144

Krug, W.; Blume, F.; Erfurth, F.: Modeling and Simulation of Automated Welding Manufacturing System 150

Voigt, G.: On the Application of Advisory Systems and Simulation Processes for the Planning and Control of Manufacturing Systems 154

Yashkov, S.F.: The Non-Stationary Distribution of Number of Calls in the M/G/1 Processor-Sharing Queue 158

2.3. Robotics

Tenreiro Machado, J.A.; Martins de Carvalho, J.L.: Robot Manipulator Systems : Analysis and Control 163

Tenreiro Machado, J.A.; Martins de Carvalho, J.L. Costa, A.M.C.; Matos, J.S.: Dedicated Computer System for Robot Manipulators 167

Miernik, J.W.; Zaremba, M.B.: Performance Approximation of Robotic Systems Using Rendezvous Networks 171

2.4. Automation of Continuous Industrial Systems

Matsevity, Yu.M.; Moultanovsky, A.V.: Solution of Parameter Optimization and Control Problems in Thermal Systems by means of a Local Adaptive Filter 175

Vajta, M.; Rijnsdorp, J.E.; van der Meulen, S.F.; Verberne, J.F.C: Optimization of Heat Delivery to Apartment Buildings Using an Improved Building Model 179

Zhou Shichang; Cao Xinming; Gao Chengrui; Yang Yanyong: Performance Measurement and Mathematical Model Identification of the Hydraulic Servo Control System of a Vacuum Electric-Arc Furnace 183

Zhou Shichang; Cao Xinming; Gao Chenrui; Yang Yanyong: Identification of the Mathematical Model of a Servo Valve by means of a Square Wave Signal 190

Piłaciński, J.: Analog Simulation of Transient Behavior of Converters with the Application of Dynamic Thyristor Models 194

Hvala, N.; Strmčnik, S.; Černetič, J.: Design and Simulation of Advanced Control for a Pulp Digester House 198

Jaworska, I.; Szulc, Z.; Kryger, G.: Modelling and Simulation of a Drive Motor System 202

Tikasz, L.; Zaymus, M.; Cseh, A.: A Simulation Model for Aluminium Electrolytic Cells Concerning on Process Control 206

Sabadí-Diaz, R.; de la Vega, E.; Fernandez, D.: A Computer System for Decision Support in the Sugar Cane and its Derivatives Industry 213

Klempous, R.; Nikodem, J.; Ulasiewicz, J.: Simulation Model of High Pressure Gas Pipeline Networks 216

Urbanik, A.: Chosen Problems of Air Traffic Control Systems Modelling 220

Wagner, M.; Hille, Ch.: Design of Multivariable Control Loops for Distillation Columns Using Digital Simulation 222

Ehrlich, H.: An Integral Look at Automation Functions in Modern Production Systems 225

3. Applications to Agricultural Systems

Dunkel, Z.; Hunkár, M.: Dynamic Simulation of Corn Growth 229

Hunkár, M.; Zemankovics, F.: Studies on the Adaptation of a Winter Wheat Model 233

Grosskopf, K.; Matthaeus, E.; Flechsig, M.: Modelling of Growth Processes in Broiler Production Using the Simulation System SONCHES 241

4. Applications to Economic Systems

Mueller, J.A. /INVITED PAPER/: Computer Aided Modelling and its Application in Economy and Ecology 245

Neck, R.: Policy Simulations and Optimal Control with a Small Econometric Model for Austria 249

Baladynowicz, H.W.: Total Pollution Effect and Total Energy Cost per Output of Different Products for Polish Industrial System 253

Hoeschel, H.-P.: A Multisectoral Econometric Input-Output-Model of the G.D.R. for Prediction and Dynamic Simulation of the National Income 258

Zeipert, S.: Statistical Methods for Supporting the Selection of Strategies in Automatization by Regard of Socio-Economic Factors 262

Cofala, J.; Ostromecki, K.: Model Analysis of Energy-Economy Interactions 265

5. Applications to Energetic Systems

Fandel, G.; Reese, J. /INVITED PAPER/: Cost Minimization in a Firm's Power Station 272

Richter, K.: Dynamic Energy Production Model 278

Mende, W.; Albrecht, K.-F.: The Growth of Electric Energy Production as an Example for the Application of the Evolon Model 282

Gacs, A.: Analysis and Simulation of the Dynamic Behaviour of Saturated Steam Turbines of PWR Nuclear Power Plants 286

Hoppe, D.; Goetz, J.: Optimal Power Control of a Nuclear Reactor by means of an Inverse Model 290

Šodan, M.; Tešnjak, S.; Marušić, A.; Tomiša, T.: Simulation Model and Microprocessor System for Digital Control of Hydraulic System in a Hydroelectric Power Plant 294

Domachowski, Z.: Power System Frequency Control Characteristics as a Function of Nuclear Power Plant Participation 298

Model, N.; Grossman, S.; Dineva, P.; Hadjikov, L.: Design of a Reduced Model for a Nuclear-Power Plant Under Earthquake Excitation 301

Haeusler, F.; Roesel, G.-St.: Mathematical Modelling and Simulation Techniques of a Complex Power Plant Pipe Network 305

6. Applications to Medical Health Care and Biological Systems

Érdi, P.: Structural Aspects of the (Natural and Artificial) Semantic Memory 309

Hauser, F.; Anděl, M.: Estimation of Drug Need From the Health Projection - The Case of Insulin 313

Pohl, Th.: Numerical Simulation of Pulsatile Flow in Arteries 317

Kotva, M.: "Balance" Segments in Multicompartmental Models 321

Shakin, V.V.; Karasev, A.A.; Przyjalkowski, V.V.: DYNAMOEXPERT: PC-Aided Simulation and Identification Tools for Biomedical Applications 325

Schierwagen, A.; Francu, J.: A New Modell for Current Flow in Dendrites of Arbitrary Geometry 329

Wuenscher, G.; Wenzel, V.; Flechsig, M.; Matthaeus, E.: Simulation System SONCHES Based Toxicokinetic Modelling as a Tool in Biological Monitoring 332

Fritzsch, Ch.: Parameter Estimation of the Blood Pressure Control System in Dogs 336

7. Applications in Other Fields

Schilling, H.: Computer Graphic for Rotational and Mirror Symmetric Fractals with m Colours and a Determined Basic Design 340

8. Late Papers

Dekker,L. /INVITED PAPER/: Simulation Environments 344

Wunsch,G. /INVITED PAPER/: Foundations of a General Process Theory 351

Kulisch,U. /ABSTRACT of an INVITED PAPER/: Fast and Accurate Arithmetic for Vector Processors 356

Sydow,A.; Bellmann,K.; Matthaeus,E.; Rudolph,P.; Straubel,R.; Wittmuess,A.: Computer-Aided Systems Analysis - Methods and Tools 357

Owsinski,J.W.; Zadrozny,S.: A Flexible System of Precedence Coefficient Aggregation and Consensus Measurement 364

Breedveld,P.C.: Systematic Procedures to Derive Block Diagrams and State Equations from Bond Graphs 371

Carl,P.: Software Engineering Aspects of Computational Systems Analysis in Physics 375

Ivanilova,T.N.: Set Deviations Modelling of Forest Fire Space State 379

Laret,L.: An Attempt for an Adapted Model Generation Methodology: Example of Application in Building Energy Modelling 383

Dubois,A.-M.: A Draft of a Component Model Library Manager, the "MODELOTHEQUE": Analysis of an Approach, and first Results 396

Kutas,T.: Models for Lake Eutrophication 404

Baškys,A.I.; Zanevitčius,D.J.; Gobis,V.J.: Ways to Reduce Time Consumption of Statistical Modelling of Integrated Circuits by the Monte Carlo Method 408

Zanevičius,D.: Mathematical Models of Non-Linear Elements in the Theory of Systems 411

Stachov,A.P.: The Medial Section as a Proportion for the Thermodynamical Equilibrium of Selgorganizing Systems 415

Singh,M.G. /INVITED PAPER/: Recent Advances in Decision Technologies for Management 417

LIST OF AUTHORS

Albrecht, K.-F. 75,282
Anděl, M. 313
Anders, S. 37,45

Balandynowicz, H.W. 253
Baškys, A.I. 408
Bellmann, K. 37,42,45,357
Bereziński, M. 69
Blume, F. 150
Bobeanu, C. 140
Braun, P. 75
Breedveld, P.C. 371

Cao Xinming 90,183,190
Çarl, P. 33,375
Černetič, J. 198
Cofala, J. 265
Costa, A.M.C. 167
Cseh, A. 206

Damrath, U. 37
Dekker, L. 344
Demetriu, S. 65
Dineva, P. 301
Dmitriev, M.G. 21
Domachowski, Z. 298
Dorrer, G.A. 21
Dubois, A.-M. 396
Dunkel, Z. 229

Ehrlich, H. 225
Érdi, P. 309
Erfurth, F. 150

Fandel, G. 272
Fernandez, D. 213
Filip, F. 140
Flechsig, M. 241,332
Francu, J. 329
Fritzsch, Ch. 336

Gacs, A. 286
Gao Chengrui 183,190
Gawłowicz, P. 29
Gerecke, D. 24
Geschke, S. 98
Gnauck, A. 79
Gobis, V.J. 408
Goetz, J. 290
Graeber, P.-W. 61
Grosskopf, K. 241
Grossman, S. 301
Gutenbaum, J. 84

Hadjikov, L. 301
Haeusler, F. 305
Hauser, F. 313
Hille, Ch. 222
Hoeschel, H.-P. 258
Hofmann, G. 37,45
Holeňa, M. 131
Hoppe, D. 290
Hunkár, M. 229,233
Hvala, N. 198

Imming, I. 37
Ivanilova, T.N. 379

Jaworska, I. 202
Jumar, U. 111

Karasev, A.A. 325
Kaschenz, H. 37
Kempe, N. 98
Klempous, R. 216
Koch, M. 136
Korbicz, J. 29
Kotva, M. 321
Krug, W. 150
Kryger, G. 202
Kulisch, U. 356
Kutas, T. 404

Laret, L. 383
Lasch, P. 42,45
Lorenz, P. 144

Mansimow, M.R. 21
Marinescu, V. 65
Martins de Carvalho, J.L. 163,167
Marušić, A. 294
Matos, J.S. 167
Matsevity, Yu.M. 175
Matthaeus, E. 241,332,357
Mende, W. 282
van der Meulen, S.F. 179
Miernik, J.W. 171
Model, N. 42,301
Moultanovsky, A.V. 175
Mueller, J.A. 245

Nachbar,M. 136
Neagu,G. 140
Neck,R. 249
Nikodem,J. 216

Ostromecki,K. 265
Ovezgeldyev,O.G. 21
Owsinski,J. 364

Page,B. 17
Parmaste,I.L. 29
Petryczek,G. 69
Piłaciński,J. 194
Płoszajski,G. 121
Pohl,Th. 317
Popescu,Th.D. 65
Przyjalkowski,V. 325
Puta,H. 24

Rathke,P. 79
Reese,J. 272
Richter,J. 51
Richter,K. 278
Rijnsdorp,J.E. 179
Roesel,G.St. 305
Rudolph,P. 75,357

Sabadí Diaz,R. 213
Schiele,K. 136
Schierwagen,A. 329
Schilling,H. 340
Schlawatzky,H. 98
Schulz,H. 45
Shakin,V.V. 325
Singer,D. 107
Şingh,M.G. 417
Šodan,M. 294
Stachov,P. 415
Stenchikov,G.L. 33
Straubel,R. 37,79,357
Strmčnik,S. 198
Sydow,A. 37,357
Szulc,Z. 202

Tenreiro Machado,J.A. 163,167
Tešnjak,S. 294
Tikasz,L. 206
Tolujev,J. 144
Tomiša,T. 294

Ulasiewicz,J. 216
Urbanik,A. 220

Vajta,M. 179
Valdés Pardo,V.G. 94
de la Vega,E. 213
Verberne,J.F.C. 179
Voigt,G. 154

Wagner,M. 222
Wenzel,V. 332
Winnicki,A. 121
Wittmuess,A. 79,357
Wrycza,St. 102
Wuenscher,G. 332
Wunsch,G. 351
Wydro,K. 125

Yang Yanyong 183,190
YaoGuiyou 90
Yashkov,S.F. 158

Zadrozny,S. 364
Zanevičius,D. 408,411
Zaremba,M.B. 171
Zaymus,M. 206
Zeipert,S. 262
Zemankovics,F. 233
Zhou Shichang 90,183,190

REVIEW OF APPLIED INFORMATICS IN ENVIRONMENTAL PROTECTION

Bernd Page [1)]

1. INTRODUCTION

An effective protection of our environment is highly dependent on precise information on environmental state and development. In this way, computer use in environmental agencies became more and more important. The state of environmental computing in practice is documented in the "Study on DP-Applications in the federal and state Environmental Offices" by the Environmental Federal Office of the F. R. G. (see /18/). It analyses the standard of environmental protection software on the basis of an empirical survey in more than 50 state agencies. The results of the study show on one hand a rapid increase of DP-activities in the environmental agencies during the last years and an intensified use of modern and powerful DP-concepts (such as interactive systems, data base systems, distributed processing with PC-networks or computer colored graphics). On the other hand, the introduction of overall concepts for the development of environmental DP-procedures in the sense of comprehensive, task oriented environmental information systems is still at the beginning. Scientific activities in environmental computing have been rather rare. Most of the environmental computer applications stem from practical projects or other scientific fields without participation of informatics. As a scientific application field for informatics, the environmental domain was only of very limited research interest, although already more than 10 years ago a first conference on this topic was held (see /10/). In the following years only few publications on this subject came out. Just recently an increasing research interest in environmental applications of informatics can be observed as shown by intensified publishing (see e.g. /17/ or /4/) and recent conferences (see /2/, /3/, /7/, /14/, /15/ or /16/). Some review articles on this subject were also published during the last years (see /13/, /19/, /2/ p. 183ff, /14/ p. 5ff, or /21/). In this paper some promising research directions for applied informatics in the environmental domain will be introduced.

2. RESEARCH SCOPE FOR APPLIED INFORMATICS IN ENVIRONMENTAL PROTECTION

In Table 1 a list of informatics methods as applied to different types of environmental computer systems is given. Only a few can be briefly addressed here.

Non-standard Data Base Applications:

Data bases and information systems are fundamental informatics concepts for environmental information processing. Data bases play the role of archives in information systems. They provide a pool with all relevant data of an application organized in a centralized and integrated manner. Conventional data base systems are mainly oriented towards classical or "standard"-data base applications typically found in the business or administrative domain. During the last years the importance of "non-commercial" computer applications - named "Non-standard" or "Non-conventional Data Base Applications" in the data base field - increased. They are mainly characterized by long, non formated, complex structured data objects, by long transactions and different versions of objects. This implies that among other things additional mechanisms for ensuring semantic integrity conditions, more powerful data models and suitable data base languages for modelling and manipulation of complex structured objects have to be developed. Application fields in the Non-standard Data Base domain with environmental reference are scientific studies (experimental and measurement data), word processing and document retrieval, process control systems, land information systems (geographical data processing) or graphical data handling. In /22/ environmental information systems are discussed in the context of Non-standard Data Base Applications. Another environment related (geographical) data base application is described in /1/.

Environmental Modelling and Simulation Software:
Modelling and simulation of environmental problems have a long scientific tradition (see e.g. /9/). In (/2/ p. 183ff) a classification of typical model applications by environmental subjects is given. The software techniques used in environmental modelling, however, are not always on a satisfactory level from the informatics point of view. Applied informatics can offer powerful concepts for supporting design, implementation and operation of models (such as modelling languages (see /17/ p. 284ff), interactive simulation systems, model base systems, decision supporting systems), which are not yet sufficiently used in environmental modelling (see /16/ p. 465ff). These software concepts provide a high degree in flexibility and user adaption in running environmental analysis methods and models. The term Method- and Model Base System stands for a system architecture concept, which is of high interest for specific environmental information system applications such as (mathematical) analysis systems or interactive operation of environmental models from a model base. By definition a Method- and Model Base System is an extendable collection of program modules representing the methods and models of an application domain - under central administration with a user and problem oriented access. The use of model based systems for operating environmental models was analysed exemplarily at the MBS, developed by the Gesellschaft für Mathematik und Datenverarbeitung (see /17/ p. 331ff). The method-/model base architecture was also taken as a basis for an analysis system for the Hamburg air pollution measurement data (see /15/ p. 173ff).

While in model base systems the administration and operation of implemented (sub-) models is of primary concern, Interactive Modelling and Simulation Systems are mainly concerned with computer support of the entire process of model building and simulation, from planning and building up to documentation of models and graphical presentation of results. With a high degree of interactivity on all system levels interactive (graphical) modelling directly at the display, interactive control of simulation flow as well as interactive preparation of simulation results is offered. The aim is a high degree of user friendliness in computer based modelling also for users with few computer experience. In /5/ the concept and implementation of an interactive simulator for System Dynamics modelling is outlined. The main asset of this modelling tool is its object oriented user interface taking advantage of modern graphical features such as windows, pull down menues, icons and mouse driven pointing devices. The user can interrupt simulation experiments at any time and alter system states. An ecological study with this interactive simulator - simulating the euthrophication of a lake - is presented in /17, p. 251ff/.

Concepts of interactive modelling and simulation are not only relevant for scientific ecological system modelling, but also for Decision Support Systems (DSS) in environmental policy decision making. Basically DSS shall provide decision aids to decision makers and their advicing staff in computer dialog as problem specific model based solutions. Typical methods and tools for DSS are statistics (descriptive, prognosis), operations research (simulation, optimization) and more recently expert system techniques. An environmental DSS approach for supporting investment decisions in pollutant emission control based on analytical models and multicriterial optimization is described in /20/. Another large-scale, model based decision support system for environmental management, namely for industrial risk assessment and management of hazardous substances, was developed at IIASA (see /8/ and /14/ p. 235ff). Finally, recent knowledge based simulation concepts should be also beneficial for environmental modelling (e.g. /14/ p. 245ff).

Expert Systems:

Artificial Intelligence and in particular expert systems are highly rated topics in modern informatics. We can understand an Expert System as a knowledge based system modelling and using the knowledge of qualified experts for a specific limited domain in order to solve complex problems of a particular type or support their solution, respectively. The applications of expert systems can be classified by three main categories:

- Interpretation of data, findings, or symptoms in order to derive diagnoses or actions (e.g. interpretation of pollution data in smog situations or ecological identification problems)
- Assistance in design and planning in complex situations (e.g. environmental impact analysis or planning)
- Knowledge transfer: Education, training, simulation (e.g. guidance in chemical toxid disturbances or accidents).

In the environmental domain only few expert system applications are known so far, although expert assistance is badly needed here due to a common lack of expertise in local environmental decision making. The Environmental Technical Information System for providing computerized decision aids in environmental impact analysis, developed in the U. S. already in the seventies (see /17/ p. 129ff), can be seen as an expert system like approach, although the knowledge representation techniques used here (cause-impact matrix, check lists) are not AI specific. Another approach for decision support in environmental impact analysis using expert system concepts is discussed in /14 p. 195/ and /23/. Additional environmental expert system applications were published for waste disposal (see /2/ p. 87ff), for ecological species determination in nature conservation /17 p. 420ff/, for identification of environmental harmful waste substances /15/, or for handling environmental regulations /11/.

3. OUTLOOK

For a long time environmental computing was a practical field mainly in environmental agencies with only few relating research input. More recently, however, the research community - including applied informatics - was paying more attention to this challenging topic as we can learn from the growing number of publications and specialized conferences. Meanwhile, an international journal "Environmental Software" is also published. Further, an own thematic section "Environmental DP and Informatics" was introduced recently in the public (German language) environmental literature data base ULIDAT run by the Federal Environmental Office (UBA) of the F. R. G.; a bibliography on this subject with about 1800 documents will be published by the office at the end of the year. Finally, research will be further pushed by an own section "Informatics in Environmental Protection" within the German Society for Informatics (GI), which was installed in 1987 with more than 100 members and will hold regular workshops and conferences.

Methods \ Systems	Information Systems	Monitoring and Control Systems	Planning and Decision Support Systems	Scientific Analysis Systems
Measurement Data Collection and -processing		X		X
Process Control Systems		X		
Computer Networks	X	X		
Data Bases	X		X	X
Modelling/ Simulation			X	X
Method-/ Model Bases	X		X	X
Computer Graphics	X	X	X	X
Picture Processing		X	X	X
Expert Systems			X	

Table 1: Informatics Methods as applied in various types of environmental Computer Systems

LITERATUR

(1) Appelrath, H.-J.: GEO - Konzept eines applikationsneutralen geographischen DB-Systems und seine Implementierung als INGRES-Fronend. In: A.Blaser, P.Pistor (Hrsg.), Datenbanksysteme für Büro, Technik und Wissenschaft. Proc. GI-Fachtagung, Karlsruhe, Informatik-Fachberichte Nr. 94, Berlin-Heidelberg 1985, p.476-486.

(2) Arbeitsgemeinschaft für Datenverarbeitung (Hrsg.): Tagung "EDV für Umweltschutz und Energiewirtschaft", Tulln, 6.-8. Nov. 1985, Wien 1985.

(3) Bossel, H., K.-H. Simon (Hrsg.): Computer und Ökologie Eine problematische Beziehung. Alternative Konzepte 54, Karlsruhe 1986.

(4) Bruenig, E. F., H. Bossel, K.-P. Elpel et al.: Ecologic-Socioeconomic System Analysis and Simulation. Inst. for World Forestry and Ecology. Hamburg 1986.

(5) Brunnstein, K., A. Häuslein, and B. Page: DYNAMIS: A System for the interactive Support of Simulation Modelling and Experimentation. Proc. 2nd. European Simulation Congress. Antwerpen, Sept. 1986, Simulation Councils, Inc. 1986, p. 297-301.

(6) Bundesforschungsanstalt für Naturschutz und Landschaftsökologie (Hrsg.): Landschaftsinformationssysteme. Natur und Landschaft 57, 1982, No. 12.

(7) Deutsches Institut für Urbanistik (Hrsg.): Kommunale Umweltinformationssysteme. Bedarf, Inhalte, Methoden, DV-Einsatz. Seminarunterlagen, Berlin 1986.

(8) Fedra, K., Weigkricht, and L. Winkelbauer: A hybrid Approach to Information and Decision Support Systems: Hazardous Substances and Industrial Risk Management. Proc. 1st Conf. on Economics and Artificial Intelligence, Aix-en-Provence, Sept. 1986, Pergamon Press 1986.

(9) Frenkiel, F. N., D. W. Goodall (eds.): Simulation Modelling of Environmental Problems. Wiley & Sons, New York 1978.

(10) Gesellschaft für Mathematik und Datenverarbeitung, Umweltbundesamt (Hrsg.): Datenverarbeitung auf dem Umweltsektor - Vorträge einer gemeinsamen Veranstaltung. München-Wien 1978.

(11) Hälker, M., V. Stahl, and F. Bubeck: Expertensystem Umweltrecht. Proc. 16. GI-Jahrestagung, Berlin, Oct. 1986, Informatik-Fachberichte 127, Bd. 2, p.436-447.

(12) Hilty, L.: Computeranwendungen im Umweltschutz - Inhalte und Methoden - Ein Überblick. Fachbereich Informatik der Universität Hamburg, Mitteilung Nr. 129, Hamburg 1984.

(13) Hilty, L., B. Page: Computeranwendungen im Umweltschutz - eine erste Analyse. Angewandte Informatik 10/85, p.409-419

(14) Jaeschke, A., B. Page (Hrsg.): Informatikanwendungen im Umweltbereich. Kolloquium des Instituts für Datenverarbeitung in der Technik des Kernforschungszentrums Karlsruhe, KfK-Bericht Nr. 4223, Karlsruhe, März 1987.

(15) Jaeschke, A., B. Page (Hrsg.): Informatikanwendungen im Umweltbereich. Proc. des 2. Symposiums, Informatik-Fachberichte, Springer Verlag 1988 (in press).

(16) Krupp, H. (Hrsg.): Beitrag der Mikroelektronik zum Umweltschutz. Proc. Workshop Ges. Mikroelektronik (GME), Hamburg 1987, vde-Verlag, Berlin 1988.

(17) Page, B. (Hrsg.): Informatik im Umweltschutz - Anwendungen und Perspektiven. München-Wien 1986.

(18) Page, B. und Mitarb.: Studie über DV-Anwendungen in den Umweltbehörden des Bundes und der Länder. Herausgegeben vom Umweltbundesamt, UBA-Texte 35/86, Berlin 1986.

(19) Pillmann, W.: Mikroelektronik um Umweltschutz. Fernerkundung - Meßtechnik - Umweltdatenverarbeitung. In: Tagung Mikroelektronik für den Menschen. Johannes Kepler Universität, Linz 1984, Bd. 2, p.1083-1097.

(20) Sydow, A.: Models and Multicriterial Control of complex Environmental Systems. Proc. 2nd European Simulation Congress, Antwerpen, Sept. 1986, Simulation Councils, Inc., 1986, p.14-16.

(21) Trauboth, H.: Was kann die Informationstechnik für den Umweltschutz tun? Automatisierungstechnik 35, 1987, Heft 11, p. 431-442.

(22) Umweltinformationssystem Dortmund: Problemanalyse/ Anforderungsdefiniton und Grobentwurf für den Bereich der Umweltverträglichkeitsprüfung. Diploma Thesis, Faculty of Informatics, University of Dortmund 1987.

(23) Brunnstein, K., A. Schwabl: Applying Expert Systems Techniques for Environmental Compatability Analysis: Approach, Realization, Problems. Proc. 7th International Workshop Expert Systems & Their Applications. Avignon, France, May 1987, Vol.2, p.1303-1322.

1) Section of Informatics at the University of Hamburg, Schlüterstr.70, D-2000 Hamburg 13

APPLICATION OF OPTIMAL CONTROL METHODS TO SIMULATION OF NATURE PROCESSES IN ARID ZONE

Dorrer G.A.[1], Ovezgeldyev O.G., Dmitriev M.G.[2], Mansimov M.R.[3]

In scopes of general system approach to the study of ecology - economical processes of the arid zone we propose the approach to simulation of the spreading processes (desertification, salination of lands etc.) under human impact.

There was considered the typical formulations of modelling problems - both direct, connected with the prediction of desertification lands boundaries under man's influence and opposite tasks, which were enclosed in the dynamical systems parameters identification by air and space photos.

I. Desertification process modelling

We consider the desertification process connected the phytocoenosis degradation as a consequence of grazing. To calculate the phytomass change model we take the linear differential equation of the first order with linear control - pressure. Here a small pressure shows the stimulating effect on the vegetation and phytomass equilibrium store reduces when antropogenic load is more than some level. As a consequence magistrals equation is defined by time-optimal problems set solution.

Supposing that the animal pressure on the phytocoenosis is proportional to the animals density we get expression for the load value $u(\rho,\omega)$ on elementary area element $d\sigma$ which is constrained by $\rho', \rho'+d\rho$ and $\omega', \omega'+d\omega$. Here (ω, ρ) are variables of curvilinear coordinated system, ω - initial direction of magistral ($0 \leq \omega \leq 2\pi$) and ρ - relative distance from a centre - water source ($0 \leq \rho \leq 1$).

The pressure u^*, for which normal phytomass volume $\bar{S}$ equals zero, is called the crytical pressure. If $u < u^*$, we have $\bar{S} > 0$ - it is possible the stable management and by $u > u^*$ we have the desertification. For our model we have also $\bar{S} \to 0$, in origin neighborhood, where $u \to \infty$.

[1] Siberian Technological Institute, Krasnoyarsk, 660049, USSR
[2] FTI AN TSSR, Ashkhabad, 744000, USSR
[3] Turkmengydromet, Ashkhabad, 744000, USSR

This fact corresponds the physics sence and observations in the areas of watering-place the plant cover is destroyed, as a rule.

2. Parameters identification of spreading processes [1].

There are often appeared the inverse problems under the models parameters estimation on the base of remote sensing side by side with the problems of direct modelling of natural processes.

The wave fronts transfer speeds are represented the greatest interestfrom the dynamical parameters. In our study the two approaches for this problem solution are proposed.

2.I Parameters estimation of wave front speeds indicatrix.

Let consider some dynamical set contours on the earth surface are fixed in sequential time moments, i.e. we have the spreading processes $X(t)$. Boundary $\partial X(t)$, which we approximate the points sequence, is the two-dimentional vectors set $\partial X_e(t)$. At first we identify the contours sequence in topographic maps. Then changed sets $\partial X_e(t)$ are used for normal speeds set construction. Further with the help of statistical analysis methods we find the formules parameters for describing of the normal speeds set.

2.2 Optimal guaranted estimates of spreading processes.

Here we use the results [2] by control systems attainability sets estimates on the base of the ellipsoids.

Let us consider the spreading process parameters estimate problem by observed fronts positions (A front is characterized by the circumscribed ellipse).

Let have the ellipse sequence $\{E(t_i)\}$, $t_i < t_{i+1}$, $i=\overline{1,m}$, $E(t)=E\{a(t), H(t)\}=\{x\in R^2: (x-a)^T H(x-a)\leq 1\}$, a - the ellipse centre, H is the matrix defining the ellipse amounts and orientation.

Our problem - find the C, f, K, where $\dot{x} = Cx + Ku + f$, $x\in R^2$.

$u \in E\{0, G\} \subset R^2,\ x(t_1) = a(t_1),\ x(t) \in E(t_i),\ t_i \le t \le t_{i+1},\ i = \overline{1,m}.$

From [2]it is known, that $x \in X(t,q) = \{x \in R^2 : (x-\bar{x})^T \sigma(t,q)(x-\bar{x}) \le 1 + \int_0^t q(\tau)d\tau\},\ \dot{\bar{x}} = C\bar{x} + f,\ \bar{x}(t_1) = a(t_1),$ $q(t) > 0$- arbitrary function,

Further we propose the next algorithm:

I. $t \in [t_1, t_2]$-calculated ellipsoid estimate X approximates to observed estimate $E(t_2)$,i.e.

$$\|\sigma(t_2,q) - H(t_2)\| \to \min_{C,K,f,G,q}$$

$$\bar{x}(t_2) = a(t_2).$$

2. $t \in [t_2, t_3]$-

$$\|\sigma(t_3,q) - H(t_3)\| \to \min_{C,K,f,G,q}$$

$$\bar{x}(t_3) = a(t_3)\ldots$$

Then constructed functions $C(t), K(t), f(t), G(t), q(t)$ we exprapolate by t and get the forecasting ellipsoidal estimates.

Numerical experiments, which were fulfilled for the dynamics predictions of the eolian land forms near the Sarykamysh lake on the basis of two set forth approaches have shown the satisfactory conformity between calculations and observations.

Literature

[I] Dorrer, G.A.: Dynamical systems parameters estimate by their attainability sets. Avtomatika and Telemechanica. (I986) I, 39-46 (In Russian)

[2] Konstantinov, G.N., G.V. Sidorenko : Outer estimates of the control systems attainability sets. Izv. AN SSSR. Techn. kybernetics. (I986) 3, 28-34 (In Russian)

A REPETITIVE WORKING IRRIGATION ALGORITHM USING ACTUAL CLIMATE DATA

Puta, H.[1] and Gerecke, D.[2]

1. INTRODUCTION

For a long time biologists and scientists in agriculture are engaged to develope growth models (see e.g. van KEULEN, 1986; AUGUSTIN, SCHMIDT, 1980). Descriptive models have the benefit to be as true as possible in detail. Control models represent as correct as possible the input-output behavior and contain the control parameters explicitely. The latter one is suitable for design of control strategies but it is more difficult to build it directly. For this end a deskriptive model can be useful as original substitute. In this paper a control model for sugar beets has been available (HOFFMANN, 1980) and also a soil water model (KOITZSCH, 1977), which enables to ascertain the water supply of the plant (called waterstress) with respect to the water amount which comes from rainfall and/or irrigation. Both models have been developed by the Academy of Agriculture of the GDR.

Using these two models it is possible to give the farmer a computer aided help for irrigation decisions taking into account the present growth state and climate data from the past.

2. MODEL DISCUSSION

The growth model of HOFFMANN (1980) can be at least described by the following nonlinear differential equations:

$$\dot{B}(t)=f_1*f_2*f_3 \tag{1}$$
$$\dot{Z}(t)=f_1*f_2 \tag{2}$$
$$\dot{W}(t)=\dot{Z}(t)-\dot{B}(t) \tag{3}$$

where $\dot{Z}$ is the total dry matter or nettophotosynthesis rate of both the body W and the leaves B. The anabolite distribution between leaves and the total dry matter describes f_3. The f_i are as following:

$$f_1=0.01*f(R,t)*B(t), \tag{4}$$

$$f_2=K_1+\sqrt{R}*(K_2+K_3*B^{2/3}+K_4*WS+K_5*(ST-2)+K_6*(ST-2)^2 +K_7*WS*(ST-2)+K_8*\ln(TFZ)+K_9*WS*\ln(TFZ)) \tag{5}$$

[1] Institute of Technology, Ilmenau, GDR

[2] Institute for Cybernetics and Information Processes, Berlin, GDR

$$f_3=a_0-a_1*B-a_2*t+a_3*ST^{1.2}-a_4*WS-a_5*T+a_6*\ln(TFZ)-a_7*t*\ln(TFZ) \qquad (6)$$

The coefficients K_i are linear interpolated within the four growth intervals: 45...63, 64...87, 88...129, 130...180th day. Waterstress WS and nitrogen content ST can be treated as the control variables, because they are directly conneted with the effects of irrigation and fertilization on the growing behaviour of the plant. Positive values of WS means water demand. Uncontrollable inputs are the radiation R and the outside temperature T. The soil quality is given by the parameter TFZ. The K_i and the constants a_i has been ascertained by regression.

The function f(R,T) gives a proper weight every day on the nettoassimilation rate in dependence of the actual courses for global radiation and temperature. Analytical models are available for R(t) and T(t) but it is also possible to use measured data.

The waterstress WS represents the state of the water supply of the plant. The soil water model of KOITZSCH (1977) can be used to elaborate the actual WS in dependence of the water supply in the past and the actual growth state of the plant. To get reference courses for WS(t) we applied a dynamic optimization procedure (ARNOLD, PUTA, 1986) with the goal to maximize the yield of dry matter of the body at harvest time. In agreement with the definition of HOFFMANN (1980) the course for waterstress results in about WS(t) = 0 without a short period between 90th and 110th day. During this period the waterstress could be a bit positiv (water demand) for better qualification of the roots. Up to the availability of a proper nitrogen soil model, nitrogen courses ST(t) from the practical experience have to be used. For given values or courses, resp., of the discussed control variables it is possible to use the growth model, eqs. (1) to (5), to predict the yield of dry matter of the body and dry matter of the leaves for any day during the growing season of about 135 days (from 165th to 300th day of the year).

3. THE IRRIGATION ALGORITHM

For application it is desirable to make control decisions (here for irrigation) with respect to the annual climate situation and adapted to the actual growth state of the plant. For this task are available the growth model and the soil water model. The latter one allows the computation of waterstress in dependence of climate date as inputs and using the growth model. The inversion of the soil water model would be desirable to compute the necessary irrigation for required values of waterstress. But this is because of the algorithmic structure of this model impossible. Therefore another strategie will be proposed.

At a given actual time t_a during the growing period we know the real climate from t_0 (initial time) up to t_a. With this information we can simulate this time interval and compute the actual growth state of the plant using both of the models. With predicted climate data for the interval from t_a to t_f (t_f is the estimated harvest time) it is now possible to give a irrigation proposal (fig. 1).

The prediction results may be applied as long as there are not too large deviations between estimated and real climate including expected rainfall. If this not hold, a new prediction is needed at a new time t_a. Therefore the algorithm should have repetitive character. A fragment of the repetitive irrigation algorithm is shown in figure 2.

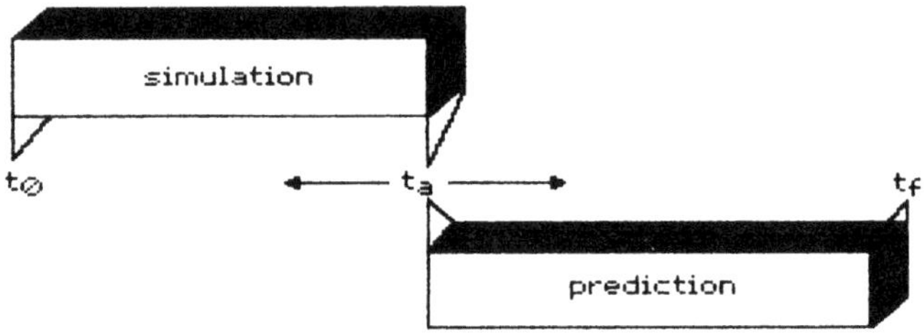

Fig. 1 Choosing t_a between t_0 and t_f

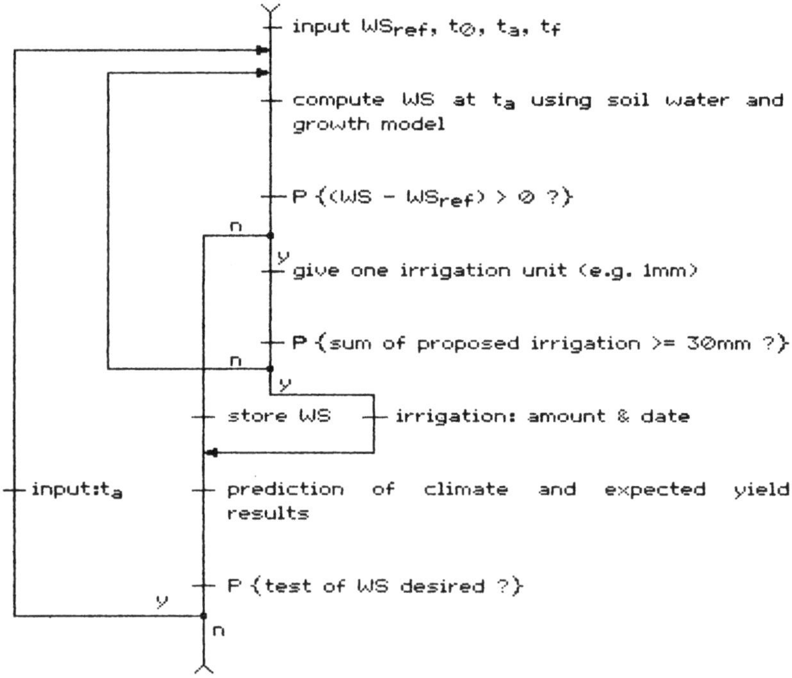

Fig. 2 The repetitive irrigation algorithm

4. SIMULATION RESULTS

The dialogoriented algorithm (GERECKE, 1988) allows the prediction of five variants of expected climate on the screen of the 8 Bit PC 1715. The program needs about 5Ø kByte of memory and is written in TURBO-PASCAL. The protocol is shown in figure 3.

Ergebnisprotokoll

	1.Variante	2.Variante	3.Variante	4.Variante	5.Variante
Blattmasse	83 dt/ha	79 dt/ha	72 dt/ha	9Ø dt/ha	41 dt/ha
Ruebenmasse	2Ø1 dt/ha	2Ø3 dt/ha	177 dt/ha	227 dt/ha	94 dt/ha
Zuckerertrag	114 dt/ha	117 dt/ha	97 dt/ha	134 dt/ha	45 dt/ha
Gesamterloes	8936.-M/ha	8947.-M/ha	7846.-M/ha	1ØØ23.-M/ha	4234.-M/ha
Transportkosten	1555.-M/ha	1538.-M/ha	1392.-M/ha	17Ø5.-M/ha	958.-M/ha
Beregnungskosten	225.-M/ha	338.-M/ha	338.-M/ha	225.-M/ha	Ø.-M/ha
>> Gewinn <<	7156.-M/ha	7Ø71.-M/ha	6117.-M/ha	8Ø93.-M/ha	3275.-M/ha
Wasserstress	-43.5	-59.9	-42.4	-59.8	-23.1
NS - Faktor	Ø mm	-5 mm	-1Ø mm	5 mm	Ø mm
Beregnungstermin und Regenmenge	9.7.:25mm 8.8.:25mm	9.7.:25mm 4.8.:25mm 9.9.:25mm	8.7.:25mm 3Ø.7.:25mm 25.8.:25mm	1Ø.7.:25mm 13.8.:25mm	keiner

Fig. 3 Protocol of five irrigation variants

In figure 4 the waterstress in the plant according to variants 1 and 5 resp. (with and without irrigation) is demonstrated. Another benefit of the dialog is that the dispatcher can also give his own irrigation proposal.

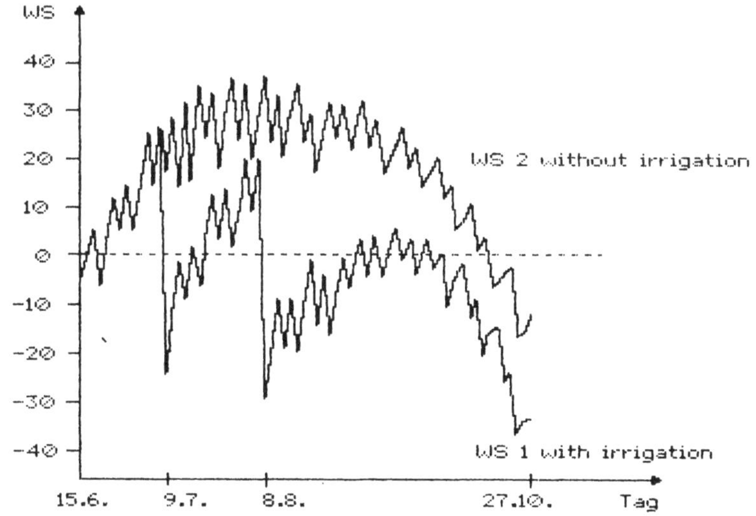

Fig. 4 Waterstress courses with and without irrigation

5 SUMMARY

In agricultural practice irrigation deliberation is already used but it is not typical to utilize models. In this paper a repetitive working dialog oriented irrigation algorithm for sugar beets using the growth model and a soil water model is proposed. Whenever the algorithm is started it works in two phases, the simulation of the past and a prediction up to the harvest time, estimating the yield with irrigation proposals, if necessary.

6. REFERENCES

Arnold, E. and Puta, H. (1986): Ermittlung günstiger Klimastrategien für Gewächshäuser - ein Beitrag zur rationellen Energieanwendung in der Landwirtschaft. III. Interdisziplinäre Konferenz "Energietechnik und Energiewirtschaft", 20.-21.2.1986, IHS Zittau

Augustin, P. and Schmidt, M. (1980): A model of yield formation as basis for optimal control of growth factors in culture of greenhouse cucumber. 1st Intern. Symp. on Systems Analysis and Simulation, Berlin, 1980

Gerecke, D. (1988): Erstellen eines Programms zur Beregnungsberatung beim Zuckerrübenanbau. Diplomarbeit, TH Ilmenau, 1988

Hoffmann, F. (1980): Entwicklung eines dynamischen Modells der Stoffproduktion von Zuckerrüben mit Hilfe der multiplen Regressionsanalyse. Biometrical J. 22 (1980), pp. 799-812

van Keulen, H. (1986): Simulation Monographs. Centre for agricultural Publishing and documentation. Wageningen, 1986, The Netherlands

Koitzsch, R. (1977): Schätzung der Bodenfeuchte aus meteorologischen Daten, Boden- und Pflanzenparametern mit einem Mehrschichtmodell. Z. Meteorol. Berlin 27 (1977) S. 302-306

SIMULATION OF URBAN AIR POLLUTANT PROCESSES AND MONITORING STATIONS LOCATION PROBLEM +

Józef KORBICZ*, I.L.PARMASTE**, Piotr GAWŁOWICZ*

1. INTRODUCTION

Investigations into air pollution problem are mainly related to modelling, predicting the pollutant concentrations and monitoring stations location and have been considered to be very important in recent years since the existence of high level concentrations for a long time is injurious to the health of the populous.

In this report, a computer model for simulation of city air pollutions and monitoring stations location problem are presented. Mathematical model describing the diffusion and transport of air pollutions from emission source into the atmosphere are expressed by stochastic advection-diffusion equation [6]. The numerical solution of this equation by using the split method and the modified finite element technique is obtained. The problem of defined minimal number of monitoring stations and their location is very important from the theoretical and practical point of view and is connected with the known problem of sensors location for distributed parameter system (DPS) [1,2,3]. Here suboptimal approach to the solution of this problem is developed based on the suboptimal state estimation theory for DPS. The computational effectivies of this algorithm is defined by using the suboptimal filtering algorithm [4], which do not require to solve the matrix Riccati equation for the filter error covariance.

Finally, the digital simulation of the air pollution processes for some region demonstrates the effectiveness of the proposed computer model and the monitoring station location algorithm.

2. DESCRIPTION OF THE PROBLEM

We consider the pollutant dispersion process in atmosphere governed by the advection-diffusion equation [7]

$$\frac{\partial q(x,t)}{\partial t} = \sum_{i=1}^{3}\left(\frac{\partial}{\partial x_i} k_i(x)\frac{\partial q(x,t)}{\partial x_i}\right) - \left(U, \operatorname{grad}\left(q(x,t)\right)\right) - \alpha q(x,t) + f(x,t) + w(x,t) \quad x\in\Omega,\ t\in(0,t_f) \tag{1}$$

with boundary condition

* Higher College of Engineering, Depart.of Applied Mathematics and Computer Science, ul.Podgórna 50, 65-246 ZIELONA GÓRA, POLAND

** Technical University of Kiev, Depart.of Computation Technique and Computer Science, prosp.Pobiedy 37, 252056 KIEV, USRR

+ This autor's research was supported by contract CPBP 01.02.

$$\Gamma(k+1|k) = R(k+1) + H(k+1)\hat{p}_{NN}(k+1|k)H^T(k+1) \tag{11}$$

prediction and filtering errors: $\delta q(x,k+1|k)$ and $\delta q(k+1|k+1)$

$$\delta q(x,k+1|k) = A_x \delta q(x,k|k) + \bar{w}(x,k) \quad x \in \Omega \tag{12}$$

$$\delta q(x,k+1|k+1) = \delta q(x,k+1|k) - K(x,k+1)\nu(k+1) \tag{13}$$

prediction error variance estimator: $\hat{p}(x,y,k+1|k)$

$$\hat{p}(x,y,k+1|k) = \hat{p}(x,y,k|k-1) + \gamma(k)\left\{\frac{1}{S_x S_y}\int_{\Omega_x}\int_{\Omega_y} \delta q(x',k+1|k) \times \right. \tag{14}$$

$$\left. \times \delta q(x'',k+1|k)\,d\Omega' d\Omega'' - \hat{p}(x,y,k|k-1)\right\}, \quad x,y \in \Omega \times \Omega$$

$$\hat{p}(x,y,k|k-1)\Big|_{k=1} = \hat{p}_0(x,y), \quad x,y \in \Omega \times \Omega$$

with corresponding boundary and initial conditions. In (8) - (14) $\hat{p}_N(x,k+1|k)$ and $p_{NN}(k+1|k)$ are the dispersion functions at measured points described in [4], $A_*[\cdot] = \mathrm{diag}[A_x j[\cdot], j=1,2,\dots,N]$, S_x and S_y are chosen constant coefficients, Ω_x and $\Omega_y \subset \Omega$ are subspaces of Ω.

3. PROBLEM SOLUTION

Assume that the priori information about the number and admissible location of sensors is known, i.e. N', $(N' \geqslant N)$, $\bar{x} = (x^1, x^2, \dots, x^N)$. But at the domain $\bar{\Omega}$ only N sensors of the N' admissible, will be located. Introduce the following parameter, defined as

$$\lambda_j = \int_{\bar{\Omega}} \delta(x-x^j)\,dx = \begin{cases} 1, & \text{if at the point } x^j \text{ the sensor is located} \\ 0, & \text{at the other measured points} \end{cases} \tag{15}$$

where $\sum_{j=1}^{N'} \lambda_j = N$, and $\Lambda = \mathrm{col}[\lambda_1, \lambda_2, \dots, \lambda_{N'}]$ is the N'-dimensional parameter vector. Taking into account the parameter (15), the observation equation (3) at the selected measured points is given by

$$Y(x^j,k) = \lambda_j\left[H(x^j,k)\,q(x^j,k) + V(x^j,k)\right], \quad j=1,2,\dots,N' \tag{16}$$

In the same way the parameter (15) may be introduced in the filter equations (8) - (14).

Now, the location problem of sensors can be reformulated as the optimal control problem, i.e., it is necessary to find an optimal signal of an artificial control λ_j^*, $j=1,2,\dots,N$, which minimizes the cost function (7) with respect to the dynamical restrictions given by the suboptimal filter equations with introduced λ_j.

The formulated optimal control problem has been solved by the use of the minimum principle for DPS and the method of switching points in the optimal bang-bang control problem. In accordance with this approach the switching function ξ_i is

$$q(x,t) = 0, \quad x\in\partial\Omega_b, \quad \partial\Omega = \partial\Omega_b \cup \partial\Omega_0 \cup \partial\Omega_H$$

$$\frac{\partial q(x,t)}{\partial x_3} = \alpha_b q(x,t), \quad x\in\partial\Omega_0, \quad \frac{\partial q(x,t)}{\partial x_3} = 0, \quad x\in\partial\Omega_H \tag{2}$$

where t is the continuous time, $q(x,t)$ is the pollutant concentration, U is the mean wind velocity vector, $k_i(x)$, α and α_b are any parameters, $f(x,t)$ is the function of emission source rate and $w(x,t)$ is the stochastic activation of the system with the known characteristics, $E[w(x,t)] = 0$ and $E[w(x,t)w(\tau,y)] = Q(x,y,t)\delta(t-\tau)$.

The measurements by monitoring station at bounded number points $x^j \in \bar{\Omega}$, $\bar{\Omega} = \Omega\cup\partial\Omega$, $j=1,2,\dots,N$ (N is the measurement points number) with sampling time Δt, $t_k = k\cdot\Delta t$, $k=0,1,\dots,K$ (K is the number of all measurements made at time) are made. Then the measurement system may be described by the following equation

$$Y(k) = H(k)q_N(k) + V(k) \tag{3}$$

where $Y(k) = [y(x^j,t_k),\ j=1,2,\dots,N]$, $q_N(k) = [q(x^j,t_k),\ j=1,2,\dots,N]$ are N-dimensional vectors, $H(k)$ is the diagonal (NxN) matrix and $V(k)$ is the N-dimensional Gaussian white noise with zero mean and the (NxN) covariance matrix $R(k)$.

The discrete-time model of the continuous system (1) may be obtained by using the Crank-Nicolson's finite difference scheme

$$q(x,k+1) = A_x q(x,k) + \bar{f}(x,k) + \bar{w}(x,k), \quad k=0,1,\dots,K \tag{4}$$

$$q(x,k+1) = 0, \quad x\in\partial\Omega_b, \quad \frac{\partial q(x,k+1)}{\partial x_3} = 0, \quad x\in\partial\Omega_H$$

$$\frac{\partial q(x,k+1)}{\partial x_3} = \alpha_b q(x,k+1), \quad x\in\partial\Omega_0, \quad q(x,k+1)\Big|_{k=0} = q_0(x) \tag{5}$$

where $\bar{f}(x,k) = \Delta t f(x,k)$, $\bar{w}(x,k) = \Delta t w(x,k)$ and

$$A_x[\cdot] = 1 + \Delta t \left[\sum_{i=1}^{3} \left(\frac{\partial}{\partial x_i} k_i(x_i) \frac{\partial[\cdot]}{\partial x_i} \right) - (U, \mathrm{grad}[\cdot]) - \alpha[\cdot] \right] \tag{6}$$

<u>Problem</u>. It is necessary to define minimal number of sensors N and fixed points of their location $x^{opt} = (x^1, x^2, \dots, x^N)$ at domain $\bar{\Omega}$, which minimizes the cost function

$$J(\bar{x}) = \sum_{k=1}^{k} \iint_{\bar{\Omega}\bar{\Omega}} \hat{p}(x,y,k|k)\,dx\cdot dy \tag{7}$$

where $\hat{p}(x,y,k|k)$ is the dispersion function estimate, with respect to restricts given by the suboptimal filtering equations [4]

<u>state estimate</u> $\hat{q}(x,k|k)$

$$\hat{q}(x,k+1|k+1) = A_x\hat{q}(x,k|k) + \bar{f}(x,k) + K(x,k+1)\nu(k+1), \quad k=0,\dots,K \quad x\in\Omega \tag{8}$$

$$K(x,k+1) = \hat{p}_N(x,k+1|k)H^T(k+1)\Gamma^{-1}(k+1|k) \tag{9}$$

$$\nu(k+1) = Y(k+1) - A_x\hat{q}_N(k|k) \tag{10}$$

$$\zeta_j = \sum_{k=1}^{K}\int_{\Omega}\Big[\hat{\phi}^*(x,k+1|k+1) - \delta\phi^*(x,k+1|k+1)\Big]\,\hat{p}^*(x,y^j,k+1|k)\,H(y^j,k+1)\times \\ \times\, q(y^j,k+1|k)\Big)\,\Gamma^{-1}(y^j,k+1|k)\Big[Y(y^j,k+1) - H(y^j,k+1)\,\hat{q}(y^j,k+1|k)\Big]dx \qquad (17)$$

where $\hat{\phi}(x,k|k)$, $\delta\phi(x,k|k)$ are conjugate functions of $\hat{q}(x,k|k)$ and $\delta q(x,k|k)$.

The minimal number of sensors choice (i.e.N) and their optimal location (i.e.choice the Λ) were determined by using the iterative procedure.

4. NUMERICAL EXPERIMENT

Air pollution process was simulated in restricted two-dimensional space $\Omega = (X_{1F}, X_{2F}) = (100,100)$ with one emission source located at $(x_1, x_2) = (20,80)$. The sampling time Δt, the coordinates division Δx, the dispersion of noise process σ_W^2, parameters α, α_b, the wind velocity $U=(u_1,u_2)$ and turbulent diffusivities $k(x) = (k_1,k_2)$ are equal to: $\Delta t=0.1$, $\Delta x_1=\Delta x_2=5$, $\sigma_W^2=0.04$, $\alpha=\alpha_b=0$, $U=(5,-5)$, $k(x)=(20,20)$, and $R(k)=\mathrm{diag}\,\sigma_v^2 = \mathrm{diag}[0.3]_{N\times N}$.

This numerical example was been studied for testing the performance or our algorithm in different cases. For instance, if estimation accurácy is equal to $\varepsilon(k)=0.75$, then the minimal number of sensors $(N_{min}=16)$ from all admissible $(N'=54)$ has been defined. On the second step of proposed algorithm the optimal location coordinates of the minimal number of sensors have been calculated. The map of all admissible measurement points and optimal places of their minimal number is the result of modelling. The dynamic and exactness of the estimation process by using the suboptimal filter algorithm with the minimal number of sensors has been shown too.

REFERENCES

[1] Kubrusly, C.S., Malebranche, H.: A survey an optimal sensors and controllers location in DPS. - IFAC 3rd Symp. "Control of Distributed Parameter Systems", Toulouse, France, (1982), 59-73.

[2] Chen,W.H., Seinfeld,J.H.: Optimal location of process measurement.- Int.J.Control. 21 (1975),1003-1014.

[3] Aidarous,S.E., Gevers,M.R., Installe,M.J.: Optimal sensors allocation strategies for a class of stochastic distributed systems. - Int.J.Control. 22 (1975),197-213.

[4] Korbicz,J.: Suboptimal state estimation algorithm for non-linear discrete-time distributed-parameter systems. - Int.J.Syst.Sci., 17 (1986), 725-234.

[5] Korbicz J., Zgurovsky,M.Z., Novikov,A.N.: Suboptimal sensors location in state estimation problem for stochastic nonlinear distributed parameter systems. - Int.J.Syst.Sci., (1988) in press .

[6] Korbicz,J., Zgurovsky,M.Z.: State and parameter estimation of stochastic distributed parameter systems on base of combined method with application to air pollution processes. - Proc.IMACS/IFAC Int. Symp. "Modelling and Simulation of Distributed Parameter Systems", Japan, Hiroshima, Oct.,(1987), 523-530.

STRUCTURAL ANALYSIS OF THE CLIMATIC RESPONSE TO A NUCLEAR WAR

P. Carl[1) and G.L. Stenchikov[2)

1. Introduction

Global and indirect consequences of nuclear warfare, which would be mediated by the atmosphere, are discussed already since the fifteeth. A definite assertion about the sign and magnitude of potential climatic effects, however, was not achieved for long. Apart from quantitative investigation, crescent warnings of a possible climate catastrophe after an all-out nuclear exchange originated in the late seventeeth from a general increase in understanding the structure and dynamics of complex systems, from the quickly developing knowledge about paleoclimatic changes, and from global energy balance studies of climate variability and stability. A serious, quantitatively based consideration of atmospheric, climatic and environmental consequences of nuclear war became available in 1982 with an international study [1], which also brought forth the startling discovery of a potential severe atmospheric impact of smoke. Due to propitious coincidence with major attention devoted to Martian dust storms, the "Nuclear Winter" theory could be quickly carved out and further developed as a well-founded scientific reasoning [2]. On the other hand, general indeterminisms (the "scenario"), poorly understood climatic feedbacks, and insufficient self-consistency and resolution of current-generation climate models span a serious range of uncertainty. Activities are worldwide in progress to narrow these margins [3]. One question is often posed in this context: What methods and what general properties of the climate system may the computational analyst rely on in this field?

2. "Nuclear Winter": Comments on its scientific base and uncertainties

Though global climate modeling is one of its major component, it should be noted first, that the unique character of the complex risk study the world's scientific community currently acts upon challenges a much broader scale of available experience. Persistent criticism however, which accompanies the scientific debate, seems doing right to concentrate on climate simulation. Parameterizations are generally used to express unresolved processes and scales by the model's variables. General Circulation Models (GCM), the most complex approach to climate modeling, have to dispense with greatest possible detail, at least for computational efficiency. So far, these models are generally destined to describe the present climate and to follow weak disturbances. In the strongly disturbed case, the theoretical and empirical base of GCM parameterizations has to be ca-

1) Central Institute for Electron Physics, GDR Academy of Sciences
Hausvogteiplatz 5-7, Berlin - 1086, GDR

2) Computing Centre, USSR Academy of Sciences
Vavilov Street 40, Moscow - 117967, USSR

refully reconsidered. Moreover, significant steps towards higher 'interactive' (self-consistent) modeling are required to allow for stronger disturbances - a work which is far from being completed today. Atmosphere/ocean and cloud/climate feedbacks, for instance, are attacked in large scientific projects to understand climate variability and man's impact on the environment. Research on climatic and environmental consequences of nuclear war (we may use the term 'Nuclear Winter' though the nature of the disturbance is far more complex than a seasonal change) profites from progress there and, in turn, contributes to solution of tasks of a more general nature. Beyond that, for present-day 'Nuclear Winter' research, traditional scientific tools like sensitivity studies and model intercomparison proved to be efficient.

We cannot yet present conclusive soothing arguments on GCM 'Nuclear Winter' studies. Some 5-6 years of experience exist, however, of an international, careful approach to understand the response of the strongly disturbed climate. Several well-accepted GCM's were used to this end. Since major differences in the short-term prognoses could be assigned to initial conditions, standard scenarios are formulated. Thus, we may turn over here to briefly comment on 'Nuclear Winter' sensitivity studies.

3. Two remarks on the use of GCM's in 'Nuclear Winter' research

Any investigation of a structurally disturbed system has to consider the nonlinear character of its response. A first structural insight may be often achieved by means of a simply-to-write, self-consistent, empirical parameterization. Sensitivity studies by parameter variation of a model that ignores feedback and instead prescribes some characteristics, on the other hand, cannot on principle properly meet our goal: Such models operate over a less structured phase space. Its topology, in fact, is largely shaped by the nonlinear part of model. Topological constraints on the system's trajectory must be considered essential after strong disturbance. Even a sophisticated self-consistent parameterization, however, is not a 'genuine' model: One free parameter seems necessary, and limitation by prescribed bounds may be reasonable to adjust the undisturbed case, the margins of possible systems evolution, and the model's sensitivity. The technique sketched here is extensively used in lower-dimensional climate modeling, preferably in energy balance models (EBM). As described below, it proved helpful in our approach to predict the structure and variability of 'the Unthinkable' [4].

There is yet another point in our understanding of the role of GCM's in 'Nuclear Winter' research: Full-scale diagnosis in space, time and mechanisms is usually required to describe a 'climate' in general; the same should hold for its complex, nonlinear response. In a broad diagnosis of the available GCM output [4a], zonally or regionally averaged time evolution plots offer a direct view of the meridional extension, the duration and the major 'carriers' of the disturbance. Different stages of response, its appearance and sensitivity in certain variables, critical phases and possible branches of system's evolution may be detected. Periods which should be more carefully diagnosed, are easily seen. Furthermore, a set of such figures under comparable conditions should be very useful for model intercomparison and reliability investigations.

4. CCAS model based analysis of the disturbed climate

The CCAS-Model [5], a coarse-resolution (12° latitude, 15° longitude) two-level tropospheric GCM, is used here in a diurnal and annual mean mode. Merits are its thermal 'ocean' and a justifiable expense for longer-term investigation. Though the coarse resolution is problematic, it seems tolerable for the task we have chosen at present - a cautious, short- to medium-term ("acute phase") structural analysis at an acceptable level of reliability. Our aerosol model has developed from stationary, optically prescribed to self-consistent transport and scavenging [4,5]. To illustrate our discussion above, we finally present results of an experiment, which may be considered "medium" in several aspects. The figure presents part of a complex diagnosis, which shows a disturbance penetrating through the system and expanding geographically. Series of numerical sensitivity experiments conducted so far [4,5] include:

- large-scale inhomogeneities of the initial smoke distribution,
- changing aerosol modeling,
- smoke amount variation,
- types of war "scenarios" (all-out / first-strike / European war).

Using the three standard "Bangkok scenarios", biologically relevant climate data ("stress matrix") are currently diagnosed [4b] to meet the international efforts to evaluate the ecological consequences of nuclear war in greater detail and in closer connection to climate modeling.

We identified in our experiments a highly nonlinear systems behaviour, with a delicate interplay between inertia and response in four channels: radiative, thermal, dynamical, and hydrological. A high sensitivity (even of the ground hydrology) was found to stratospheric smoke modeling. Though most of our experiments followed a branch of systems evolution that is characterized by decreasing dynamical energy, different regimes seem possible [4b]. A further branching is governed by competition of equator-near aerosol lofting and washout. Impact on the Southern Hemisphere is dependent upon the rate at which the equatorial circulation barrier is suppressed. We have seen in either case a serious impact on tropical zones, especially with respect to insolation and water cycle. The strongly nonlinear climatic response drives the system far from normal. Sound indication of a quick return was not found - we have observed instead the potential of a dramatic structural change. Stratospheric smoke load, atmosphere/ocean feedback and biosphere response are crucial for the "chronic" phase of disturbance - and for sweeping conclusions on the threat of nuclear war for biosphere and man's civilization.

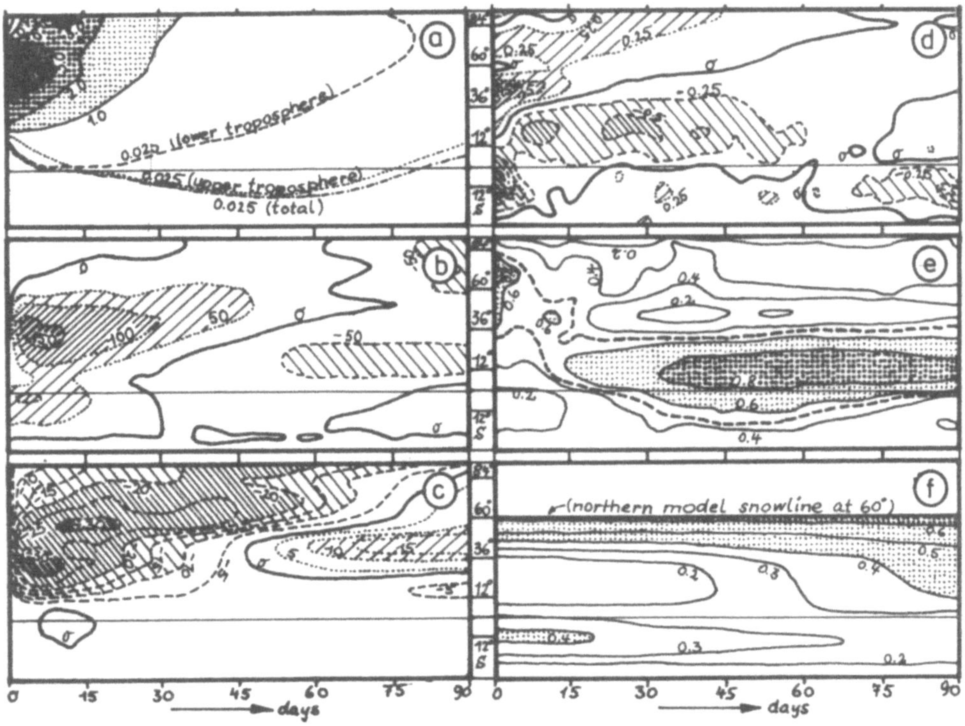

Figure: Zonally averaged evolution of a war-generated optical disturbance and the climatic response:
a - vertical opacity of the model atmosphere (hemispherically averaged initial value: 2.2),
b - radiation balance (in W m^{-2}; negative: net back radiation),
c - surface air temperature change over land (in K),
d - vertical mass flux (in kg m s^{-1}; negative: rising air),
e - cloudness (0 - cloudfree, 1 - overcast),
f - ground wetness (0 - dry, 1 - saturated).

5. References

[1] Royal Swedish Academy of Sciences: "Nuclear War: The Aftermath", AMBIO 11(1982) No. 2/3;

[2] PETERSON, J.: "Scientific Studies of the Unthinkable - The Physical and Biological Effects of Nuclear War", AMBIO 15(1986)60;

[3] GOLITSYN, G.S. and M.C. MacCRACKEN: "Atmospheric and Climatic Consequences of a Major Nuclear War: Results of Recent Research", preprint UCRL-96597, LLNL, Livermore, September 1987, - to be issued by WMO/ICSU;

[4] STENCHIKOV, G.L. and CARL, P.: Manuscripts BA.45/46.87, MO.26.88,
(a) SCOPE-ENUWAR Workshop, Bangkok, Feb. 1987;
(b) SCOPE-ENUWAR Workshop, Moscow, March 1988;

[5] STENCHIKOV, G.L. in: "The Night After ... Climatic and Biological Consequences of Nuclear War" (Ye. Velikhov, Ed.), Mir Publishers, Moscow, 1985, 165p.

PEMU - AN IMPACT MODEL BASED ENVIRONMENT PROTECTION DECISION SUPPORT SYSTEM

A. Sydow[1], K. Bellmann[1], R. Straubel[1], I. Imming[2], H. Kaschenz[2], U. Damrath[3], G. Hofmann[4], S. Anders[4]

Summary: PEMU is being developed as a user friendly decision support system for environmental conservation an a national level. It supplies connections between pollutant emission of point sources, immission (conc./depos.) at receptor points or grid elements and environmental consequences. The user is provided with a tool for evaluation emission reduction strategies by scenario technique and by optimization facilities in order to calculate optimal timing of control strategies and minimal needed immission reduction to meet future goal states of environmental systems. Presently, PEMU is based on SO_2 and pine stand decline. Inclusions of dust, nitrogenium species, summer/winter immission and spruce stand impacts are being prepared.

1. Introduction

PEMU is being developed at ZKI as a collaborative work mainly together with Centre of Environmental Management of the GDR, Meteorological Service of the GDR, Institute of Forest Sciences/Eberswalde, Section Forestry/Tharandt (TU Dresden). PEMU is used in designing and evaluating strategy options of pollutant control with respect to pine stand productivity and stability dynamics. Beside this, PEMU helps to identify gaps in understanding pollutant and stress driven loss of pine stand productivity in given areas.

2. Implemented Basis Software for Modelling and Optimization

In PEMU is included a branch specific simulation system (SONCHES) for building validation and use of ecological models. By a hierarchy of menues, the user is leaded in designing the model by an appropriate input language and aided in model validation and scenario performances

---

1) Central Institute of Cybernetics and Information Proc./Acad. Sci. GDR, Kurstraße 33, Berlin 1086

2) Centre of Environmental Management of the GDR, Schnellerstraße 140, Berlin 1190

3) Meteorological Service of the GDR, Michendorfer Chaussee 23, Potsdam 1561

4) Institute of Forest Sci., Alfred-Möller-Straße, Eberswalde-Finow 1300

Fig. 1 PEMU/Air - A National Environmental Protection Decision Support System

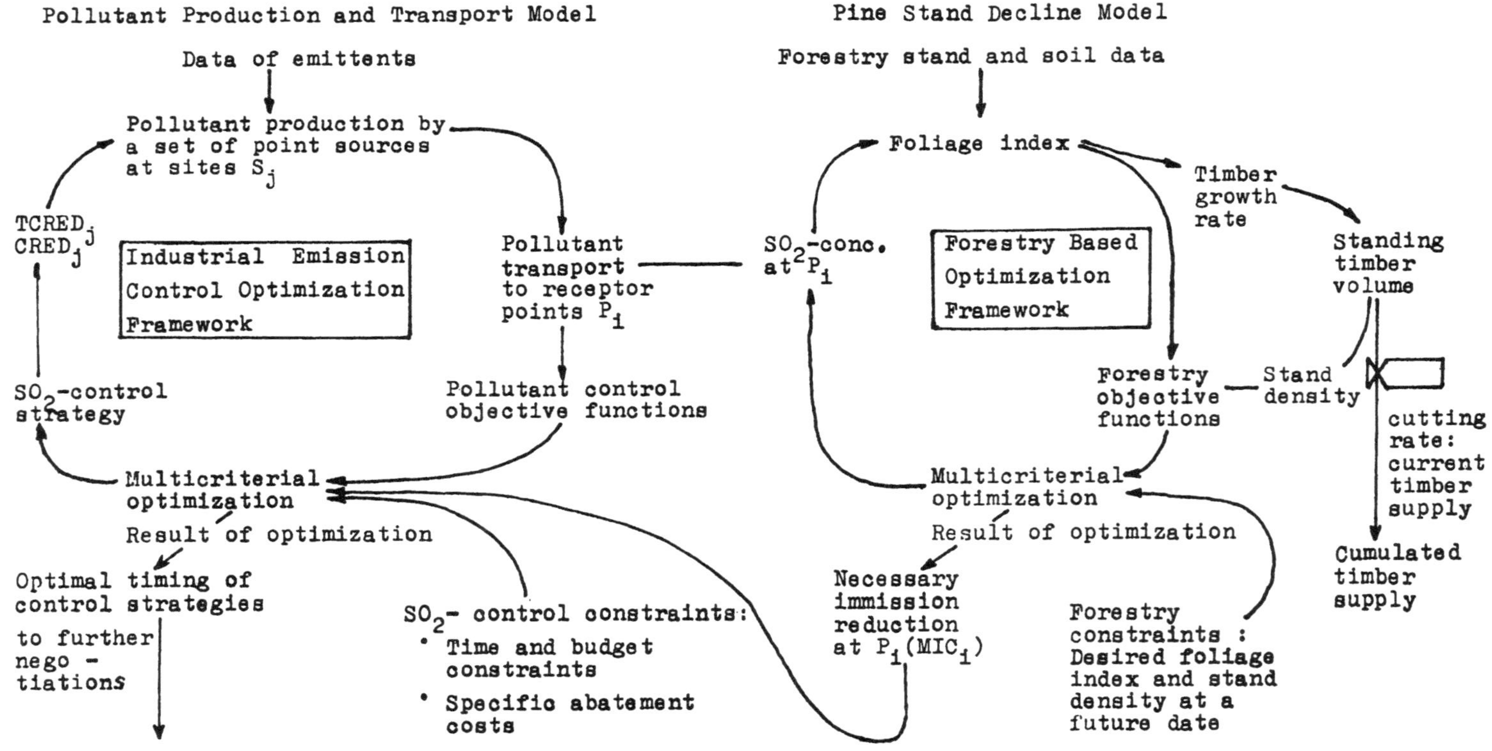

$TCRED_j$, $CRED_j$ - time and amount of emission reduction at the j-th source

by flexible facilities in sensitivity analysis, behaviour analysis, Monte Carlo studies, parameter optimization, result processing and a variety of display options. The pine stand decline model is implemented in SONCHES. Secondly, the user is flexibly supported in preparation of decision alternatives by the multicriterial optimization system REH which yields Pareto-optimal solutions in the case of non-convex and non-connected domains of control variables, using gradient procedures or an evolution strategy based optimization package.

3. PEMU Submodels

PEMU is of modular construction. It consists of the submodel 'Pollutant Production and Transport' and 'Forest Decline' (see Fig. 1); it is open for further impact models (spruce stand decline, human impact in preparation).

3.1 Pollutant Production and Transport Model

Inputs are:
Point source data set of emittents S_j: geographical position, real and effective stack height, time point of appearence of a new stack source, time point of emission reduction and its amount, actual emission of S_j.
Receptor point data set: geographical position, index set over i describing subregions which are intended to be considered.
For immission calculations normalized immission matrices of SO_2-conc. and dry and wet deposition (annual average) for reference emittents (differing in height, geographical position) are used. These matrices were computed by The Meteorological Service of the GDR using a three-dimensional Euler-transport-model [1].

The immission calculation can be performed by using optionally the grid mode (10 x 10 km, 16 x 16 km) or the Receptor point mode with freely chosen P_i utilizing non linear spline interpolations. - In order to identify the main sources being responsible for the immission at a given point P_i a facility for Responsibility Analysis is implemented. After setting a threshold parameter one gets a list of mainly responsible stacks with their individual contributions to the total immission at P_i. For further informations see [2].

Outputs are:
Pollutant concentration, wet and dry deposition for any future time point (Grid mode) or as time series (Receptor point mode), responsibility list for any point or grid element and future time point.

3.2 Pine Stand Decline Model

Inputs are:

Receptor point data: Acidity of immission (bark analysis) and soil, soil nutrient and moisture level, approx. retrospective immission conc. time series for decline interval, intended future fertilization and liming.

Forest stand data: present foliage index, age class distribution at P_i, area which is represented by P_i, initial values of the main yield table variables (e. g. present standing volume, stem nbr.).

The pine decline model is based on the well established yield table equations, additional functions describing pollutant caused damage and fertilization effects and the stand state depending needle growth rates. All functions were provided by the Institute of Forest Sciences/Eberswalde. For further information see [3].

Fig. 2 PEMU Pine Stand Decline Model

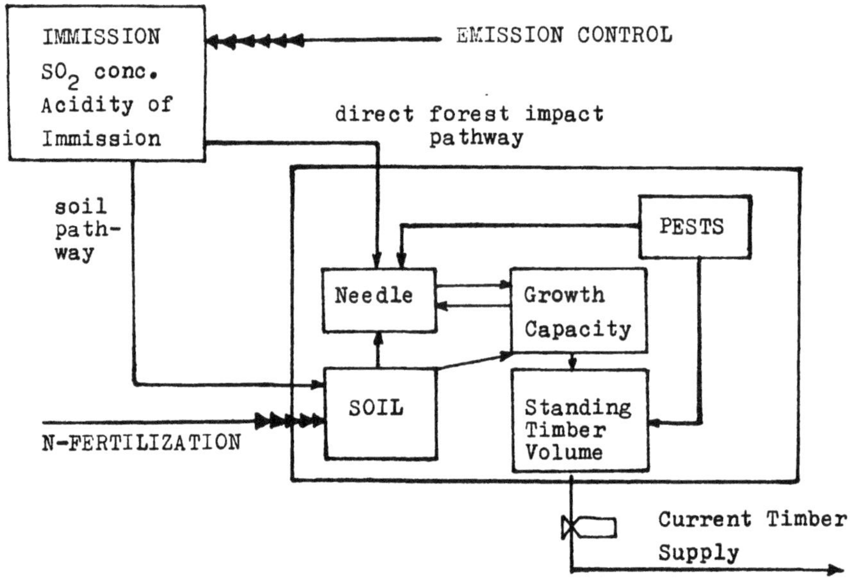

Outputs are:
Time series of the forest stand variables of productivity (e. g. timber growth rate, standing volume, current and cumulated timber supply), stability (stem nbr., density), and foliage (nbr. of neddle age classes, biomass, necrosity) per stand or as average over a set of stands at different P_i (see Fig. 2).

4. Performances of PEMU

4.1 Scenarios yield pollutant immission inventory insights by regional classifications (risk areas) and clarifications of connections between pollutant control options and forest decline dynamics.

4.2 Optimizations can be performed within two frameworks: Forestry optimizations and Industrial pollutant control optimizations.

Industrial pollutant control optimizations

By using REH the user is aided in solving the problem how to get the optimal control device and its optimal timing within a limited future time horizon in such a way, that for a maximal nbr. of receptor points P_i the i-specific necessary immission reduction due to forestry and human demands for a given future date are met, under time and budget constraints and specific abatement costs. For this task forestry based MIC_i-values are used which were optimized within the other optimization frame.

Forestry optimizations

The user is supported by the system in calculation of forest-MIC's. The task is to find for a given forest stand at P_i the necessary immission reduction so that at a given future time point variables (e. g. foliage index, stand density, cumulated timber supply) set by the user as his minimal future demands are met.

References

[1] Discher, H.-J., U. Damrath: Das operative regionale Ausbreitungsmodell der DDR zur Bestimmung grenzüberschreitender Schadstoffe; In: Abhandlungen des Meteorologischen Dienstes der DDR, 1986

[2] Lasch, P., N. Model, K. Bellmann: The PEMU Transport Model Based on Emittent-Receptor Point Transmission Calculation (in this volume)

[3] Bellmann, K., P. Lasch, S. Anders: The PEMU Forest Impact Model FORSTK - a Pine Stand Decline and Wood Production Model (in this volume).

THE PEMU/AIR POLLUTANT TRANSPORT MODEL BASED ON EMITTENT-RECEPTOR-POINT-TRANSMISSION CALCULATION

P. Lasch, N. Model, K. Bellmann[1]

1. Introduction

PEMU (Prognosis and Decision Support Model for Environmental Protection) has been designed to be especially useful to decision makers involved in SO_2 - emission control in the GDR. PEMU deals with a) pollution generation from point sources (substances : SO_2, in preparation nitrogen, dust) b) pollutant transport processes and c) environmental impacts (forest decline and productivity). The general PEMU - concept is described in [1], an overview about the forest decline model is given in [2] in this volume. In this contribution we describe the implementation and use of part a) and b).

2. PEMU Transport Model

For model calculations transfer matrices are used elaborated for PEMU by MD (Meteorological Service of the GDR). This transfer matrices are based on a three-dimensional Euler-model for medium range transport of air pollutants. For fixed SO_2 - emission (1000kt/y) SO_2 air concentrations and sulfur depositions were calculated for different weather classes (classification of wind, precipitation and other meteorological data) and were superposed using the probabilities of the weather classes, because the processes are assumed to be linear with respect to concentration and deposition [3]. This calculations are performed for so called reference emittents (19) for both different stack height (50,75,100,150,200,300 m) and their geographical position on the territory of the GDR. For each reference emittent a geographical reference area is defined. A real emittent's distribution is calculated by use of the transfer matrix of its corresponding reference emittent. According with this the real emittents must be classified corresponding to their stack height and geographical position. The reference matrices are given for the (25 x 25 km) grid of the MD. The time scale for calculation is one year (in the future half a year : summer, winter). Two possibilities of calculation are prepared : Grid mode, Receptor point mode.

---

[1]Central Institute of Cybernetics and Information Processes, Academy of Sciences of the GDR, 1086 Berlin, Kurstr. 33

2.1. Grid Calculation

Calculations of concentrations and depositions are possible for different grid square sizes :

- 10 x 10 km especially for the Centre of Environmental Management of the GDR
- 16 x 16 km ECE grid

Using a modified two-dimensional spline interpolation method for each real emittent normalized concentration (deposition) values are calculated for each grid element. These normalized values are multiplied by the real emission figure (per year) of the emittent. This absolute concentrations (depositions) are summed up per grid element (superposition) and one gets the total concentration (deposition) for each grid element. The two-dimensional spline interpolation is described in [4]. If the stack height of a real emittent does not meet the height of any reference emittent a height interpolation with cubic splines is used (see [4]).

2.2. Receptor Point Calculation

A (real valued) receptor point on the territory of GDR is given by its geographical position in the north-south-direction and east-west-direction (km). In the same way as described above for grid calculations normalized concentration (deposition) values are calculated for each considered receptor point (by means of spline interpolation); they are multiplied by the emission of the emittents and summed up over all emittents. One gets the total concentration (deposition) for this receptor point. These calculations are possible for a future time period chosen by the user. The resulting time series for receptor points are inputs for impact models in PEMU , e.g. the SONCHES implemented pine stand decline and productivity model FORSTK (see [2]).

3. Use of the Emission and Transport Model in PEMU

The way of using PEMU for scenario analysis is shown in table 1. In the first column , called Scenario preparation, the user has to decide 7 options ,corresponding to his task. After this the user specifies Calculation options(second column). In order to design emission reduction scenarios the user is supported by responsibility analysis , with this it is possible to find the main emittents which are responsible for x % of concentration (deposition) at a given grid element (receptor point). In the third column Output mode options are offered (tables,maps). Grafic display options are in preparation.

Table 1 : Emission - Transport - Block

<table>
<tr><td>1. Scenario preparation</td><td>2. Calculation specification</td></tr>
<tr><td rowspan="3">User options are :

- <u>Working mode</u>: grid,receptor point calculation

- <u>Grid</u>: 16 x 16 km ECE
10 x 10 km grid

- <u>Calculation mode</u>:
annual average, summer or winter average

- <u>Substances</u>: SO_2-concentration,S -deposition (wet and dry)

- <u>Input of emission data</u>:
set of emittents with geogr. position, stack height, SO_2-emission

- <u>Input of reduction parameters</u>: degree of stack specific emission reduction and time point of reduction

- <u>Input of receptor point data</u>: geogr. position of receptor points on the territory of GDR</td><td>User options are :

- <u>Calculation</u> of concentration/deposition for each grid element for a given future time point (grid mode) or time series (receptor point mode)

- <u>Responsibility analysis</u> for a given set of grid elements or a set of receptor points

- <u>Calculation</u> of an emission reduction scenario</td></tr>
<tr><td>3. Output specification</td></tr>
<tr><td>User options are :

- <u>Tables</u> of emittent and receptor data

- <u>Maps</u> of concentration/deposition grid data

- <u>Tables</u> of concentration/deposition time series

- <u>Tables</u> with results of responsibility analysis</td></tr>
</table>

4. References

[1] Sydow,A., Bellmann,K., Straubel,R., Imming,I., Kaschenz,H., Damrath, U., Hofmann,G., Anders,S.: PEMU - an impact model based environmental protection decision support system. in this volume

[2] Bellmann,K., Lasch,P., Hofmann,G., Anders,S., Schulz,H.: The PEMU - forest impact model FORSTK - a pine stand decline and wood supply model.in this volume

[3] Discher,H.-J., Damrath,U.: Das operative regionale Ausbreitungsmodell der DDR zur Bestimmung grenzueberschreitender Schadstoffe. Abhandlungen des Meteorologischen Dienstes der DDR,Nr.136,Bd.18 (1986),S.63-72

[4] Model,N., Born,J. : Calculation of air pollution immission values by using of two-dimensional spline approximations. Syst. Anal. Model. Simul.6(1989)1, in press

The PEMU Forest - Impact - Model FORSTK

A Pine Stand Decline and Wood Supply Model

K. Bellmann[1], *P. Lasch*[1], *G. Hofmann*[2], *S. Anders*[2], *H. Schulz*[3]

Summary:
The simulation model FORSTK is an impact component of the national Decision Support System for Environmental Protection (PEMU/Air) in order to test consequences of SO2-immission reduction strategies for the future dynamics of pine forest stands with respect to productivity and stability variables by means of scenario and optimization calculations.

A Short Model Description

FORSTK is being developed at ZKI since 1983 as an interdisciplinary work. It is based on broad field observations at a set of test sites along an immission gradient, gas chamber experiments, chemical analyses of needles, soil and bark, existing yield table equations and new ones. Both the direct and soil impact are mapped. Driving forces are N-fertilization and SO_2-conc. time series, calculated by the PEMU - SO_2 - transport model. BOX 1 describes the model structure. FORSTK yields future time series of foliage index, stem wood growth rate and its average, standing volume, timber supply rate, totally grown stem wood, diameter, height, stem number, stand density, depending on age, site quality, acidity of soil and immission, soil nutrient and moisture level, a.o. Nearly all necessary initial values and parameters are measurable. The foliage damage dynamics is based on the concept of site specific compensation ability (BOX 2). For regional considerations all output variables can be aggregated over a set of sites and/or age classes (BOX 3). FORSTK is implemented by use of the object- and task-oriented simulation system SONCHES for ecosystem modelling. It runs on 32-bit microcomputers and 16-bit PC.

Some Results

In Fig.1 and 2 the effects of immission reductions with respect to totally grown stem wood and foliage index are demonstrated. For each site and age class SO_2-immission intervals for future behaviour modes can be calculated (restoring, delayed decline, break down). Within PEMU, FORSTK is linked with the decision support system REH for optimization problems. Given desired future stability and productivity demands (stem

[1]Central Institut of Cybernetics and Information Processes, Academy of Sciences of the GDR, Kurstr. 33 Berlin 1086
[2]Institute of Forest Sciences, A.-Moeller-Str. Eberswalde 1300
[3]Institute of Landscape Research and Nature Conversation, Academy of Agric. Sciences of the GDR, Neuwerk 4 Halle/Saale 4020

BOX 1 FORSTK-The PEMU-Pine-Stand Decline & Wood Supply Model (SONCHES-implemented)

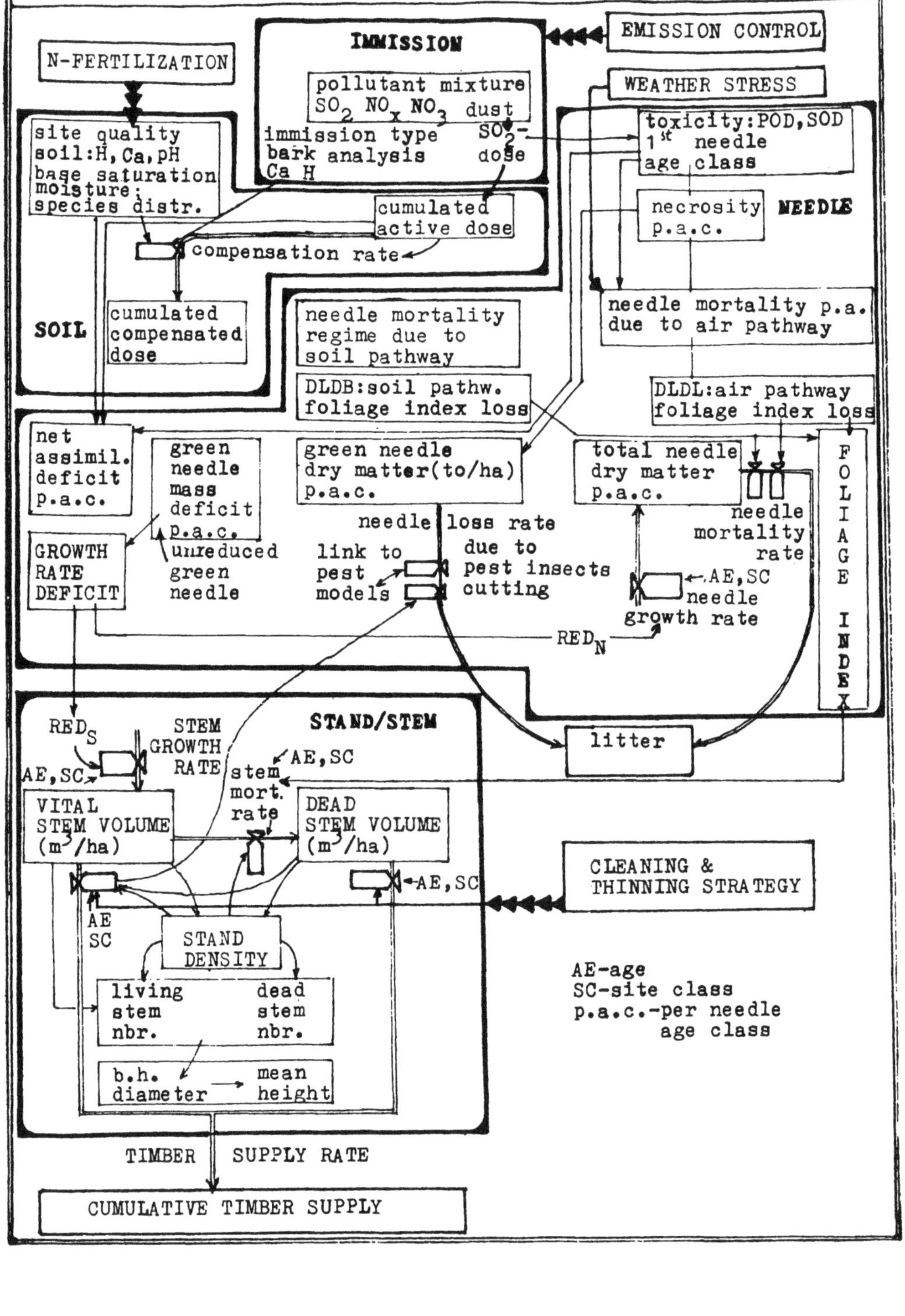

BOX 2 Compensation Rate

The compensation rate KR_t used in FORSTK is the fraction of $CIMS_t$ which remains without damage effect with respect to foliage index loss $DLD_t = DLDB_t + DLDL_t$

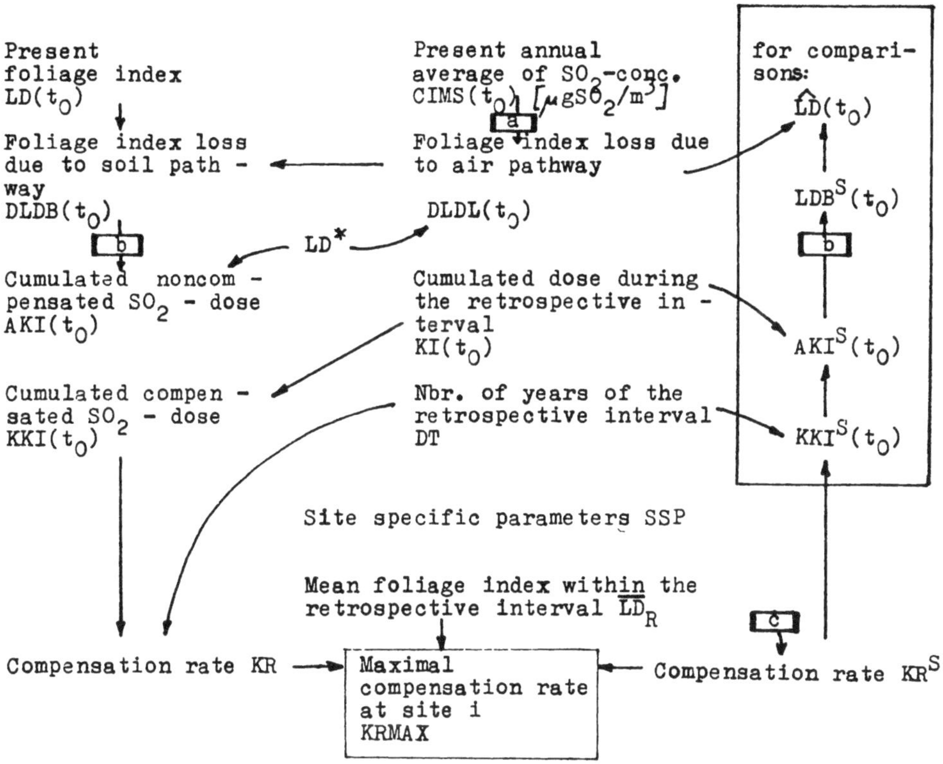

<u>SSP - parameter vector</u> from

(1) <u>Soil analysis</u> :
H^+, Ca^{2+}, p_H, base saturation
macro-, micro - nutrient level
moisture

(2) <u>Bark analysis</u> :
H^+, Ca^{2+}

(3) Natural vegetation analysis :
species ratio vector

<u>Observables</u> are :

$LD(t_0)$, $CIMS(t_0)$, $KI(t_0)$, DT, SSP, $\overline{LD}_R$

[a] $DLDL(CIMS; LD^*, b_L)$

[b] $AKI(LDB; LD^*, b_B)$

[c] $KR^S(F(SSP), b_S)$

LD^* - maximal foliage index at site i

b_L, b_B - lin.regress. coeff. (soil and air pathway, resp.) <u>foliage index loss</u> on <u>SO_2-conc.</u> (b_B) and on <u>AKI</u> (b_L) resp.

b_S - lin. regression coeff. KR^S on F(SSP)

BOX 3

FORSTK Result Aggregation

Region under test	R			
Set of representative sites RS_i within R	$RS_1 \ldots RS_i \ldots RS_n$			
Area per site RS_i (ha)	$A_1 \ldots A_i \ldots A_n$			
Stand & soil parameters and initial values per site RS_i	$I_1 \ldots I_i \ldots I_n$			
Control at each site RS_i: fertilization, cleaning & thinning strategy	$\underline{u}_1 \ldots \underline{u}_i \ldots \underline{u}_n$	sum	mean	over sites per age class
m=4 optional age classes AC_1		$\underline{Y}_{.1}$	$\underline{y}_{.1}$	
AC_j	$\underline{y}_{ij}$	$\underline{Y}_{.j}$	$\underline{y}_{.j}$	
AC_m		$\underline{Y}_{.m}$	$\underline{y}_{.m}$	
sum	$\underline{Y}_{1.} \ldots \underline{Y}_{i.} \ldots \underline{Y}_{n.}$	$\underline{Y}_{..}$		total sum
mean	$\underline{y}_{1.} \ldots \underline{y}_{i.} \ldots \underline{y}_{n.}$		$\underline{y}_{..}$	total mean
	over age classes per site			

Fig. 1 Totally Grown Stem Wood (TGSW)

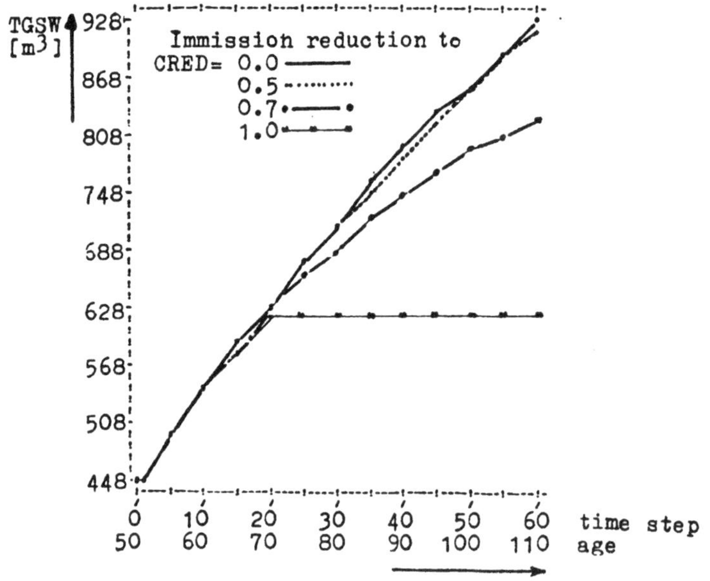

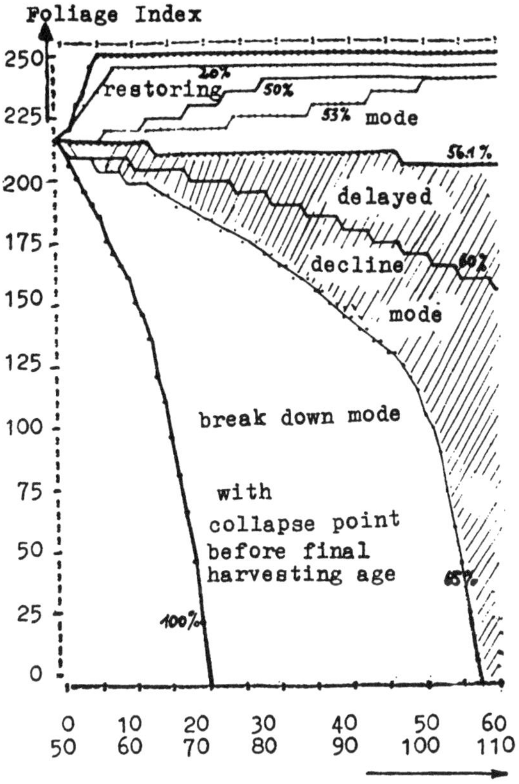

Fig. 2

Foliage Index Time Series of Immission Reduction Scenarios with varying Reduction Parameter CRED

(Time point of reduction: t_2)

CRED= 0.0
0.2
0.53
0.561
0.6
0.65
1.0

$CIMS(t_2)=CIMS(t_o) * CRED$

CIMS- SO_2-conc.[$\mu g/m^3$]

Fig. 3

Optimization of Minimal Needed Immission Reduction Depending on Target Foliage and Reduction Date

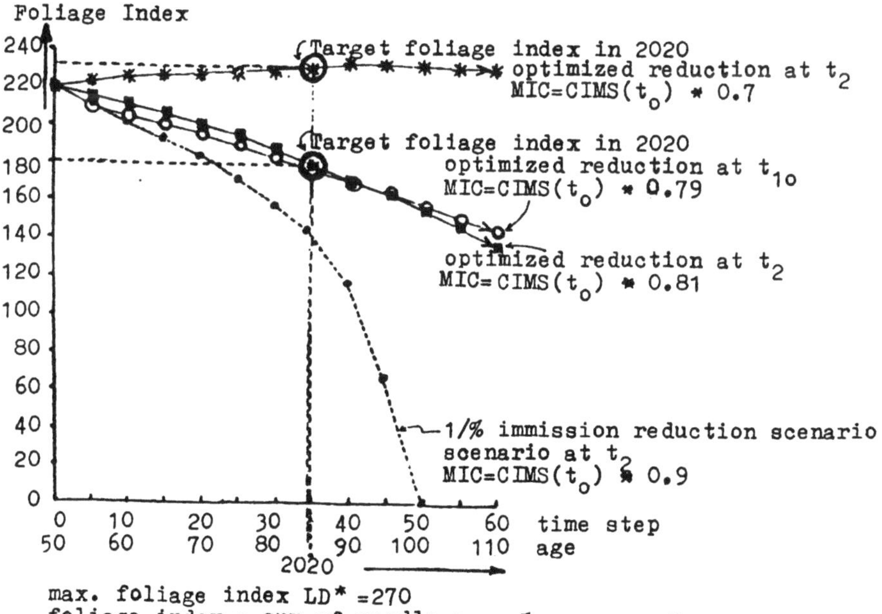

max. foliage index LD* =270
foliage index = sum of needle-mass-% per age class over all a.c.

density, foliage) for a given age class and site class by optimization the necessary SO_2-immission reduction at this site is calculable: one gets MIC - values (minimal immission concentration) based on forestry demands. The MICs are used in optimization functionals for optimal industrial pollutant reduction strategies (see [1]); they depend on present foliage index, on the target foliage and its time point and on the reduction time point (Fig.3). If for age 85 (year 2020) a target foliage index 180 is desired, it is necessary to reduce the SO_2-conc. from CIMS(t_0) to 79% or 81% for reduction date t_{10} or t_2 resp. With this the system is shifted from break down mode to delayed decline mode. For a desired foliage index 230 a reduction to 70% is needed (at this specific site). Forest decline is to be considered under different aspects [2]. One important is the ecological one. In [3] and [4] some of such problems were discussed.

References

[1] Sydow, A., K. Bellmann, R. Straubel, I. Imming, H. Kaschenz, U. Damrath, G. Hofmann, S. Anders: PEMU - An Impact Model Based Environment Pro - tection Decision Support System. (in this volume)

[2] Kairiukstis, L., S. Nilsson, A. Straszak (Eds.): Forest Decline and Reproduction : Regional and Global Consequences. Proc. Workshop Kraków ,1987. IIASA WP-87-75 (Sept.)1987

[3] Bellmann, K., P. Lasch : SONCHES-Modelling of Dynamics of Growth and Damage of Pine Exposed to SO_2-Immission. SiSy Prague. 1986

[4] Lasch, P., K. Bellmann :Simulation of Environmental Impact Assessments (Scenarios in Pine Forest Productivity). SiSy Prague. 1986

WATER AND MATTER DYNAMICS IN AGRO-ECOSYSTEMS
- A SPECIAL COOPERATIVE PROJECT OF THE GERMAN RESEARCH FOUNDATION AT THE TECHNICAL UNIVERSITY OF BRAUNSCHWEIG -

J. Richter *

The Objectives: The main objective of this major program is the analysis and forecasting of effects which farming and natural properties of specific sites have on energy fluxes, water fluxes, matter dynamics, microclimate and soil organisms. The differentiation of water and matter dynamics in space and time is analyzed. Special emphasis lies on the investigation of the long-term changes in agro-ecosystems. The atmosphere, the soil surface, the saturated and the unsaturated zones of the soil are the main compartments to be analyzed. Therefore all investigations are focussing on the soil.

All questions analyzed are related to actual world-wide problems of agro-ecosystems, their cultivation and deterioration, i.e.:

- changes of the physical properties of soils as a result of soil cultivation, crop rotation, ploughing in of straw, organic fertilization, splash, overland flow, accumulation of eroded material;
- acceleration of the soil erosion depending on the vegetation cover, soil cultivation, size of the drainage area, soil properties, etc.;
- changes of the chemical properties of soils depending on fertilization and deposition of pollutants;
- changes of the biological soil properties;
- changes of the evapotranspiration and other agrometeorologic parameters as a consequence, for example, of changing landuse;
- changes of the vulnerability of plants;
- pollution of the surface water and groundwater
- behaviour of herbicides in the soil.

The Methods: Physically and chemically based models shall be developed and applied not only to analyze data from field and laboratory experiments, but also to connect the various research projects. The models will describe and forecast the ecosystem. Water fluxes, matter dynamics, canopy development as well as physical, chemical, and to some extent the biological conditions of the system and its changes are to be described by dynamic models differentiated in space and time. The various submodels combine to a general model concept.

A transfer of the results from the research sites to larger landscape units (catchments, etc.) is planned by combining regionalization procedures with distributed models. The data base for regionalization is

---

* Institut für Bodenkunde der Universität Hannover, FRG

composed of remote sensing data, relief data, and soil mapping data which will finally be combined in a Geographical Information System with a resolution down to single fields.

The Structure of the Special Cooperative Program: Eight institutes of the Technical University of Braunschweig, the Federal Biological Research Centre for Agriculture and Forestry, the Central Research Institute of Agricultural Meteorology of the German Weather Service, the Geological State Survey of Lower Saxony, and one Institute of the University of Hannover are participating in the research program. The Department of Physical Geography and Landscape Ecology at the Technical University of Braunschweig is in charge of all the central tasks.

Division A of the project (Modeling System "Water and Matter Dynamics in Agro-Ecosystems") is the focus of all research activities. It comprises a number of sub-projects concerned with the analysis and modeling of processes in the soil and in the atmosphere, i.e. with energy fluxes (A4), with water fluxes in the soil (A6) as well as in plants and in the atmosphere above the soil surface (A4), the transport of matter by water and wind (Soil Erosion, A5), nutrients transported in the soil and taken up by plants as well as the growth of biomass (A7) and the dynamics of herbicides (A9). Nutrient dynamics at different sites are analyzed by the sub-projects A4, A6 and A7, while the sub-projects A3 and A7 are concerned with their modeling. The sub-project A1 integrates the results of various sub-projects into models describing the water and matter dynamics at specific sites and in whole catchments.

To be able to make statements representative of a catchment, extensive preliminary work is necessary. This is the task of division B of the program (Regionalization and Remote Sensing). The sub-projects concentrate on investigating and processing relief and geomorphologic data (B1), soil data (B2), data of unconsolidated sediments (B1, E2), and landuse data (B3, B1) which will be processed for model application by sub-group B1 in cooperation with sub-group Z1. The sub-group B4 will develop data structures suitable for the Geographical Information System which has to be generated. It is planned to simplify the model application extremely and to evaluate the knowledge at hand comprehensively (Expert System).

The division C of the project integrates the physical and chemical research on basic processes and structures which the projects of the divisions A and B require.

The division A of the program will be assisted by the results of division D which is concerned with biological research.

Significant species groups and plant diseases which have effects on the biomass development of agricultural crops are analyzed. The biological

soil condition will be characterized in its effects on the soil structure and on transformation processes induced by microorganisms.
The division E (Runoff and Groundwater) focusses on runoff processes and the material transport into the groundwater and into flowing water. Water balances are being modeled which are indispensible for verification of the dynamic models.
Finally, division Z of the program is responsible for central services and includes computer facilities for data collection and data organization, mini- and microcomputers with an digital image analysis system, the central laboratory, the secretarial office, and the central information bureau for scientific exchange.

DIVISION A: MODELING SYSTEM WATER AND MATTER DYNAMICS OF AGRO-ECOSYSTEMS

Subproject A1: Integrated Site and Catchment Models of the Dynamics of Water and Matter

Within the Special Cooperative Program 179 the subproject A1 is working at site and catchment models of a high degree of integration.
The other subprojects are concerned with analytical and experimental investigations into single processes of energy, water, and matter dynamics in agro-ecosystems. Their objective is a mathematical description of these processes by individual dynamic models which should be as deterministic (physically, chemically and biologically based) as possible.
The task of the subproject A1 is to combine the individual models to model systems. Integrated site models will be used for the most detailed description of the agro-ecological process-system. They refer to small areas with little spatial variation in the plant cover, the microclimate, and the soil properties.
In close cooperation with the other subprojects the integrated site models will be used for the analysis of agro-ecosystems and for predictive simulations of selected scenarios. The investigation will focus on short, medium, and long-term effects of climatc changes and anthropogenic interventions on the turn-over rates of matter of agro-ecosystems. This depends especially on agricultural practices and intensities wich influence the turn-over rates by, for example, the accumulation or reduction of organic matter content, the soil loosening or campaction, soil erosion or leaching, and surface sealing.
The consequences of environmental or anthropogenic effects in respect of the turn-over rates of catchment areas or of whole landscapes will be studied by integrated catchment models.
The models are designed to bridge the gap between the ecological research at individual sites and catchment hydrology.

Subproject A3: Models of Nutrient Transport and Chemodynamics

The aim of this subproject is the quantitative description of the nutrient ion balance in agricultural soils as a function of cropping practices and intensities predominantly by means of deterministic model approaches. In particular the physical interactions and immobilisation and mobilisation processes as well as the ion uptake by plant roots are to be considered. Single ion models, which will first be developed for the one-dimensional case with a horizon-related vertical differentiation, will be combined to a comprehensive model approach under the condition of a balanced charge in depth and time. Furtheron, it is planned to develop an approach for calculating the H^+ and HCO_3^- production in the soil based on a complete estimation of the ion and matter fluxes between the compartments of the ecosystem. The experimental part of the subproject includes the measurements of the parameters and a preliminary calibration of the ion-models by means of undisturbed soil column and field experiments.

Subproject A4: Investigation and Modeling of Agrometeorological Structures and Processes which Influence the Bugdet of Water and Matter

The contribution of the agrometeorological subproject to the research program "Water and Matter Dynamics in Agro-ecosystems" is above all the specification of evapotranspiration and rainfall interception. Evaporation and transpiration must be determined separately in order to differentiate the water and material transport to the plant roots as well as to the soil surface. The close coupling of the water and heat budgets makes it possible to include the energy budgets of the canopy and the soil in the models.

The main objectives will be reached by developing the existing models; in particular, the determination of the micrometeorological conditions inside different canopies including the soil temperature and a wider range of different soils and crops. Moreover, the possibilities of extrapolating the meteorological boundary conditions for the models have to be expanded, including advective effects.

Statistical analyses of the existing climatological data should result in further extrapolations of the meteorological boundary conditions and enable the evaluation of ecologically effective combinations of meteorological parameters.

Subproject A5: Analysis and Modeling of the Processes of Soil Erosion and the Matter Transport of Surface Runoff

The soil erosion leads to a considerable decline of the crops, high nutrient losses, and a far-reaching pollution of the surface water. To

minimize erosion damages successfully, it is necessary to obtain sufficient knowledge about soil erosion processes. The main objects of the research in sub-project A5, therefore, are a) the analysis of soil erosion processes as the detachment, transport and accumulation, b) the description of processes using transferable physical and chemical algorithms, and c) the validation and application of the models developed.
Special emphasis lies on the verification of the dynamic models for the description of the aggregate stability, the soil particles removed by splash and surface runoff, the transport and accumulation of eroded material, the changes in the soil surface as a result of those processes as well as the changes in the transport of nutrients and pollutants due to soil erosion.
The boundary conditions, the system variables and the model parameters are determined by measurements in the field and in the laboratory.

Subproject A6: Analysis and Modeling of Infiltration, Soil Water, Soil Air, and Hydrodynamic Dispersion

Subproject A6 is developing sub-models of water and matter transport in soil and groundwater. The parameters relevant to the transport processes of water and dissolved components in porous media (K,pF,D) will be determined by field and laboratory experiments. The temporal development of the state variables soil moisture potential, water content, and - in cooporation with sub-project A7 - the ion concentrations will be measured quasi continuously at a number of representative sites.
Various phenomena occuring in agricultural soils, e.g. temporal change in pore volume, retardation of infiltration by surface crusting, or soil compaction as well as the influences of hysteresis, soil air compression or inclosure, inhibition of wetting, etc., on the transport processes will be analyzed and used to improve the existing models.
To enable the transfer of site-related data to a catchment, the spatial variability of soil-physical properties will be studied and methods of parameterization will be developed. Comparative laboratory and field investigations are planned to obtain suitable algorithms for the derivation of soil-physical parameters from basic soil data in cooperation with subproject B2.

Subproject A7: Nutrient Budgets of Agricultural Catchments

The subproject A7 "Nutrient Budgets of Agricultural Catchments" intends to evaluate the input (by air, precipation and fertilization), the transport and turnover (by fixation, translocation, and plant uptake) and the output (by plant removals and in dissolved or gaseous forms) of the most important mineral nutrients depending on location and cropping intensity. This requires an extensive sampling and analytical program

the results of which are supposed to be representative for the various parameters and variables which are essential in modeling agro-ecological situations.
For nitrogen, being the ecologically most critical nutrient, we have developed the corresponding models. They must be improved to quantify the management-induced losses by leaching and denitrification. With regard to nitrogen and phosporus this also includes an investigation of the influence of farming on the surface water quality. For this purpose measurements of the nutrient exports at the outlets are to be supplemented by corresponding surveys at the conjunction of individual drains from differently cropped sub-areas and of the nutrient turnover within the water.

Subproject A9: The Dynamics of Herbicides in Soils
The activity of the herbicides applied for weed control remains for a certain period depending on environmental conditions. Under unfavourable circumstances residual herbicide activity can cause side effects to microorganisms, soil fauna, following crops e.g. in an intensive crop rotation. To assess a particular residue situation in the soil correctly it is necessary to have sufficient knowledge of the dynamics of herbicides in the soil. Sorption, microbial and abiotic degradation, volatilization, leaching, as well as uptake and breakdown by plants and other organisms are important processes, which can occur simultaneously and influence each other more or less intensively. Within a long-term research program the influence of different cropping intensities on residue situations should be investigated in regard to soil and groundwater protection.

DIVISION B: REGIONALIZATION/REMOTE SENSING
Subproject B1: Central Working Group Regionalization/Remote Sensing
It is the task of this subproject to gain parameters required for catchment models. The sampling methods have to be automated in the long run. The data of the spatially differentiated boundary conditions and parameters will be obtained by regionalization and remote sensing techniques. The data will be used in dynamic models of the water and nutrient fluxes. It will be organized in a Geographical Information System.
In cooperation with the related subprojects subproject B1 will develop interpolation algorithms to regionalize the wind pattern and numerous other parameters in soil physics, soil chemistry and biology. A Digital Terrain Model and a Digital Sediment Model will be worked out which will be used for correlative interpolations.
Many parameters, for example LAI, radiation temperatures, soil moisture,

areas of erosion and accumulation are planned to be mapped using remote sensing techniques which work operationally (e.g. Landsat) as well as experimentally (e.g. radar-flights).

Subproject B2: Soil Properties, their Lateral and Vertical Variability, which Influence the Water and Matter Dynamics in Soils

Subproject B2 is concentrating on the investigation and processing of basic pedological data (field and laboratory data) in order to recognize and describe vertical and lateral distribution patterns. The results will be based on field and laboratory data and will consider the typical variation of the different pedological parameters, to achieve a more objective and systematic method of describing soil units and defining the boundaries between them.

These basic soil data will provide regionalization parameters for the cartographic presentation of the results which have been derived with help of models within the framework of the Special Cooperative Program.

Subproject B3: Segmentation of Remote Sensing Images by Means of Texture Analysis

Digitalized aerial images are investigated by means of texture analysis and region growing methods, some of which are taken from classical image processing techniques, and some of which have been newly developed. The aim is to extract the contours of terrain regions featuring a homogeneous structure, as for instance: soil erosion, accumulation, damaged fields. Depending upon the image content, statistical texture models or region growing techniques are used. Texture analysis aims at extracting the borders of regions characterized by a statistical local homogeneity feature. On the other hand, the region growing approach is more effective in cases where differences between regions are more related to the structure of their primitive components than to the grey level distributions. These primitives can be merged by means of a region growth process, followed by a simple contour extraction algorithm.

Subproject B4: Expert Systems and Land Information Systems

The project will develop expert systems based on a deductive data base for earthscience applications.

The expert systems will contain, divided in 3 layers: facts, knowledge and rules for deduction and will provide operations for insertion, updating and querying. Facts, gathered within the Special Cooperative Program, will be stored in a data base; knowledge, known and new found relationships, will be kept by a knowledge base and rules for application and deduction of knowledge and facts will be managed by a metarule base. In addition to maintenance of knowlegde, special components for knowledge based problem solving, generation and verification of complex models and qualitative and quantitative data analysis will be included in the expert systems.

DIVISION C: BASIC STRUCTURES AND PROCESSES

Subproject C1: Soil Structure and Transport Properties: Soils as Correlated Random Media

The transport of water as well as the ensuring "Hydrodynamic Dispersion" of solutes in soil depends on soil structure. Starting from existing information on soil genesis and measureable soil properties, we set out to develop a statistical description of the network of connected soil pores on different length scales ("Scaling Model" of pore diameter and pore connectivity correlations). From such a description, the transport properties can be understood, categorized and calculated by known methods of Statistical Physics, including numerical simulation techniques. This method of typization should also have an impact on the design and optimization of experimental "strategies" (to obtain good estimates of the model parameters at minimum cost).

Subproject C4: Gasexchange at the Soil/Air-Interface

The dynamics of matter in agro-ecosystems are partly defined by the vapour phase transport at the soil/air-interfaces.This project centers around research on the dynamics of N_2O. In cooperation with project A7 we want to investigate the influence of soil parameters such as soil moisture, soil temperature and soil structure on the chemodynamics and the transport of N_2O.
For this purpose we developed a chamber by which it is possible to measure the gasexchange in undisturbed soil columns in the laboratory. Furtheron, we plan to extend our research on other gases relevant to agro-ecosystems (CO, CO_2, NO_x,...).Our measuring system allows also to investigate the deposition velocities of these gases.

DIVISION D: BIOGENIC MATTER DYNAMICS AND BIOINDICATORS

Subproject D1: Investigations of the Influence of Different Pathotops as well as the Variability of Susceptibility of Cereals on the Spread of Powdery Mildew and Rust in Agro-Ecosystems

With the aid of powdery mildew and rust on cereals the pathotop-hypothesis is checked. In these experiments the conditions are investigated that are important for the development of the mentioned pathogens within special habitats. The relations to microclimate, the water and nutrient budget of the soil and the host plants are considered. In addition it is investigated in which way the forecasting with models concerning the spreading of the pathogens in agro-ecosystems can be made more accurately by incorporating data of the disposition of their hosts.

Subproject D2: Soil Biology - Earthworms and Collembola

Among soil animals earthworms and Collembola are choosen, since they are important for turnover processes in the soil. Important physical and chemical parameters are modified in order to screen their influences on the population dynamics of both taxa. It shall be tested, wether special species can be used as bio-indicators for soil humidity and soil pollution. Further, we want to investigate other important problemes, such as: influence of earthworms on decomposition of organic material in the soil; the importance and function of bio-pores; testing of toxic substances in the laboratory.

Subproject D3: Influences of Different Agricultural Cropping Intensities on Microorganisms in Soils and their Activities

Microorganisms contribute decisively to important transformation processes in the soil and also to the soil ecosystem. Within a special long-term research program at two ecologically different sites the influence of different cropping intensities (e.g. fertilization, pesticide use, plowing) on soil microorganisms shall be investigated. Of special importance are the reactions of the microbial biomass (ATP content, respiration) and some microbial activities (straw decomposition, respiration, dehydrogenase, phosphatase) to agriculture. During the whole growing season the upper soil layer (0-30cm) must be investigated regularly.

DIVISION E: RUNOFF AND GROUNDWATER

Subproject E1: Analysis and Modeling of Runoff Processes in Rural Catchments and Total Water Balances

The project comprises the hydrologic components precipitation and runoff and the water balance of the catchments. Further the runoff processes and influences of man on the runoff are under study.

Important objectives of the hydrological investigations are the generation, the progress and variation of runoff as a result of agricultural land reclamation and channel improvement and its dependency from soil, topography and weather. Considering the relationship between the quantitiy and quality of water, the research program includes measurements of runoff and precipitation in the catchment areas, experimental investigations on plots with rainfall simulators, analysis of data and further development of continuous, space-time differentiated models for surface runoff, subsurface runoff from drains or interflow respectively and the channel flow in agro-ecosystems.

Subproject E2: Investigation of Chemodynamics and Transport Processes in the Groundwater of Agro-Ecosystems

The agricultural use of land influences the groundwater resources and its chemical state, depending upon the conditions present. The interactions between the transport and dilution processes and the turn-over of nutrients, especially nitrogen in its various combinations, are relatively well known in the upper unsaturated zone. On the other hand, reaction mechanisms and nutrient decomposition processes in the deeper unsaturated zone, the capillary seam and the groundwater itself have, up to now, hardly ever been the object of extensive examinations. In the course of this project, these conversion and material transport processes are to be examined in pleistocene pore aquifers which lie below intensively fertilized fields.

AUTOMATICAL CONTROL STRATEGY OF NON-TECHNICAL SYSTEMS IN THE GROUNDWATER ZONE

Gräber, P.-W. [1)]

In Central Europe, in particular in the German Democratic Republic, soil and groundwater zone are intensive and repeatedly used.

Today about 80 per cent of the trinking water is pumped out the groundwater, this are about 1,5 billion cubicmeters per age.
To pump about six cubicmeters per tonne lignite in the open-cast lignite mines of the G.D.R. are necessary. That are 1,5 billion cubicmeters per age and this is more than the water-delivery of the watermanagement in the G.D.R. About 17 per cent of the costs and 30 per cent of the energy demand of the open-cast mine are needed for mine drainage.

Also the industrial farming and animal production prodece complicated conditions for the groundwater. In this connection is to be mentioned the large water removal out the circuit of the repeated using by the evapotranspiration. This consumption amounts to about 1,5 billion cubicmeter per age.
Usally hierarchly system conception is applyed to the research of the processes in the soil and groundwater zone. This hierarchy is constructed by the territorial pattern. The technical devices, e.g. wells drainages and so on generate the lowest level. Next steps in the hierarchy are the objects, the regions, the territorials, the countries and the continent.
This hierarchy demonstrates very good, that each system of a level is an element of level following next and each element of a level is a system of the lower level.
This hierarchy system is characterized by, that the grade of the detail and the grade of the determination of the information decrease from the lower to the higher level.
The aggregation, the complexitation and the time horizont increase from the lower to the higher level. In the same way the number of destination functions and the number of users, which influence on the same system are increasing. That means, that the human influence also is increased from the lower to the higher level.
Different software methods are used from the adequate different levels. In the higher level the games and the expert systems dominat. Near by the processes the computer aided monitoring and the computer aided manufactoring are used.

[1)] Dresden University of Technology, Water Science Devision

The non-technical systemes of the soil and groundwater zone are characterized

- by the complexity and the strong non-linearity as the "large-scale systems"
- by the bad condition and by the very different time constant, as the "stiff systems"
- by the non-exactly determination, as the "fuzzy systems" and
- by the reference input of the social, economical, ecological, biological or political type.

Nevertheless a closed loop by the automatical control is found for the soil and groundwater zone. Here the roll of the man is not important. The type of the loop can be on-line or off-line.
This control loop consists of the process in the soil and groundwater zone, of the data base, of the computer simulation with included animated pictures and of the control devices.
The reference inputs may be

- of physical kind, e.g. water level, flow, temperatur
- of chemical kind, e.g. conductivity, pollution, concentration, pH-value
- of social, economical, ecological and political kind, e.g. environment, cost of energy.

If in a hierarchy level several different rivaled reference inputs exist, than the expert have to find a optimum, because this is not automatical to be found.
Meteorological phenomenons as well as effects of changes the groundwater zone appears as a load variable. By means of the large-scale response time the load variable must be measured before they strike at the groundwater zone. This feedforward control improves the conduct of the control.
The data base consists of the historical and of the actual datas of the groundwater layers. The historical datas are storaged in a data bank and the actual datas of the state variable and of the parameters are measured. The parameters have non-static character, especially in mining area. On the one hand, continuous variation of the data by a continuous improvement of the model by geological explorations is necessarily, on the other hand the mining process itself couses a continuous change in the soil parameters (native rock is changed into dumps). Therefore new automatical measurement methods are used in laboratorities and in field. In the Dresden University of Technology, Water Science Devision a new type of laboratory devices is built. This concept consists on a two-level local area computer network. In the high level a 16-bit-PC exists, which is used for the experiment planing, monitoring and interpretation. In the low level several different stand-alone devices are arranged.

Such devices are e.g. particel analysor, batch devices, retention measurement. In these devices a 8-bit-real-time-processor is integrated. This makes the automatical control of the measurement processes and the handing over of the measure values on the high level.
Also a know-ledge base and the reference inputs are contained in the data base. The know-ledge representations are assembled by the huma experience and by the technical laws, so-called standards. The representation take place by means a relational data base concept.
Adequate to the territorial hierarchy and to the different legal claim of this area many different data bases exist. The problem specifical data base is built.
For the computer simulation very different methods are used. The palette reaches from the determinate to the stochastical methods, from the continuous to the discontinuous methods. Subsequent methods are used: analytical solution, integral transformations, electro-analogous models, fininte-element-method, finite-difference-method, discrete integral transformation and the boundary-elements-method. Each computer simulation method requires an adaption from the mathematical model of the original to the mathematical model of simulator. The mathematical model of the groundwater flow and of the migration process, the quality (head, pollution, conductivity) dispersion and convection, consist of two coupled non-linear partial differential equations of parabolic or elliptic types.
The technology of the animated pictures is coupled on the computer simulation and on the data base. Here the display of the pre-processing, e.g. the graphic data input, the numerical input or the choice by means the menue, the selection of representative values and the coupling of the data base are realized by means the graphic methods.
The two- and tridimensal pattern of the water levels, of the water-time-diagram and of the geohydrological cross-sections are the content of the post-processing. The operator at the monitor obtains the possibility to change these pictures, to choice and to move the segments. This new created picture is the base of new input data and of a new simulation run.
The practical realization of the automatical control strategy use a large palette of the computers hard- and software. The scale begins at the main-frame computers with a special array precessor, go over the 32-bit-mini-computers over the 32-bit-supermicro-computers to the 16-bit-workstations and 16-bit-personal-computers. The peripherical devises play the most decisive roll, indepentend of the central processor unit. For the realization of the animated pictures technology the graphic devices, e.g. graphic workstation, digitizer, plotters and the lokal area networks with computers of different power are of especally signification.

On the area of the software engineering, especally the softwaretools from the software-houses and the application software, which is made by groundwater engineers, are used. The application software is written for the arithmetical parts by means of the programming language FORTRAN 77, for the graphic parts by means of the graphic tools GKS (graphic kernel system). The softwaretool dBASE III or IV realizes parts of the data input and the relational data management. The users quide is written by the methods of the expert systems by means of the language PROLOG.

References

/1/ Gräber, P.-W.: Einsatzmöglichkeiten von Mikrorechnern bei der Informationsgewinnung und -verarbeitung im System "Unterirdisches Wasser". Wiss. Zeitschr. der TU Dresden 30 (1981) 1, S.105-109.

/2/ Gutt, B., F. Reichel, and P.-W. Gräber: Computer-aided modelling of groundwater dynamics in mining areas. Int. Symposium on Groundwater Monitoring and Management, Proc. Complex V, Dresden 1987.

/3/ Beims, U., und P.-W. Gräber: CAD-Arbeitsplatz "Grundwasserabsenkung". Int. Symposium on Groundwater Monitoring and Management, Proc. Complex E, Dresden 1987.

/4/ Luckner, L., und W. Schestakow: Simulation der Geofiltration. VEB Deutscher Verlag für Grundstoffindustrie, Leipzig 1975.

/5/ Luckner, L., und W. Schestakow: Migrationsprozesse im Boden- und Grundwasserbereich. VEB Deutscher Verlag für Grundstoffindustrie, Leipzig 1986.

/6/ Kaden, St., et all: Decision Support Systems for the Analysis of Regional Water Policies. Working Paper, IIASA, Laxenburg 1986.

MODELLING AND FORECASTING OF RIVER FLOWS
USING STOCHASTIC MODELS

Th.D. Popescu [1], S. Demetriu [2], V. Marinescu [2]

1. Introduction

The river flow process is a very complex process that involves a number of primary as well as auxiliary variables such as rainfall, groundwater and geomorphologic characteristics of the basin. Because of the complicated nature of the process, it is hard to quantify the contribution of different factors on the flow. In this situation modelling of the river flow process by a stochastic model is recommended.

The paper presents several aspects of modelling and forecasting for river flow series by stochastic difference equations. The first part of the paper is dedicated to the fitting of multiplicative seasonal stochastic model ARIMA for the monthly flows of the river Bistriţa, using a fully interactive software package offering complete single - equation Box-Jenkins time series modelling and forecasting capabilities. The last part considers single and multiple - step - ahead forecasting for the river flow sequence analysed. Such forecasts are needed in the operation of reservoirs, agricultural planning, and the like.

2. The Box-Jenkins modelling procedure

Consider a seasonal time series realisation of length N, denoted by z_t. According to Box and Jenkins [1] we can explain its movements - in the most fundamental way - by a combination of past movements in the series itself and a white noise, i.e. a series of identically distributed uncorrelated deviates :

$$(1-B^s)^D(1-B)^d(1-\varphi_1 B - \dots - \varphi_p B^p)(1-\Phi_1 B^s - \dots - \Phi_P B^{Ps})\, z_t =$$
$$= (1-\theta_1 B - \dots - \theta_q B^q)(1-\Theta_1 B^s - \dots - \Theta_Q B^{Qs})\, a_t + \theta_0 \qquad (1)$$

The model (1) is referred as the general multiplicative seasonal Autoregressive Integrated Moving Average (ARIMA) model of order $(p,d,q)(P,D,Q)_s$ in the regular and seasonal terms, respectively.

Box and Jenkins propose a practical three-stage procedure for finding a such of model: identification, estimation and diagnostic checking. A good model includes the smallest number of estimated parameters needed to adequately fit the patterns in the available data.

In the identification stage, using an estimated autocorrelation function (acf) and an estimated partial autocorrelation function (pacf) is choose one or more models as candidates (values of s, p, d, q, P, D and Q in equation (1) and preliminary estimates for model parameters; the default values are set to 0.1 for all parameters). In the second phase, the parameters of the model (s), chosen in the first phase, are estimated by means of non-linear least squares method discussed in [1]. Finally, in the third phase, the candidate model is checked for adequacy. The estimation and diagnostic - checking stages provide warning signals telling us when, and how, a model should be reformulated. We continue to reidentify, reestimate, and recheck until we find a model that is satisfactory according to several criteria. This iterative application of the three stages does not guarantee that we will finally arrived at the best possible model, but it stacks the cards in our favor.

---

1) Research Institute for Computer Technique and Informatics, Bd. Miciurin 8-10, 71316 Bucharest, Romania.

2) Faculty of Civil Engineering, Bd. Lacul Tei 124, 72302 Bucharest, Romania.

3. Case study

In this case we study the Bistrița river flows (m^3/s), using 120 monthly observations covering the years 1971-1980, a nonstationary and seasonal time series. All data analysis presented in this case study was carried out using the TSYST program package, a new version of AUTOB&J program package [2]. It is implemented on a minicomputer PDP11/34 under RSX operating system.

Identification. The first step in identification is to look at the plot of data (see Fig.1). Two features of this series stand out: (i) the overall level modifies in time, suggesting that the mean is not stationary and that nonseasonal differencing is required; (ii) there is an obvious seasonal pattern within each year.

Fig.2 is the estimated acf and pacf of the data. The slow decay of the acf along the first lags shows that the mean of the data is nonstationary; calculating nonseasonal first differences is appropiate. The slow decay of the autocorrelations at lags 12 and 24 suggest that seasonal differencing is needed.

Fig.3 is the estimated acf and pacf after both seasonal and nonseasonal first differencing. It is now fairly easy to identify an appropiate model. The spike of acf at lag 1 suggests a nonseasonal MA term, and the spike at lag 12 calls for a seasonal MA coefficient. The spike of pacf at lag 1 suggests a nonseasonal AR term (in the first step we ignore the spike at lag 3), and the spike at lag 12 calls for a seasonal AR coefficient. In this case resulted a model $(1,1,1)(1,1,1)_{12}$ for the original time series :

$$(1-B)(1-B^{12})(1-\varphi_1 B)(1-\Phi_1 B^{12}) = (1-\theta_1 B)(1-\Theta_1 B^{12}) \qquad (2)$$

Estimation. Estimation results for default values of model parameters are shown in Table 1.

Table 1. Estimation results for model (1)

Coefficient	Estimate	Std.Error	T - Value
Phi 1	0.405	0.032	12.706
Phi*12	-0.158	0.128	-1.230
Theta 1	0.982	0.025	39.280
Theta*12	0.656	0.102	6.444

Adjusted RMSE = 790.331

Correlations

	1	2	3	4
1	1.000			
2	-0.460	1.000		
3	-0.826	-0.112	1.000	
4	-0.155	0.542	-0.1582	1.000

Three estimated coefficients have absolute t - values exceeding 2.0 and stationarity and invertibility conditions are satisfied. Also, the correlation matrix point out evidence that the coefficient estimates are not highly correlated, therefore, the estimated model is satisfactory in this regard [3].

Diagnostic checking. The residual acf for the above model appears in Fig.4. The residual acf point out a significant spike at lag 3. It suggests that the model is not adequate because the residuals are significantly correlated. This call for a return to the identification stage.

Further identification. We return to the acf and pacf of the differenced original series (Fig.3) and choose for the original data a model of the form :

$$(1-B)(1-B^{12})(1-\varphi_1 B-\varphi_3 B^3)(1-\Phi_1 B^{12}) = (1-\theta_1 B)(1-\Theta_1 B^{12}) \qquad (3)$$

Further estimation and checking. Estimation results for model (3) are shown in Table 2.

Table 2. Estimation results for model (3)

Coefficient	Estimate	Std.Error	T - Value
Phi 1	0.215	0.103	2.082
Phi 3	-0.227	0.040	-5.709
Phi*12	-0.322	0.143	-2.261
Theta 1	0.985	0.010	98.543
Theta*12	0.478	0.138	3.465

Adjusted RMSE = 769.845

Correlations

	1	2	3	4	5
1	1.000				
2	-0.437	1.000			
3	0.185	-0.261	1.000		
4	-0.043	-0.822	-0.147	1.000	
5	0.207	-0.121	0.685	-0.201	1.000

All estimated coefficients have absolute t - values exceeding 2.0 and stationarity and invertibility conditions are satisfied. Also, estimated coefficients are not too highly correlated as shown by the correlation matrix. The residual autocorrelation at lag 3 in Fig.5 is now much smaller than it was in Fig.4. We conclude that the final model is a good model for the analysed time series.

Forecasting. Forecasts for lead times l = 1,2,...,12 from origin t = 120(december 1980) appear in Table 3.

Table 3. Forecasts from model (3)

Time		Forecast Values	95 % Confidence Limits	
			Lower	Upper
1981	1	29.6478	- (*	84.7032
	2	30.2020	-	88.6040
	3	49.7581	-	108.3788
	4	99.2733	40.6474	157.8993
	5	115.1540	56.4432	173.8648
	6	96.4222	37.5673	155.2772
	7	81.5605	22.5333	140.5877
	8	81.0633	21.8521	140.2746
	9	61.6431	2.2433	121.0429
	10	44.9447	-	104.5343
	11	38.8996	-	98.6795
	12	30.9579	-	90.9277

* negative values.

Model (3) forecasts rather well. The forecasts faithfully track the strong seasonal pattern.

4. Literature

[1]. Box, G.E.P., and G.M. Jenkins: Time Series Analysis: Forecasting and Control, Holden Day, San Francisco 1976.

[2]. Popescu, Th.D.: AUTOB&J - A Computer Aided Procedure for Time Series Analysis and Forecasting, Proc. 3 rd IFAC Symp. on Computer Aided Design in Control and Engineering Systems, July 31 - August 2, Lyngby 1985.

[3]. Pankratz, A.: Forecasting with Univariate Box-Jenkins Models, John Wiley & Sons, New York 1983.

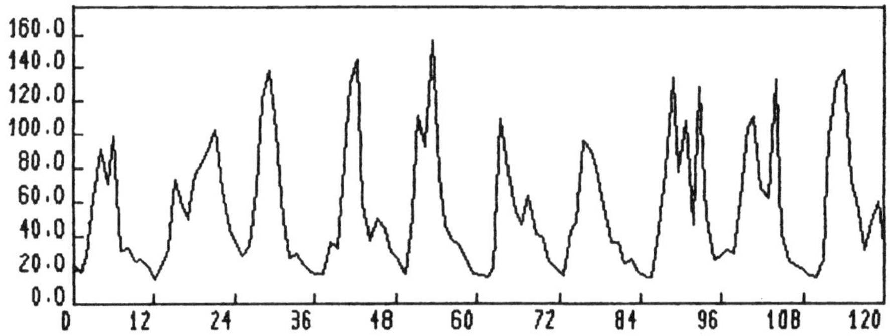

Fig. 1 Bistrita river flows (m^3/s) 1971-1980

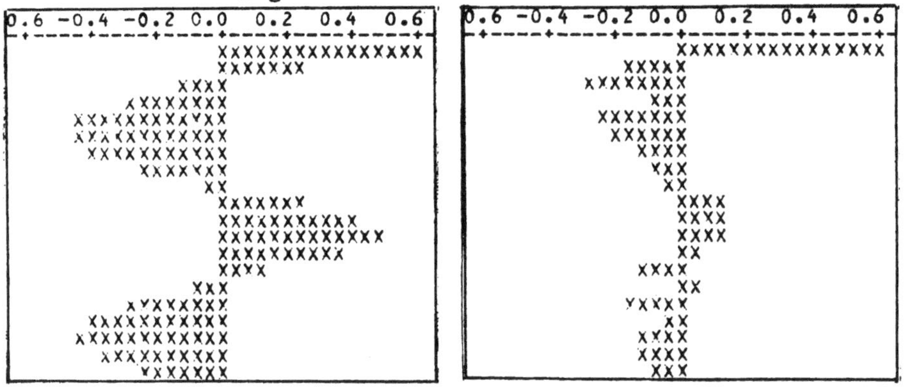

Fig. 2 Estimated acf and pacf for original series

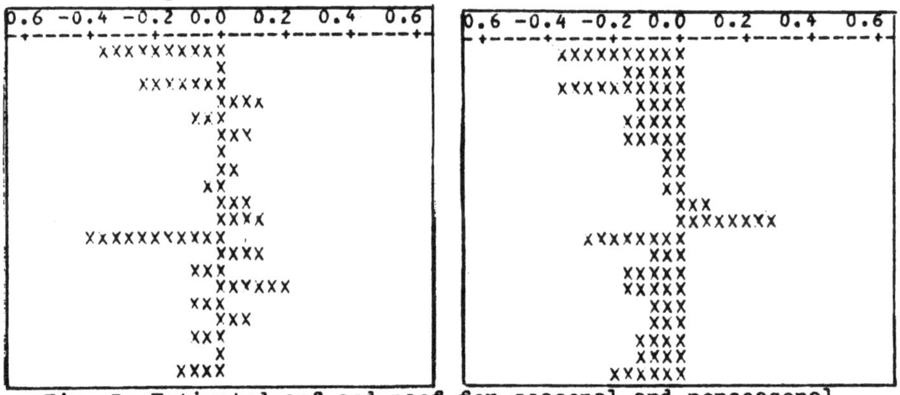

Fig. 3 Estimated acf and pacf for seasonal and nonseasonal first differencing

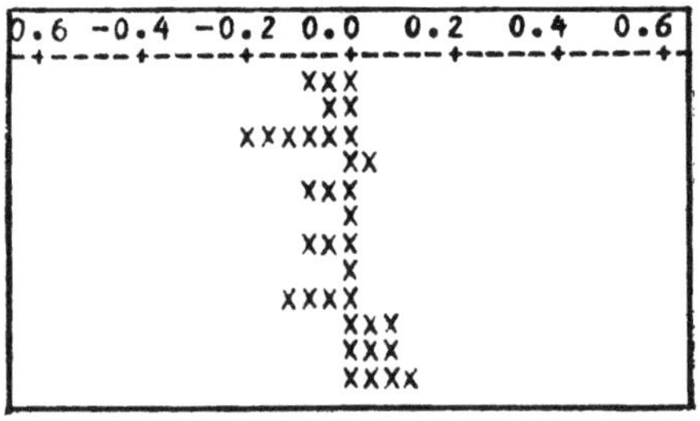

Fig. 4 Residual acf for model (2)

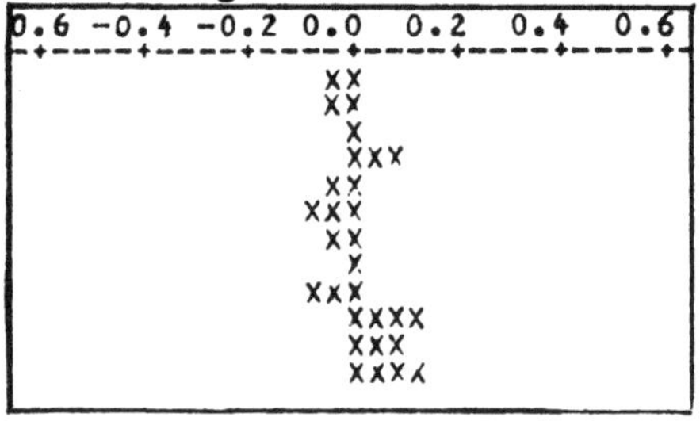

Fig. 5 Residual acf for model (3)

APPLICATION OF TRAJECTORY DECOMPOSITION METHOD OF THE WATER QUALITY CONTROL PROBLEM

Mirosław Bereziński, Grażyna Petryczek

Systems Research Institute
Polish Academy of Sciences
Newelska 6, 01 - 447 Warszawa, Poland

This paper deals with the problem of determining optimal location of waste discharge points along the whole length of the river and with optimization of waste load quantities discharged into the river. Optimization is made in order to ensure a required quality of river-water.

The steady-state conditions are assumed, i.e. both model variables and model parameters are considered to be constant in a given period of time although they can change their values down the river. Here we present a model describing the water quality control problem. The model is formulated in terms of dynamic optimization and has two-level hierarchical structure. Coordination of local solutions corresponding to the particular river sections consists in prediction of the quantities of pollutant loads discharging at the initial point of each river section. The model is solved by means of both modified and ordinary trajectory decomposition method.

1. SIMPLIFIED MODEL OF WATER QUALITY CONTROL PROBLEM

The underlying equations, which are common to most mathematical models used to solve water quality control problems are those expressing mass balance and continuity principles. In fact, the typical model which describes both transportation and dilution processes as well as biochemical reactions affecting the pollutants in rivers-water has the following form (see, for instance, Biskwas, A.K. (1981), Thomann R.V. (1972), Young, Beck (1974)

$$\frac{\partial C}{\partial x} = \frac{\partial}{\partial x}\left(E\frac{\partial C}{\partial x}\right) - \frac{\partial}{\partial x}\left(\frac{Q}{A}C\right) + f(C) + \sum_k S_k \tag{1}$$

$$C(0,t) = C_o(t)$$

where: C - pollutant concentration $[kg/m^3]$

E - longitudinal dispersion coefficient $[m^2/s]$

Q - flow $[m^3/s]$

A - cross-sectional area at location x $[m^2]$

S_k - sum of sinks and sources of pollutants $[kg/m^3s]$

x - downstream distance as measured from discharge point $[m]$

t - time $[s]$

$f(C)$ - function describing chamical and biochemical processes.

Equation (1) constitutes one-dimensional steady-state model in which only longitudinal diffusion is reflected. In order to obtain the model better fitted to reality and more suitable for solving water quality control problems let us make the following assumptions:

A1) The influence of longitudinal dispersion is relatively small and may be neglected.

A2) A river can be devided into a set of N various-length sections.

The section initial-points are situated into:

- the pollutant discharge points,
- inlets and outlets of the river,
- the stations of water intake, and
- into the points in which hydraulic parameters of the river may change.

A3) Biochemical reactions are of the first type, i.e.

$$f(C) = k_i C_i, \tag{2}$$

where k_i - biochemical reaction coefficient.

A4) Both superficial flow and ground-waters are considered to be pollutant sources and to be distributed down each section.

A5) Model corresponds to the steady-state situation.

A6) For the i-th section, i=1,...,N, both cross-sectional area A_i and first-order biochemical reaction coefficient k_i are constant.

Under assumptions A1-A6, the model describing pollutant distribution into the i-th river-section may be written in the form:

$$\frac{dC_i}{dx} = -\frac{C_i}{Q_i}\frac{dQ_i}{dx} - \frac{A_i}{Q_i}k_i C_i + \frac{S_i}{Q_i}\left(\frac{dQ_i}{dx}\right), \tag{3.a}$$

$$x \in [x_{oi}, x_{fi}]; \; i=1,\ldots,N; \quad x_{fi} - x_{oi} = L_i,$$

where: S_i - concentration of pollutants originating from superfical flow,
x_{oi} - initial point of the i-th section,
x_{fi} - final point of the i-th section,
L_i - length of the i-th section,
N - number of sections, i.e. i=1,...,N.

The boundary conditions for the i-th section are given by the following water balance equation:

$$Q_i(x_{oi})C_i(x_{oi}) = Q_{i-1}(x_{f,i-1})C_{i-1}(x_{f,i-1}) + \zeta_i, \; i=1,\ldots,N \tag{3.b}$$

where ζ_i - waste load discharged at the initial points of the i-th section.

The model is completed with the equation:

$$x_{oi} = x_{f,i-1}, \quad i=1,\ldots,N. \tag{3.c}$$

For the pollutant decay model, given by the equations (3.a)-(3.c), we formulate the following problem. Let ζ_i (quantity of the waste load discharged at the point i) and S_i (distributed waste loads) be decision variables of the model and consider the problem of finding such their values which ensure desirable quality of the river-water. In this case, because location of the waste discharge points is know the whole river is divided into segments and $x_{oi}=x_{i-1}$, $x_{fi}=x_i$. The problem can be formulated in an equivalent mathematical form as follows (Problem 1):

Find the values of decision variables S and ζ minimizing the quality functional:

$$J(S,\zeta,C) = \frac{1}{2}\int_{x_1}^{x_N} \|(C_i(x)-\bar{C})_+\|^2 dx + \kappa(S,\zeta) \qquad (4.a)$$

subject to equations (3.a)-(3.c).

Equation (3.b) implies that the concentration trajectory can be decomposed into a set of arcs. In this case we can obtain the functional $J(.,.,.)$ in a slightly different form

$$J(S,\zeta,C) = \sum_{i=1}^{N} \{ \frac{1}{2} \int_{x_{i-1}}^{x_i} \|(C_i(x)-\bar{C})_+\|^2 dx + \kappa_i(S_i,\zeta_i)\} \qquad (4.b)$$

$$x \in [x_{i-1}, x_i]$$

where: x_1, x_N - initial and final points of the river under consideration, i.e. $L = x_f - x_o$ - length of the river,

x_{i-1} - initial points of the i-th segments,

x_i - final points of the i-th segments,

$\bar{C}$ - permissible level of concentration of the given sort of wase.

Consider now the situation when location of the waste discharge points is not known in advance. In this case one should find both expected quantities of waste loads and location of the waste discharge points. It results in the following Problem 2:

$$\min\{\tilde{J}(S,\zeta,C) = \sum_{i=1}^{N} \{ \int_{x_{oi}}^{x_{fi}} \frac{1}{2} \|(C_i(x)-\bar{C})_+\|_F^2 dx + \kappa_i(S_i,\zeta_i)\}\} \qquad (5.a)$$

subject to equations (3.a)-(3.c). In this problem the value of x_{oi} are not known.

2. APPLICATION OF THE TRAJECTORY DECOMPOSITION METHOD TO PROBLEMS 1 AND 2

It follows from the above considerations that the model of pollutant distribution in a river can be subdivided into N independent submodels corresponding to particular river section by introducing the interconnection variables given by the mass balance equation (3.b) and interconnection equation (3.c). On the basis of this decomposition, the two-level algorithm of trajectory decomposition (Bauman J.E. (1971)) can be applied to Problem 2.

Let us assume that the number of waste discharge points is known and that one should find optimal allocation of these points. Functional (5.a) represents costs related to ensuring an appropriate quality of water. For given x_{oi} and ζ_i and for each particular river-section $[x_{oi}, x_{fi}]$, we solve the following local problem:

$$\min_{C_i,S_i} \{J_i(S_i C_i) = \frac{1}{2} \int_{x_{oi}}^{x_{fi}} \|(C_i(x)-\bar{C})_+ {}^2 dx + \kappa_i(S_i,\zeta_i), \quad i=1,\dots,N, \qquad (6.a)$$

subject to

$$\frac{dC_i}{dx} = -\frac{\nu_i}{Q_i} C_i - \frac{A_i}{Q_i} k_i C_i + \frac{\nu_i}{Q_i} S_i \tag{6.b}$$

$$\nu_i = \frac{dQ_i}{dx}$$

$$Q_i(x_{fi}) C_i(x_{fi}) = Q_{i+1,o} C_{i+1}(x_{o,i+1}) - \zeta_{i+1} \tag{6.c}$$

$$x_{o,i+1} = x_{fi} \tag{6.d}$$

Here S_i, i.e. ditributed waste load, is treated as control variable, whereas C_i, i.e. pollutant concentration in the i-th section, occurs as state variable. The task of the second level is to predict quantities of waste loads ζ_i diacharged at partition points as well as to find location of partition points (i.e. to find x_{oi}). The procedure for calculating ζ_i and x_{oi} is as follows:

$$(x_{o,i+1})^{l+1} = (x_{o,i+1})^{l} - k\delta x^{l}_{o,i+1} \tag{7.a}$$

$$(\zeta_{i+1})^{l+1} = (\zeta_{i+1})^{l} - k\delta\zeta^{l}_{i+1} \tag{7.b}$$

$$\delta x_{o,i+1} = -\frac{\nu_i}{Q_i(x_{fi})} \cdot \psi_i(x_{fi}) \cdot C_i(x_{fi}) - H_i\Big|_{x=x_{fi}} + H_{i+1}\Big|_{x=x_o,i+1} \tag{7.c}$$

$$\delta\zeta_{i+1} = -\frac{Q_{o,i+1}}{Q_i(x_{fi})} \cdot \psi_i(x_{fi}) \tag{7.d}$$

l - index of the second-level iteration
where

$$H_i = -J_i(C_i,S_i) - \psi_i\left(\frac{\nu_i}{Q_{if}} C_i + \frac{A_i}{Q_i} k_i C_i + \frac{\nu_i}{Q_i} S_i\right), \quad Q_{if} = Q_i(x_{fi})$$

and ψ_i - conjugate variable connected with equation (6.b). Notation $H_i|_{x=x_{fi}}$ and $H_{i+1}|_{x=x_{o,i+1}}$ means that Hamiltonian values are calculated for $x=x_{fi}$ and $x=x_{o,i+1}$ correspondingly. According to the above-presented model, each section is considered independently of the others on the first level of control structure. On the other hand, it is the task of the second level to find such values of the variables ζ_i and x_{oi} that allow to meet all the requirements concerning water quality along the whole length of the river. If the location of waste discharge points is known, the problem becomes simplified and on the second level only quantities ζ_i of waste loads at discharge points are calculated.

Because in the Problem 1 we deal with a fixed partition of the river into a set of sections and since locations of the partition points are known, one can apply modified method of trajectory decomposition. It results in the following two-level structure:

(1) The lower level is related to particular sections $[x_{i-1}, x_i]$, $i=1,\ldots,N$, and solve the following local problems:

$$\min_{S_i} \tilde{L}_i(S_i,\zeta_i,\pi,\rho) \tag{8.a}$$

subject to equation (3.a), where

$$L_i(S_i,\zeta_i,\pi,\xi) = J_i(S_i,C_i)+\pi_i\left[\frac{Q_i(x_i)}{Q_{i+1}(x_i)}C_i(x_i)+\frac{1}{Q_{i+1}(x_i)}\zeta_i - -C_{i+1}(x_i)\right]+\frac{1}{2}\rho\left\|\frac{Q_i(x_i)}{Q_{i+1}(x_i)}C_i(x_i)+\frac{1}{Q_{i+1}(x_i)}\zeta_{i+1}-C_{i+1}(x_i)\right\|^2 \tag{8.b}$$

The task of the lower level is to find (for each section) an optimal value of control variable S_i and corresponding to it value of state variable C_i.

(2) Overall optimal solution one obtains as a result of coordination of local solutions corresponding to particular river-sections. Multipliers π in equation (8.b) and the quantities ζ_i of waste loads discharged at partition points are treated as coordination variables.

The second level determines the optimal values of multiplier π and variables ζ according to the following iteration schemes:

$$\zeta_{i+1}^{l+1}(x_i) = C_{i+1}^{l}(x_i) - \frac{Q_i}{Q_{i+1}}C_i^{l}(x_i) \tag{9.a}$$

$$\pi_i^{l+1} = -\frac{\partial H_{i+1}}{\partial \zeta_{i+1}} - \rho\left(\frac{Q_i}{Q_{i+1}}C_i(x_i)+\frac{1}{Q_{i+1}}\zeta_{i+1}(x_i)-C_{i+1}(x_i)\right) \tag{9.b}$$

where l - index of the second-level iteration.

The iterations (9.a)-(9.b) are obtained from stationarity conditions.

3. CONCLUSIONS

Application of trajectory decomposition method to the problem of finding optimal values of waste loads and locations of discharge points under steady-state conditions leads to two-level problem and gives a clear physical interpretation of the problem.

One should remark, that the waste load quantities ζ (decision variables) are determine on the upper level and are treated as coordination variable. Distributed waste loads S_i and point-type loads ζ_i are assumed to be control variables. The values of control variables are determined on two various levels. Quantities of waste loads S_i connected for instance, with agricultural pollutants are determines independently for each river-section (on the lower level) whereas quantities of waste loads ζ_i connected with industrial and urban sources as well as their locations are calculated on the coordination level.

They have to ensure obtaining nominal level of water quality for the whole river (global task).

The above approach can be used in situations, when the river under consideration belongs to an agricultural region with intensively polluted ground-water. In this case, influence of agricultural pollutants can not be omitted in the model. Given quantity of chemical substances used in agriculture one may control (locally) their concentration in the particular river-section. Given quantity of production generated by particular industrial plants one can look for the optimal policy of pollutant discharging.

REFERENCES

Bauman, J.E.: Trajectory decomposition in D.A. Wismer ed.: Optimization Methods for Large-Scale Systems with Applications. McGraw-Hill, New York 1971.

Biskwas, A.K.: Models for Water Quality Management. McGraw-Hill, Series in: Water Resources and Environmental Engineering (1981).

Gutenbaum, J., A. Łodziński: Dynamic polyoptimal of multistage processes. Proceedings of an IIASA Task Force Meeting (1981).

Petriczek, G., Z. Uhrynowski: Methods for Pollutant Decay Coefficients Determination in River in Decision Models. Proceedings of the ISEM Symposium, Warsaw 1986.

Systems Analysis for Water Quality Management, Proceedings of a Seminar Sponsored by the Regional Office for Europe of the World Health Organization. Budapest 1975.

Thomann, R.V.: Systems Analysis and Water Quality Management. McGraw-Hill, New York 1972.

Young, P., B. Beck: The modelling and control of water quality in a river system. Automatica 10 (1974) 3, 455-460.

COMPUTER AIDED DECISION FOR WATER QUALITY MANAGEMENT

P. Braun[1)]

P. Rudolph[2)] K.-F. Albrecht[2)]

Abstract

The complex system analysis of water quality models is introduced with software support by the system CANDYS. The model is a system of ordinary deterministic or stochastic differential equations with a great number of free parameters. The quantitative validation of the model (parameter fitting) , the qualitative validation (validation of stability, limit cycles, bifurcation etc) and statistical analysis of the stochastic model (time dependent state distribution, statistical parameters, significance of statements etc) is done.

Introduction

For the management of water resources the problems of water quality are of exceptional importance. It is neccessary to describe the qualitative behaviour of water quality processes to make computer studies of the long range influence of control strategies. For modelling such aquatic ecosystems one has to take into account that the real process is an energetically open and stochastically disturbed system of complex non-linear dynamic. This is the reason that only a model with extremly reduced number of states can be designed (the influence of some states is neglegted of others aggregated). But nevertheless the model has to reflect the main qualitative behaviour of the process. Because there is no reasonable theoretic law which guarantees a unique model parameter fluctuation and the influence of internal and external noise can give more reliability to the accuracy of the model. The aim of the complex system analysis in such cases is to design an adequate model with enough free parameters which reflects the true dynamic of the cooperation of the complex biological,chemical and physical processes. This means that the model should reflect

-the quantitative behaviour (it has to approximate measured dates)

-the long range qualitative behaviour (stability or instability constellations etc) and

-the stochastic character of the real process and should serve as well

-statistically significant statements for control strategies.

The software instrumentarium CANDYS is a computer support for this purpose.

---

[1]Institut für Wasserwirtschaft in der WWD-Berlin/Außenstelle Potsdam DDR-1500 Potsdam,Schloßstraße 1

[2]Zentralinstitut für Kybernetik und Informationsprozesse Akademie der Wissenschaften der DDR,DDR-1086 Berlin,Kurstraße 33

The Model ERNA

The model ERNA describes the dynamic of eutrophication processes in slowly flowing rivers. The local mass balance equation is described under suitable conditions as a parabolic partial differential equation. With a number of heuristic arguments (Braun [1]) the partial differential equation can be reduced to a system of ordinary nonlinear differential equations

$$\frac{dc}{dt} = f(t,c,p) + \frac{Q}{V}(c_{input} - c)$$

which describes now a homogeneously stirred river segment of volumina V and flow rate Q. The first term of the right hand side reflects the autochthonuous metabolism and growing processes and the second term the outer mass balance of the river segment. 3 groups of alga (diatomeen, green and blue alga), death material and 4 nutrients (silicate, ammonium, nitrate, phosphate) represents the 8 state variables of the system.
A typical nonlinear dependence for growth rate for a group of alga is

$$\mu_{eff,k} = \mu_{max,k} \prod_j f_{jk}(N_j)\, F(T)\, F(L)$$

c_{input}	:inflow
k	:k-th group of alga
$\mu_{eff,k}$	:effective growth rate
$\mu_{max,k}$	:maximal growth rate
f_{jk}	:dependence of the growth rate on the j-th nutrients (Monod kinetic)
$F(T)$	:dependence of the growth rate on water temperature
$F(L)$	:dependence of the growth rate on light intensity

The balance equation for the k-th group of alga is for instance

$$\dot{c}_k = \text{assimilation} - (\text{respiration} + \text{mortality} + \text{excretion}) - \\ - \text{feeding} - \text{sedimentation}$$

with the typical growth term assimilation = $\mu_{eff,k}\, c_k$ and a summarized loss term

$$(\text{respiration} + \text{mortality} + \text{excretion}) = p_{\varrho k} \exp\left[\frac{T - p_{\sigma k}}{p_{8k}}\right]^2$$

The Analysis of the Model

A lot of different models for describing the dynamic of eutrophication exists which all reproduces nearly the same behaviour for a sample of state variables. This is because of the incomplete knowledge of the complex ecological, chemical and physical processes and their cooperation in the real system. On the other hand not all states which really influence the water quality are measurable. Thus modelling with the inclusion of parameter fluctuation and the influence of external and internal noise can give more reliability to the accuracy of the model. Diffusion processes form a class of stochastic processes, which take into account these facts. The stochastic model is a multidimensional stochastic differential equation in the Ito-sense

$$dc = a(t,c,p)\ dt + b(t,c)\ \Sigma\ dw$$

a is the right hand side of the deterministic system and b is a matrix function. The positive definite matrix Σ has additional parameters.

The analysis of such complex models described as a system of ordinary parameter dependent deterministic or stochastic differential equations can be done with the help of the software system CANDYS.

CANDYS/CM	software support for developing a parameter dependent system of deterministic or stochastic differential equations (Funke [2])
CANDYS/QA	software support for analysing the qualitativ behaviour (Jansen, Feudel [3])
CANDYS/SA	software support for stochastic analysis (Simulation of stochastic differential equations,parameter estimation,statistical analysis)

A neccessary step to simplificate and validate the model in comparison with other ones is to know its qualitative attributes (limit points, limit cycles, stability, bifurcation etc). The software system CANDYS/QA analyzing the qualitative behaviour is described in detail in [3].

CANDYS/SA is a software support for analyzing systems of stochastic differential equations. The simulation of such systems is realized with derivatefree and other integration procedures with step size control of the form

$$P\left[\left| c^{2h} - c^{h}\right| < \varepsilon\right] > 1-\alpha$$

the probability of the difference of 2h- and h-step size approximations beeing not significant. The Bayesian estimates for "quasi"linear parameters and linear covariance parameters are included. The parameter estimation of the nonlinear parameters for the water quality model ERNA has been done with different functions of accuracy for 50 parameters by the help of a MINUIT-based ([4]) software system.

References

[1] Braun ,P.:Das Eutrophierungsmodell ERNA, Dissertation ,TU Dresden, 1984

[2] Funke ,R.:CANDYS/CM-A Dialogue System for Modelling Continuous Dynamical Systems with Chain Structure by Differential Equations this Volume

[3] Jansen, W.,Feudel, U.: CANDYS/QA- A Software System for Qualitative Analysis of the Behaviour of the Solutions of Nonlinear Dynamical Systems, this Volume

[4] James,F.,Roos,M.: MINUIT,Computer Physics Communication 10(1975)

PARETO-OPTIMAL COST DIVISION FOR THE DESIGN OF SEWAGE WATER TREATMENT PLANTS BY MEANS OF THE DSS REH

A. Gnauck[1)], P. Rathke[2)], R. Straubel[2)], and A. Wittmüss[2)]

INTRODUCTION

The achievment of regional water quality goals often involves substantial capital investments and changes in public attitudes concerning resource management. The economic impacts may include not only the costs of facilities designed to reduce the discharge of contaminants into natural surface waters or to improve the quality of waste-receiving waters, but also any limitations on economic development in a particular river basin. Those responsible for the formulation and approval of water quality plans or management policies must have means of estimating and evaluating the temporal and spatial economic and environmental or ecological impacts of these plans and policies. This need has stimulated the development and application of a wide range of mathematical modeling and optimisation techniques for predicting various physical, chemical, biological and economic impacts of alternative pollution control schemes.

WATER QUALITY SIMULATION MODELS FOR RIVERS: THE SPROX-FAMILY

Most of the water quality simulation models in actual use today are extensions of the equations proposed by STREETER and PHELPS (1925) for predicting the biochemical oxygen demand (BOD) and dissolved oxygen (DO) concentrations or deficits resulting from point and non-point discharges of easily biodegradable organic waste into a river system. According to BECK and YOUNG (1976) a family of selfpurification models, called SPROX, were developed for describing dynamic interactions between physical, biological and biochemical constituents of river water. For all cases the water quality relations are expressed by systems of coupled linear differential equations. The SPROX models simulate the longitudinal and temporal variations of DO and BOD levels caused by point and non-point sources and sinks of organic loads. In dependence of the main biochemical processes the basic water quality model covers DO-BOD interactions due to decay and reaeration, where these processes are represented by first-order kinetics.

For modelling the river stretch under consideration is divided into segments of different length'. Organic load inputs were located at the beginning of the segments (if any). Each segment is considered as

---

1)Water Authority of Berlin, Institute for Water Management, Schnellerstrasse 140, Berlin, 1190, DDR. Present address: Academy of Medicine, Institute for General and Public Hygienics, Fetscherstrasse 74, Dresden, 8019, DDR.

2)Academy of Sciences of GDR, Central Institute for Cybernetics and Information Processes, Kurstrasse 33, Berlin, 1086, DDR.

a CSTR approximately. Water quality observations are available at the beginning and at the end of the segments, where the output of the i-th segment is equal to the input of the segment i+1 (fig. 1.). For water quality management different hydrologic situations are taken into account. The results presented in this paper referred to mean flow conditions.

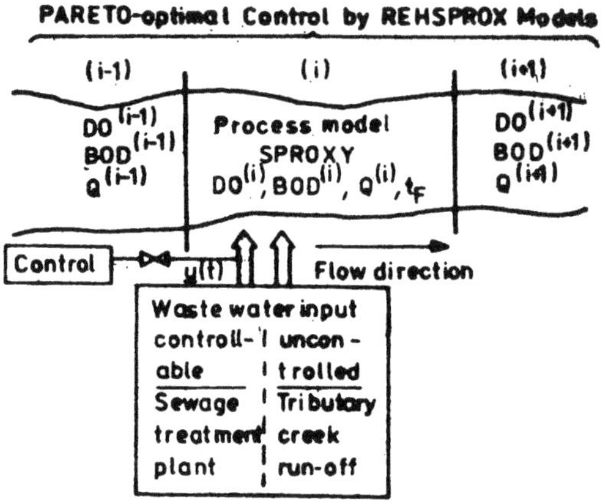

Fig. 1. Schematic representation of the proposed methodology by river segmentation. The organic waste load of river water is divided in controlled and in uncontrolled sources. In flow direction each segment influences the following segment after a time-delay depending on water velocity inside of the segment characterised by the flowing time t_F.

According to the main ecological processes in river water three modified STREETER-PHELPS-type models are used for simulation tasks: The first (basic) one (called SPROXY) describes the selfpurification process in a river mainly influenced by organic waste loads caused by point sources. The second model (SPROXN) takes into account the nitrification process and DO losses caused by respiration, while the third model (SPROXE) reflects DO-BOD interactions in a eutrophic water body. The basic equations are valid for all models, where the input terms are multiplied by a time-varying function $\underline{u}(t)$ to get any control scheme for waste load input along the river. To the models SPROXN and SPROXE equations for describing nitrification and eutrophication processes are added. In order to reduce computation times for solving differential equations simulation runs start on the basis of analytical solutions, where some integrals approximated numerically (GNAUCK et al., 1989). The following basic equations are valid:

DO - equation:

$$\dot{c}_1^{(i)} = -(K_2(t)+A^{(i)})c_1^{(i)} - K_1(t)c_2^{(i)} + A^{(i)}c_1^{(i-1)}(t-t_F^{(i)}) + K_2(t)c_{1S}(t) + B_1^{(i)}u^{(i)}(t) \qquad (1)$$

BOD - equation:

$$\dot{c}_2^{(i)} = -(K_1(t)+A^{(i)})c_2^{(i)} + A^{(i)}c_2^{(i-1)}(t-t_F^{(i)}) + B_2^{(i)}u^{(i)}(t) \qquad (2)$$

where the dot notation is referred to the time derivation and c_1-DO concentration, c_2-BOD concentration, c_{1S}-(temperature-dependent) DO saturation concentration, K_1-BOD decay rate constant, K_2-reaeration rate constant, $A^{(i)} = Q^{(i)}/V^{(i)}$ is constant for each segment with Q is the mean volumetric flow rate and V is the mean volume of a segment. u(t) is a control variable with $0 = u(t) = 1$.

The DO saturation concentration within a segment is given by a third order polynomial equation (THOMANN, 1972):

$$c_{1S}(TW) = 14.65 - 0.41022\,TW + 0.007991\,TW^2 - 0.0000474\,TW^3 , \qquad (3)$$

where the annual time course of water temperature TW is modeled by a sinusoidal function (GNAUCK et al., 1987):

$$TW(t) = 13.16 + 10.23\cos(2\pi(t-213)/365). \qquad (4)$$

The coefficients B_j $(j=1,2)$ are defined as follows:

$B_j^{(i)} = QE_{max}^{(i)}/V^{(i)} c_{jE}^{(i)}$, with QE denotes the additional volumetric flow rate, c_{jE} denotes additional inputs of DO and BOD concentrations due to waste water and/or tributaries, respectively. i is the number of segment.

RIVER WATER QUALITY MANAGEMENT AND CONTROL BY DSS REH: THE REHSPROX MODELS

To get PARETO-optimal decisions (PARETO, 1896) for water quality management and control the SPROX models were implemented into the interactive, hierarchically structured DSS REH developed by STRAUBEL and WITTMÜSS (1983, 1986). The DSS REH works in an dialogue form. No special requirements are necessary for the type of process equations and for the mathematical formulations of the objectives. Nonlinear process equations with time delay of any order, non-convex or non-concave goal functionals, and also disconnected reaches of control variables are accepted for computation. Time-dependent restrictions of the control variables in form of lower and upper bounds and other implicite formulated restrictions between state and control variables are taken into account additionally.

According to the choosen model structures the objectives, related to different river stretches, are referred to the state variables of the main water quality processes:

$$\max_u \sum_i \sum_t DO^{(i)}(t); \ \min_u \sum_i \sum_t BOD^{(i)}(t); \ \min_u \sum_i \sum_t NIT^{(i)}(t);$$
$$\max_u \sum_i QE^{(i)}; \ \min_u \sum_i \sum_t PHOS^{(i)}(t). \qquad (5)$$

Solving the planning problem for extension of sewage treatment plants the QE-objective is replaced by a cost function with discrete values in the solution space: $\min_u \sum_i COSTS^{(i)}(t)$. NIT, PHOS and COSTS denote the ammonia-nitrogen input, the phosphate-phosphorus input and the investment costs, respectively.

APPLICATION OF DSS REH TO RIVER WATER QUALITY MANAGEMENT

As a basis for planning decisions the DSS REH with implemented SPROXY process model was used to get PARETO-optimal solutions for investment cost division of sewage treatment plant extensions. The practical problem arises from the socio-economic development of municipalities and industries in the region under consideration. Replacing the QE-objective by a cost objective, the compromise strategy is shown in fig. 2. This model version is called SPROXEC, a part of the REHSPROX models.

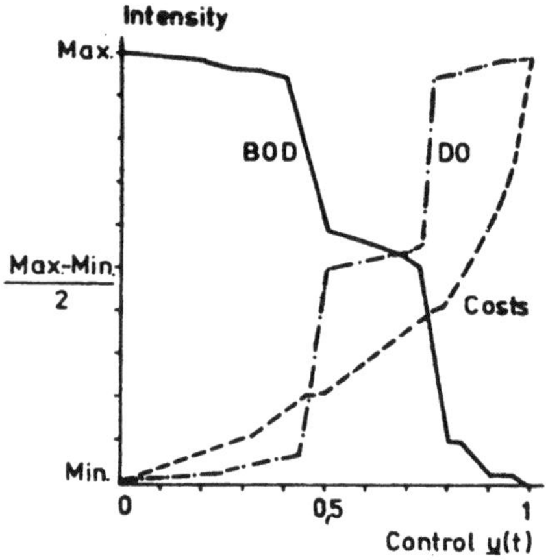

Fig. 2. Schematic representation of the compromise strategy for solving design problems of planned extensions of sewage water treatment plants.

Considering the valid water quality standard and under the assumption of a 85%-degree of BOD removal with fixed design costs, a set of planning scenarios were investigated. For the observed river stretch segments no. 3, 7 and 9 are characterised by high waste water input. Tab. 1. contains the QE-values corresponding to the investment costs.

Table 1. Waste water input and investment costs

Segment	QE (m3/d)	Costs (Mio.Mark)
3	0	0.00
	3000	8.25
	6000	16.50
7	0	0.00
	1000	1.00
	2000	2.00
	4500	4.25
	4850	4.50
	6000	5.50
	7000	6.50
9	0	0.00
	1200	1.00
	2000	1.60
	3600	3.00
	5000	4.00
	6750	5.60
	25000	20.80

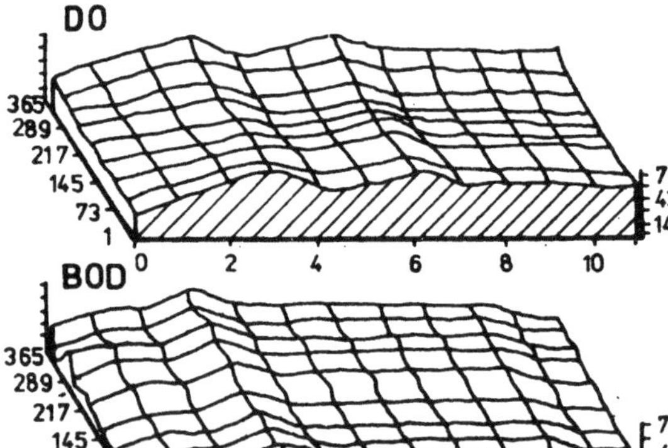

Fig. 3. Optimisation results for DO and BOD by DSS REH with implemented SPROXY process model.

In fig. 3. the influence of waste water input in segment 3 on the DO level along the observed river stretch can clearly be seen. While the DO curves (DO surface) show a more or less acceptable time behaviour, critical BOD variations exist in the lower part of the river stretch during summer and autumn period. Fig. 4. shows a consequence of sewage treatment plant extension for segments 7 and 9 corresponding to a higher degree of BOD removal.

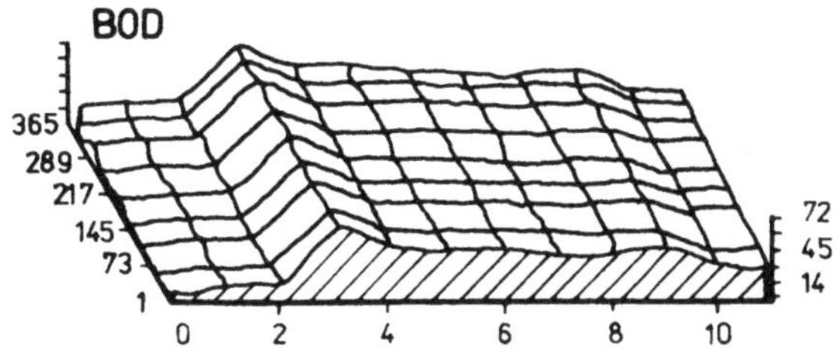

Fig. 4. Effect of sewage treatment plant extension on water quality level, expressed as BOD, in a river basin.

CONCLUSIONS

The optimal valuations of the monetary effectiveness of sewage treatment plant extensions give a basis for decision makers to get objective decisions for cost division. Furthermore, the results obtained by DSS REH in combination with water quality process models lay out new directions for further research in water quality management.

REFERENCES

BECK, M.B., and P. YOUNG: Systematic identification of DO-BOD model structure. J. Environm. Eng. Div., Proc. ASCE 102(1976)EE5,909-927.

GNAUCK, A.H., W. WINKLER, K. GROTH, F. KOMOSS, and E. SIX: Modellierung und Simulation des Phytoplanktongehaltes in einem eutrophen Fliessgewässer. Acta hydrochim. hydrobiol. 15(1987)2,148-159.

- , R. STRAUBEL, and A. WITTMÜSS: Mehrkriterialer Steuerungsentwurf der Abwassereinleitung für ein Flussgebiet. Acta hydrochim. hydrobiol. 17(1989)5, in press.

PARETO, V.: Cours d'Economie Politique. Pouge. Lausanne 1896.

STRAUBEL, R., and A. WITTMÜSS: Ein interaktives Entscheidungsmodell zur optimalen Steuerung eines Systems mit mehreren Zielfunktionen. Messen-Steuern-Regeln 26(1983)1,2-4.

- - : Das Programmsystem REH zur rechnergestützten Entscheidungshilfe. ZKI-Informationen 2/86. Akademie der Wissenschaften, Berlin, 1986.

STREETER, H.W., and B. PHELPS: A study of the pollution and natural purification of the Ohio River. Publ. Health Bull. 146(1925),1-25.

THOMANN, R.V.: Systems Analysis and Water Quality Management. McGraw Hill, New York 1972.

TWO-LEVEL HEURISTIC ALGORITHM FOR PARALELL MACHINES SCHEDULING

Jakub Gutenbaum [1]

1. *INTRODUCTION*

In automated manufacturing systems we often try to minimize work-in-progress, the main part of which consists of in-process inventories necessary for normal functioning of the system.

In this paper we consider the problem of manufacturing a certain product to the common storage by a set of machines with different production rates. The output from the storage is determined by the production plan.

The problem consists in scheduling machines for a given time horizon in such a way that the cost function involving the sum of inventories is minimized under constraints on the storage capacity.

The problem formulated as nonlinear integer programming problem can be reduced to an equivalent linear binary programming problem and solved by library procedures. However these procedures can not solve problems with more than several hundred variables.

In the paper a two-level heuristic algorithm for solving the problem is proposed: on the first level an order, in which machines are switched on in each time interval, is established; on the second level the number of working machines in each time interval is determined.

2. *PROBLEM FOMULATION*

We have at our disposal the set of K machines with production rates Q_i^j, where $i \in \{1,2,\dots,N\}$ - time interval, N - planning horizon, $j \in \{1,2,\dots,K\}$ - machine index.

Fof a given sequence of produkt demand a_i, $(i=1,\dots,N)$, we have to determine the subset of working machines in each time interval, that minimize the cost function:

$$J = \sum_{i=1}^{N} (a_i - \sum_{j \in I_i} Q_i^j)^2 \qquad (1)$$

where: I_i - the subset of working machines in the i-th interval,

Q_i^j, a_i - integers,

under constraints on the storage level s_i:

$$S^{min} \le s_i \le S^{max} \ ; \ \forall \ i \in \{1,2,\dots,N\} \qquad (2)$$

where:

$$s_i = s_{i-1} - a_i + \sum_{j \in I_i} Q_i^j \qquad (3)$$

The problem (1) - (3) is NP - complete; it can be reduced to the

[1]) Polish Academy of Sciences, Systems Research Institute

following binary programming problem [2]:

-the cost function

$$J(x_i^j,y_i^j)= \sum_{i=1}^{N} [a_i^2 + \sum_{j=1}^{K} Q_i^j(Q_i^j-2a_i)x_i^j + 2 \sum_{j=2}^{K} \sum_{l=1}^{j-1} Q_i^j Q_i^l y_i^{jl}] \quad (4)$$

should be minimized with respekct to the binary variables (x_i^j, y_i^j),

where:

$$x_i^j = \begin{cases} 1 & \text{- if the j-th machine works in the i-th time interval} \\ 0 & \text{- in the opposite case} \end{cases} \quad (5)$$

$$y_i^{jl} = x_i^j x_i^l \quad (6)$$

under constraints for $\forall i \in \{1,\dots,N\}$:

$$S^{min} \le s_{i-1} - a_i + \sum_{j=1}^{K} Q_i^j x_i^j \le S^{max} \quad (7)$$

$$x_i^j + x_i^l -1 \le y_i^{jl} \le (x_i^j + x_i^l)/2; \quad \forall (j,l) \in \{1,2,\dots,K\} \quad (8)$$

A heuristic algorithm is necessary for solving the problem (4) - (8) even for small values N and K.

3. *THE HEURISTIC ALGORITHM*

A two-level heuristic algorithm is proposed. On the first level the order, that explicitly determines the sequence in which machines are switched on, is established independently for each time interval; on the second level only the number of working machines in each time interval is evaluated considering constraints (2).

3.1. *Ordering machines for given i*

We want to order the machines in such a way that those of the smallest production rate are switched on or switched off when the correction imposed by constraints is necessary.

Initially the machines are ordered according to nonincreasing production rate value (fig.1):

$$Q_i^{j_i} \ge Q_i^{j_i+1} \quad ; \ j_i = 1,\dots,K-1 \quad (9)$$

where; j_i - the initial index of the machine in the i-th interval.

The final ordering of machines k_i is established in accordance with the following rule.

For i, in which a_i is greater or equal than the sum of the production rates for all machines - the final ordering is identical with the initial ordering:

$$k_i = j_i \quad \text{for } i: \ a_i \ge \sum_{j_i=1}^{K} Q_i^{j_i} \quad (10)$$

For i, in which the sum of the production rates for all machines is greater than the production demand a_i :

$$i : a_i < \sum_{j_i=1}^{K} Q_i^{j_i} \tag{11}$$

-the number $K_i \leq K$ is evaluated from the equation

$$K_i = \arg\min_{0 \leq r_i \leq K} \left(\sum_{j_i=1}^{r_i} Q_i^{j_i} - a_i \eta_i \right)^2 \tag{12}$$

$$\text{where: } \eta_i = a_i / \sum_{j_i=1}^{K} Q_i^{j_i} \tag{13}$$

Then for:

-$K_i = 0$ the initial ordering is reversed

$$k_i = K+1-j_i, \text{ for } j_i = 1, \ldots, K \tag{14}$$

-for $0 < K_i \leq K$ the following rule is used:

-for $0 < j_i \leq K_i$ the final ordering is identical with the initial ordering

$$k_i = j_i, \quad \text{for } 0 < j_i \leq K_i \tag{15}$$

-for $K_i < j_i \leq K$ the order is established in accordance with the table given below (fig. 2):

k_i	j_i	
	$K - K_i$ even	$K - K_i$ odd
$\frac{2K+K_i+1-j_i}{2}$	$K_i+1, K_i+3, \ldots, K-1$	$K_i+1, K_i+3, \ldots, K$
$\frac{K_i+j_i}{2}$	$K_i+2, K_i+4, \ldots, K$	$K_i+2, K_i+4, \ldots, K-1$

Value K_i can be changed in such a way that the machine with the least production rate ($j_i = K$) has the final index R_i (fig. 3),

$$R_i = \arg\min_{1 \leq r_i \leq K} \left(\sum_{k_i=1}^{r_i} Q_i^{k_i} - a_i \right)^2 \tag{16}$$

3.2. Evaluation of the number of working machines in each time interval for the ordered set of machines

The order imposed in *3.1.* determines the sequence in which the machines are switched on that in terms of binary variables x_i defined by (5) is expressed by

$$x_i^{k_i} = 1 \text{ implies } x_i^1 = \ldots = x_i^{k_i-1} = 1 \tag{17}$$

$$x_i^{k_i} = 0 \text{ implies } x_i^{k_i+1} = \ldots = x_i^K = 0 \tag{18}$$

By virtue of (16) - (17) the cost function (1) can be expressed as a linear form of binary variables $x_i^{k_i}$ [1]

$$J = \sum_{i=1}^{N} (a_i^2 + \sum_{k_i=1}^{K} b_i^{k_i} x_i^{k_i}) \qquad (19)$$

where: for $i\in\{1,\dots,N\}$

$$b_i^1 = Q_i^1 (Q_i^1 - 2a_i)$$

$$b_i^{k_i} = Q_i^{k_i} [Q_i^{k_i} - 2 (a_i - \sum_{l=1}^{k_i-1} Q_i^l)] ; k_i = 2,\dots,K \qquad (20)$$

However, the following additional constraints on the binary variables are necessary after ordering the machines:

$$x_i^{k_i} - x_i^{k_i-1} \le 0 ; i = 1,\dots,N; k_i = 1,\dots,K \qquad (21)$$

In this way the problem on the second level is to minimize the linear binery function (19) under the constraints (7) and (21). This problem is easier than the initial problem (4) - (8), because we do not need the additional variables y_i^{jl} defined by (6).

The problem on the second level can be formulated as the integer programming problem if the number of working machines in each time interval r_i is introduced as the integer variable. We know that if r_i machines are working in the i-th time interval then their indexes are $k_i=1,2,\dots,r_i$.

The cost function (1) can be expressed as the following function of r_i:

$$J = \sum_{i=1}^{N} (a_i - \sum_{k_i=0}^{r_i} Q_i^{k_i})^2 \qquad (22)$$

where: $Q_i^0 = 0$.

The problem is to minimize (22) with respect to the integer variables $r_i\in\{1,\dots,K\}$, under the constraints (2) where

$$s_i = s_{i-1} - a_i + \sum_{k_i=0}^{r_i} Q_i^{k_i} ; \quad i\in\{1,\dots,N\} \qquad (23)$$

We can represent the problem (22),(2),(23) on a tree. The nods of the tree represent the storage levels. The storage level s_i is determined by the storage level in the previous time interval s_{i-1} and the number of the working machines in the i-th time interval:

$$s_i = s_i(s_{i-1},r_i) \qquad (24)$$

The nod s_0 is the root of the tree and the nods $s_N(s_{N-1},r_N)$ are the leafs of the tree. For the nods on the available path the constraints (2) should be fulfilled. The number of all path is equal to $(K+1)^N$.

Each arc of the tree is characterized by the value of the cost function in the i-th time interval with $r_i\in\{0,\dots,K\}$ machines switched on

$$J_i(r_i) = (a_i - \sum_{k_i=0}^{r_i} Q_i^{k_i})^2 \qquad (25)$$

For solving this problem a heuristic algorithm was elaborated and tested on a IBM PC [1].

4. *CONCLUDING REMARKS*

The further research are going in two directions. The first direction is to develop programs for big problems and to obtain the exact solutions for these problems using linear binary programming library procedure for the tasks of ordered set of machines. These solutions will be used to evaluate the heuristic algorithm performance. The second direction is to implement the algorithm for ordering the set of machines and to check the adopted rule of ordering.

This research was supported by the grant from RPBR - RP.I.02/5.1 problem (The theory of control and optimization of continuous dynamic systems and discrete processes).

References

[1] Gutenbaum J., Olinger W.: Równoległe wykonywanie zadań produkcyjnych przez urządzenia o zadanym priorytecie. Sprawozdanie ZTS IBS PAN, ZTS - 20/70/RPBR - RP.I.02/5.1/87

[2] Walukiewicz S.: Programowanie dyskretne. PWN, Warszawa 1986.

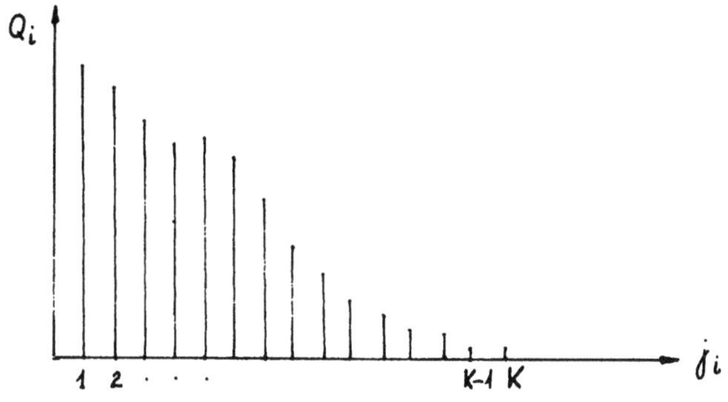

Fig. 1

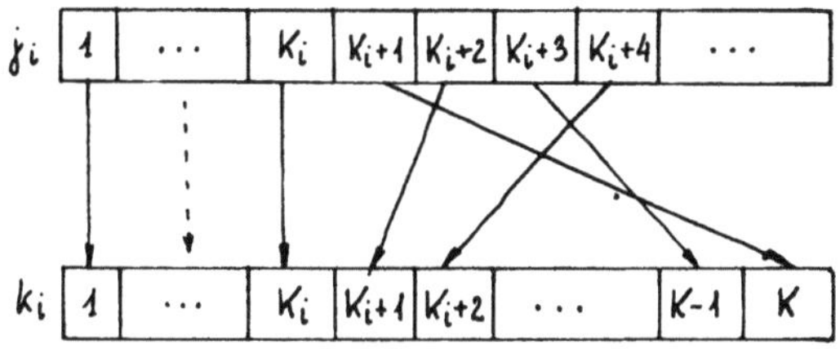

Fig. 2

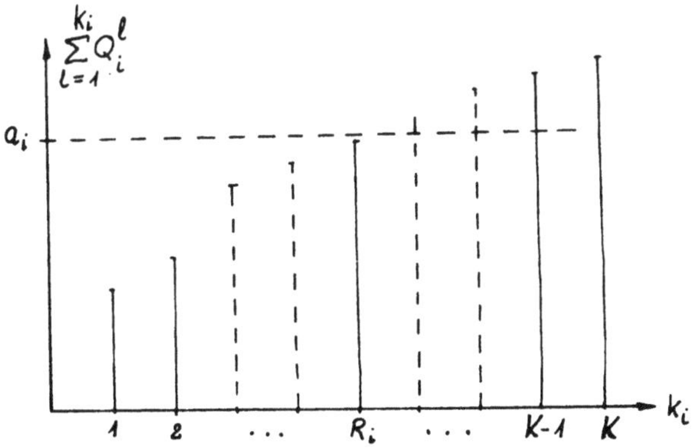

Fig 3

THE STATE VARIABLE DIAGRAM OF A HYDRAULIC CONTROL SYSTEM

Zhou Shichang professor
Cao Xinming professor
Yao Guiyou lecturer

Department of Mechanical Engineering
Northeast University of Technology
Shenyang, Liaoning, P. R. China

ABSTRACT

This paper suggests a new method of pictorially describing the mathematical model of a hydraulic system by combining the advantages of block diagram and bond graph. The fundamental components are equal flow junction, equal effort junction, power transmission line, control line and various elements. The relationships between the variables are written in the block of every element . The differential relationship is used, so that is convenient to list the state equations from the diagram. A valve controlled motor system is used as an example so as to illustrate the procedures of drawing the state variable diagram.

INTRODUCTION

In the pictorial representation of a hydraulic control system, both the block diagram and the bond graph are usually used. The block diagram was developed first. It is usually used in the frequency domain analysis with transfer functions. The relationship between variables is explicit and visual. But the use of the block diagram is convenient mostly in single-input-single-output linear system[1]. The bond graph is a power flow diagram, it is summary and can be described by a few standard elements, but it is a rather abstract diagram. Although its power flow direction is clear, the numerical transfer relationship between variables is not so explicit as in the block diagram. Taking the advantages of both the block diagram and bond graph, the two methods were combined into one and a new pictorial representation of the mathematical model was developed. It is named the "state variable diagram".

FUNDAMENTAL COMPONENTS OF STATE VARIABLE DIAGRAM

The idea of power flow exchange is also used to represent the relationship between the components of an engineering system. The power is equal to the product of the effort and flow variables. To form a state variable diagram, the following fundamental components can be used:

1. Junction

The junction may be classified into three kinds:
(1) Equal effort junction: Every power flow into or out of this junction has an equal effort variable and the sum of the flow variables around this junction is equal to zero. Around every equal effort junction, only one effort variable may be treated as the input, all other efforts are output. The direction of the input effort is oriented towards the junction and the output is led away from the junction. The equal effort junction is represented by a solid line circle, the symbol of the effort is written in the circle, such as (P_1), all of the efforts around this junction have the same effort as P_1 .
(2) Equal flow junction: Every power flow into or out of this junction has an equal flow variable and the sum of effort variables around this junction is equal to zero. Around every equal flow junction, only one flow variable may be treated as the input, all other flow variables are output. The direction of the input flow is oriented towards the junction. The equal flow junction is represented by a dash line circle, the symbol of flow is written in the circle, such as (Q_1) . All the flow variables around this junction are equal to Q_1 .
(3) Comparing junction: The control variables (signals) are compared at this junction. It is represented by a small circle, such as o.

2. Transmission lines

There are two kinds of transmission lines.
(1) Power transmission line: Power is the product of the effort and flow variables, so the power transmission line is represented by two lines: the effort transmission line and the flow transmission line. The solid line represents the transmission of effort and the dash line represents the transmission of flow. The arrow represents the direction of the transmission. The directions of every pair of effort and flow transmission lines are contrary. The small arrow between the two transmission lines represents the direction of power flow.
(2) Control line: The transmission of control signal is represented by a thin solid line. Only one variable (effort or flow) is transmitted in a control line, therefore, only signal (but no power) is transfered.

3. Element: Every element is represented by a block. The relationship between variables (including state variables and control variables) determined by the physical laws is written in the block. The relationships may be proportional, differential or nonlinear. Though in most bond graph representations the integral relationship is used to simplify the computer calculation[2,3], the differential relationship is used to simplify the procedure of listing state equations. The state variable diagram is a visual mathematical model, it can be easily drawn according to the hydraulic circuit. Because the relationship between the variables are described in the diagram, the transforming relationship between the variables can be clearly seen from the diagram. The symbol in the circle can clearly reflect the variable of that junction. Through some simplification procedures the state variable diagram can be transformed into a block diagram. An example will be taken to illustrate the procedures of drawing the state variable diagram.

EXAMPLE OF DRAWING STATE VARIABLE DIAGRAM

Fig. 1 shows an equivalent circuit of a valve controlled motor system.

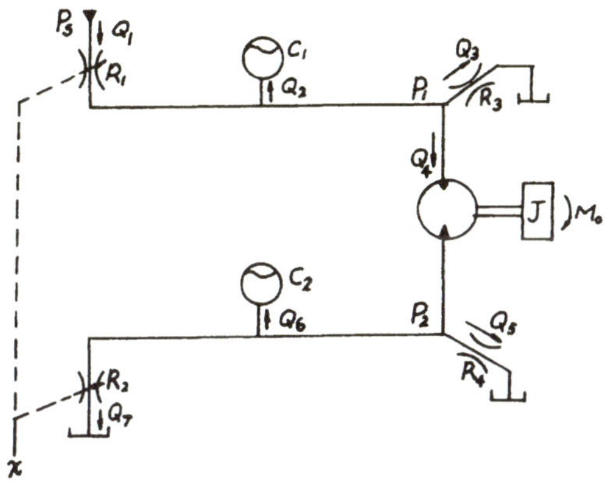

Fig.1 Equivalent circuit of a valve controlled motor system

R_1 and R_2 represent the hydraulic resistances of the two valve ports. C_1 and C_2 represent the hydraulic capacitances before and after the motor (including the capacitances of the motor itself). R_3 and R_4 represent the hydraulic resistances of leakage flow. P_1 and P_2 are the pressures before and after the motor. P_s is the constant source pressure. The return line is connected to a tank at zero pressure. The motor has a torque load M_0 and an inertia load J .

The procedures to set up state variable diagram may be as follows: There are two pressures P_1 and P_2 (not including the constant source pressure P_s and the zero pressure of the tank), so there are two equal effort junctions (P) and (P) . Between P_1 and P_s , the flow through R_1 is constant and equal to Q_1 ,so there is an equal flow junction (Q) . Between P_1 and P_2, the flow Q_4 through motor is constant, so there is an equal flow junction (Q_4). After the junctions have been determined, the elements should be determined. There are two restrictions at R_1 and R_2. The flow equation of the orifice should be satisfied there. i.e.

$$Q_1 = CA\sqrt{2(P_s - P_1)/\rho}$$

$$Q_7 = CA\sqrt{2\ P_2/\rho}$$

where Q_1 and Q_7 are the flow rate through R_1 and R_2,respectively; C is the flow coefficient of valve port ; A is the area of valve port. If the above equations are linearized,

$$Q_1 = Z_1(P_1 - P_2)$$

$$Q_7 = Z_2 P_2$$

where Z_1 and Z_2 are the flow conductances of the valve ports.

The two hydraulic capacitances C_1 and C_2 should satisfy the following equations, respectively.

$$\dot{P}_1 = \frac{\beta_e}{V_1} Q_2$$

$$\dot{P}_2 = \frac{\beta_e}{V_2} Q_6$$

where β_e is the bulk modulus of oil, V_1 and V_2 the effective volumes of C_1 and C_2. The two hydraulic resistances of the leakage should satisfy

$$Q_3 = Z_3 P_1$$

$$Q_5 = Z_4 P_2$$

where Z_3 and Z_4 are the conductances of leakage resistances.

There is a proportional element in the motor to transform the hydraulic energy into mechanical energy, the transformation ratio is equal to the volumetric displacement of motor D_m. The mechanical energy is used to overcome the load M_o and J .

$$M_o = M_2$$

$$\dot{\omega} = \frac{1}{J} M_1$$

where ω is the angular velocity of the motor. The two loads have the same angular velocity, therefore, there is an equal flow junction (ω).

The overall state variable diagram of the system can be drawn as Fig.2. Where the R_1 and R_2 are represented by linear functions.

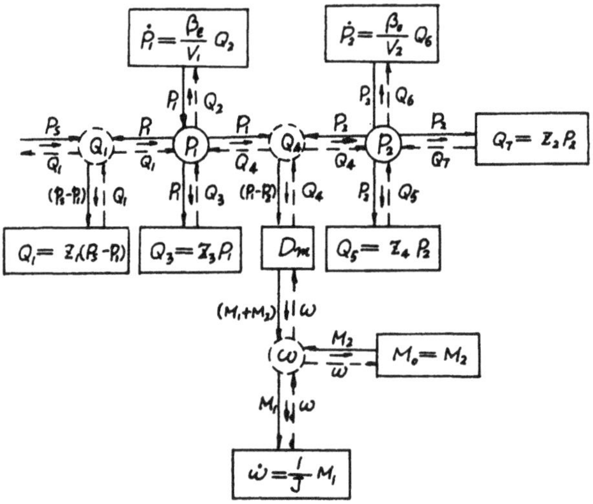

Fig.2 State variable diagram of the system

For all the elements which store energy, the differential relationship is chosen, so that the state equations can be easily listed. According to the principle that only one effort can be treated as the input of an equal effort junction, we can determine the directions of all the efforts and flows.

There are three differential elements, so there are three state equations. After simple analysis, these can be directly written in state equations as

$$\dot{\omega} = \frac{1}{J} M_1 = \frac{1}{J}[(P_1 - P_2)D_m - M_o] \tag{1}$$

$$\dot{P}_2 = \frac{\beta_e}{V_2}(Q_4 - Q_7 - Q_5) = \frac{\beta_e}{V_2}[\omega D_m - Z_2 P_2 - Z_4 P_2] \tag{2}$$

$$\dot{P}_1 = \frac{\beta_e}{V_1}(Q_1 - Q_4) = \frac{\beta_e}{V_1}[Z_1(P_s - P_1) - \omega D_m] \tag{3}$$

Eqs. (1), (2) and (3) compose the state equation of system.

CONCLUSION

1. A state variable diagram has the adventages of a block diagram in so much as the relationship between the variables is visual. On the other hand, it also avoids the inadequacies of the bond graph in that a complex procedure is needed for deriving the state equations from a bond graph. A state variable diagram can be used for the analysis of a linear or non-linear, single-input-single-output or multi-input and multi-output system.
2. Because the state equations have a differential form, it is benificial to express the elements of state variable diagram with differential relationship.
3. The state variable diagram is used to analyze a hydraulic system in this paper, but the principles of the state variable diagram can also be used in electrical, mechanical and even thermal systems.

REFERENCES

1. Katsuhiko Ogata " Modern Control Engineering " Printice-Hall, Inc, Englewood Cliffs, N. J. 1970
2. P. Dransfield " Hydraulic Control Systems -- Design and Analysis of Their Dynamics " Springer-Verlag, 1981
3. J. V. Thoma " Introduction to Bond Graphs and Their Applications " Pergamon Press, 1975

DECISION-MAKING SIMULATIONS WITH THE **Micro-CED** PROGRAMMING SYSTEM.

Prof. Victor Giraldo Valdés Pardo[1]

INTRODUCTION

In agreement with the economical and social directives for the years 1986-90, analyzed and approved of in the Third Congress of the Communist Party of Cuba, a considerable number of microcomputers has been installed in different highschools and higher education centers in our country, with the objective of providing access to computer techniques to instructors and students, as well as to improve the efficiency and quality of the educational process.

Some of the major ways for utilization of computers as instructional media embrace the use of program packages and/or programming systems to aid in class contents review, evaluation of acquired knowledge, or learning of new subjects, by means of computer assisted instruction (C.A.I.) techniques, all of which represent an important complement to traditional teaching activities. Other ways are the use of didactical games and simulations. Through modelling and simulation, students can be trained on inductive reasoning, in view of mastering a methodology.

The didactic programming system **Micro-CED**, developed at the Central University of Las Villas [2], is designed to facilitate the production and delivery of personal computer courseware, aimed at different disciplines and educational levels. Among other possibilities, the system allows for the composing and editing of textual and graphical frames, which can be logically sequenced according to a dialog-oriented pedagogical strategy.

The objective of this paper is to discuss different approaches for implementing didactical simulations, and the corresponding facilities offered by the **Micro-CED** system for creating learning programs based on decision-making simulations.

DIDACTIC SIMULATIONS

Simulation techniques can be valuable in an educational environment, to teach about some aspects of reality through imitation and replica-

---

[1] Dept. of Electronics, Central University of Las Villas, Cuba.

tion. A proposed model for computer assisted instruction [1] defines four main phases to be implemented: presenting the student with information, guiding the student in acquiring the information or skills, providing practice to enhance retention, and assessing learning. In contrast with interactive tutorials, didactic simulations may be used for any of these four phases. Their major modalities are:

Experimental simulation: By means of the dialogue, the student can conduct several experiences with the model of a real phenomenum, expressed in the form of a program. He can point out the elements needed (materials, equipment, etc), the parameters which are to be set (initial value, rate of increment or decrement, units, etc) and the computer gives him the results of experiences, helping to draw conclusions, which are the basis for learning.

Methodological simulation: The computer helps to master a methodology. As the student progresses in its reasoning, the system amends his errors. Two variants are possible: imposed strategy, in which the program guides the student step by step, so that he can not depart significatively from the trajectory the instructor wants to be followed, or free strategy, where the program solely intervenes in the formation of solution steps and gives corrective elements only at the end of the student's trajectory.

Instrumental simulation: This usually requires a graphic display and can assume two forms: the student learns to build an apparatus (based, for example, on a set of parts which are showed, and in case of error he receives corrective messages) or he learns to operate a model (on the screen appears, for example, a representation of the command board of the equipment actually used) and he can observe the effects of the orders given to the model.

Decision-making simulation: In this approach, the student is presented with a set of interrelated problems, each one having a number of possible solution alternatives. This situation is also referred as an "action maze", where the student must make successive decisions to move through the maze towards a predefined target state. This method may serve as a basis for didactic games, in which the system simulates different conditions as a function of the decisions made at each step, provides guidance and evaluates student's performance in connection with the problem-like conditions being analyzed. This method should be valuable, for example, in order to develop mastery in technical system fault-finding or in medical diagnosis. A general structure model for didactic simulations is shown in Fig. 1.

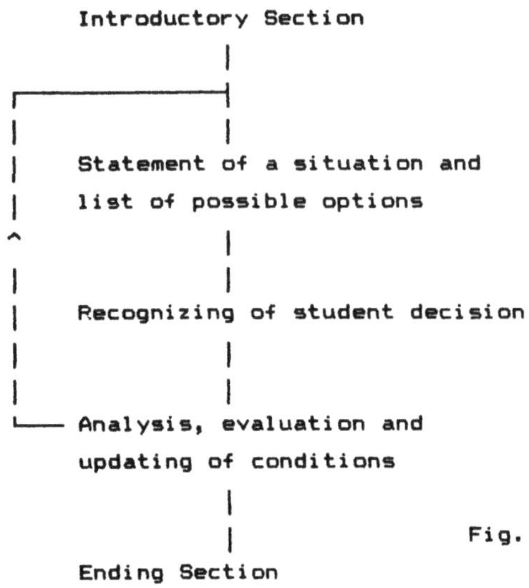

Fig. 1. General structure model for didactic simulations.

The sequencing of frames in decision-making simulations is driven by the selection of alternatives presented to the student in different situations that he confronts. There could be two kinds of selection: single and multiple. The first is used to establish main paths through the action maze, transferring from a section (scenario) to another. Multiple selection is used when data retrieval is required to clarify the conditions upon which the decision is to be made. The pedagogical design of the simulation could start from an optimal path to arrive from the initial state to the solution of the problem, and there could also be other less optimum paths, which have an associated cost relative to the initial condition impairment, with the possibility of arriving to other solutions or break points to establish and revise the causes of failure.

CHARACTERISTICS OF THE Micro-CED SYSTEM

The system has been designed with the following objectives in mind:

1- To facilitate the use of microcomputers available in highschools and higher education centers as instructional media, by means of an author language which could be easily mastered.

2- To decrease time and efforts required for programming and implementation of courseware.

As general working possibilities, **Micro-CED** offers the following:

1- Elaboration of data structures and programs to support C.A.I. activities (text and graphics edition, program translation).

2- Execution of the didactical programs.

3- Revision, by the instructor, of accumulated data about student's performance with the program, to improve it and increase its didactic effectiveness.

Sentences of the **Micro-CED** author language can be classified, from a functional point of view, into five categories:

a) Information displaying on screen, which can be textual, graphical, or both.
b) Answer gathering, which could be expressed in written form (strings of characters) or be associated with the location of a cursor on the screen.
c) Answer analysis and branching, according to the pedagogical strategy programmed.
d) State updating and registration of performance results, for later revision by the instructor.
e) General control of the program, with optional transfering to different software modules.

As final remarks we would say that the use of microcomputers as instructional resources, properly combined with traditional methods and media, permits to actively involve the student in the learning process and promotes its intensification. The **Micro-CED** system provides a set of informatic tools specially suited to the creation and delivery of C.A.I. materials for different instructional designs: tutorial, drill, and simulation. A number of didactic packages are in course of elaboration at present to assist the study of different subjects in technical, medical, and social sciences, by cuban universitary students.

REFERENCES

[1] Alessi, S.M. and S.R. Trollip: Computer Based Instruction. Methods and Development. Prentice-Hall Inc. 1985.
[2] Valdés Pardo, V.G.: Investigation and elaboration of systems for Computer Assisted Instruction. Ph.D. Thesis. Central University of Las Villas, Santa Clara, Cuba, 1987.

APPLICATION OF A HIERARCHICAL COMPUTER SYSTEM FOR PICOSECOND TECHNOLOGY

S. Geschke, N. Kempe, H. Schlawatzky 1)

1. INTRODUCTION

The unique properties of pico- and especially femtosecond laser light pulses allow for the first time a deep insight into the nature, into the temporal evolution of some of the most fundamental processes in matter, such as the free decay of molecular vibrations, electronic relaxation processes in small and large molecules as well as in solids, phonon decay and energy migration in solids, charge transfer processes and other non-radiative transfer processes, elementary processes of photosynthesis and of other biological phenomena. Moreover, picosecond technology has provided new possibilities for the manipulation of photophysical and photochemical processes and it is necessary for a number of modern technologies such as the heterogeneous catalysis, making extremly fast electrooptical components, e. g. switches, modulators and receivers, for different kinds of modern electronics, optical computing technique and for biological technology and medicine. This situation makes high demands on the scientific instruments in this field, especially because the majority of the users of these instruments are no "laser scientists". On the other hand, a great number of such instruments is the basis for the investigation of phenomena, not yet fully investigated. This report is to show how we in the Centre of Scientific Instruments approached the problem of optimized equipment for various applications with very high flexibility and convenience by means of modern microprocessor electronics.

2. MAIN PROPERTIES OF OUR MODULAR SYSTEM

We are building a fully modular system consisting of a number of several modules such as laser systems, generating the tunable cw or ps pulse laser light, beam manipulators, receivers for optical signals and some control modules. Most of them contain their own microprocessor control electronics, but there are no control elements at the front panel of the devices. Each module has to be connected by means of a standard interface with the customer's computer and is controlled only by this computer in a strong hierarchical master-slave manner [1,2].

---

1) Academy of Sciences of the GDR, Center of Scientific Instruments, GDR - Rudower Chaussee 6, Berlin, 1199

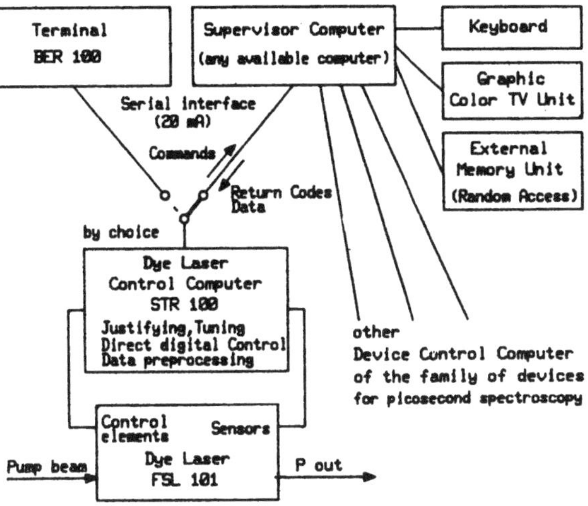

Fig. 1.
Microcomputer architecture of the ps-system

The features of the system are:

- the modular character of our set, allowing to be fitted together in various combinations according to the specific application
- separate operation of single units for other purposes
- uniform structure in all variants of spectrometers for control and data processing (interactive operation, standard interfaces, possibility for modular software etc.
- real-time software system in separately controllable modular units
- strictly nierarchical microprocessor structure in the system
- possibility to build expert systems for different picosecond laser light pulses' applications

In principle all all-purpose computers are suitable to work as a master computer in this system. The only hardware precondition is a standard serial interface. Often it is necessary to have more standard serial interface possibilities than an all-purpose computer has. Then the use of a so-called interface channel multiplexer which contains an own microprocessor and is therefore programmable for several baud rates and protocols seems to be a good solution.

The slaves are dedicated computers optimized for their special function. If possible the slaves are realized as built-in computers, single-board and back-plane free. They don't have data processing peripherals. The conception of a modular microprocessor system here allows to extend or to add single slave computers, to applicate new processors or other components, which are available.

The software of the slave computers is very specific according to the device function, and in most cases it is separated into a real-time operating system to solve the problem-independent elements and a task model and interrupt routines for the problem-related elements. We always use the real-time operating system PEST, developed in the Institut of Energetics in Leipzig [3].

3. HARDWARE AND SOFTWARE OF THE DYE LASER FSL 101 COMPUTER

As an example we want to present you more detailed the dye laser slave computer. Hardware and software are designed as a dedicated solution for the dye laser control. This solution has to meet the general demands on the ps-system.

In this sense the hardware is the result of the necessary electronics for sensors' and actors' control, demands of measurement data processing and the planned dye laser's behaviour in the hierarchical structured distributed ps-computer-system.

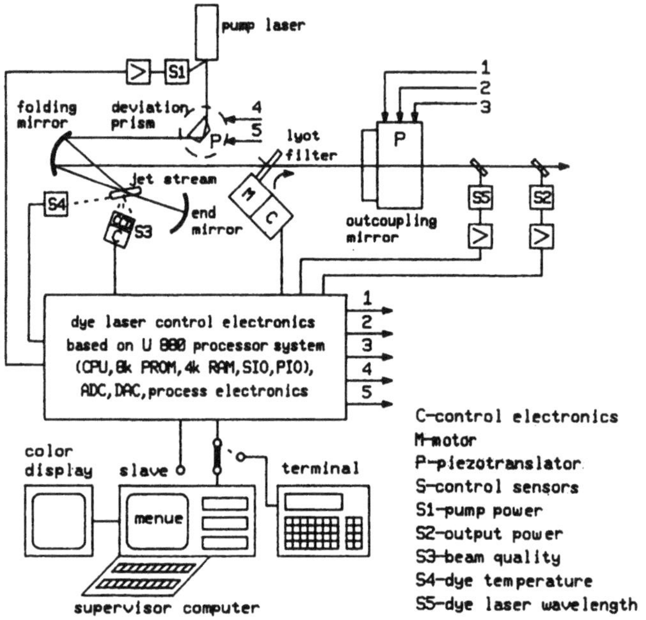

Fig. 2 Automatically controlled ps laser FSL 101

The dye laser computer contains an U 880 system (CPU, 8k PROM, 4k RAM, PIOs, SIOs). Values of five sensors are registered with an ADC. By means of DACs and high voltage amplifiers the piezotranslators are controlled.

The software of the dye laser computer is firmware. That means the user doesn't know it. The only contact of the user with this software is the command dialog to execute functions and/or to get measuring results via a serial interface. All the dye laser's facilities are completely described by the set of commands. The commands are executed in a return-on-completion-regime, that means one command at a time will be executed.

The design of the command set is a decisive creative step in the development process. It has to be done in close cooperation of all people involved in the slave's definition and development. In general this design takes time and a stepwise refinement is necessary.

The dye laser computer software consists of a very compact real-time operating system (less than 2k PROM) and a task model (5 - 6k PROM).
There are two important steps related to a task model :
- The comprehensive device function is to divide and to assign to tasks
- The tasks running must be synchronized.

The task model we used is very simple. It consists of the following tasks: self-inspection, communication, interpretation of commands and execution control, direct digital control of power and wavelength, cyclic sending of information, wavelength tuning, control of subordinated CCD-module. The operating system causes only few restrictions in the microprocessor system's facilities for the user tasks. Thanks to another remarkable feature - the so-called conditioned processor's deprivation - it is very convenient to synchronize the tasks and to manage the resources.

The hardware and software principles presented here were conceptualized for our dye laser module. In the other modules of our ps-system we use the same principles, even in modules where the physical background is quite different and other processors and electronics are used.

4. CONCLUSIONS

The presented hierarchical system solves the difficulties mentioned in the introduction. Furthermore it allows independence of the changing hardware situation. This flexible system allows the permanent improvement of the master computer's software according to the results of the measurements. In this way the necessary problem-oriented software in device complexes can be developed in a simple language like BASIC with the support of the users (biologists, physicians etc.). After qualification this may be the basis for an expert system.

At present we are preparing our device system for two applications:
- measurement of fluorescence decay curves of photosensibilisators for the photo-radio-therapy with ps excitation and simultaneous visual observation of the changes of the investigated cell (together with Humboldt University Berlin)
- photo-acoustic spectrometer for the investigation of special polymeres for the printing trade (together with TH Leipzig)

Other applications are in preparation in cooperation with the users.

LITERATURE

[1] Kempe, N., Langhoff, N., Schlawatzky, H.: Scientific Instrumentation, 1 (1986), 3, 53-62

[2] Geschke, S., Schlawatzky, H.: A Distributed System for Ultrashort-Time Spectroscopy, Proceedings of the MMPS 87, Eisenach, September 7 - 11, 1987

[3] Simon, S.:Das Betriebssystem PEST, Informationsheft des WTZ MRT/ME des MKE, 22.10.1984, 20-25

SOME COMMENTS ON CONSTITUENTS OF CONTEMPORARY INFORMATION SYSTEMS DEVELOPMENT METHODOLOGIES

Stanisław Wrycza [1/]

ABSTRACT

Development of complex information processing systems in business organizations by interdisciplinary designing team requires specific guidelines - the methodologies, enabling the system developers to transform user's needs into operating system accomplishing the assumed criteria. In the last two decades numerous information systems development methodologies ($ISDM_s$) has been proposed. Their heterogeneity inclines one to comparisons suggesting the proper selection of tools for practitioners.

The limited scope of this paper allows only to signal certain chosen questions on this subject. After choracterization of the methodological diversity prerequisites, the general classification of methodologies and specification of their constituents are presented. According to this pattern the introductory specification of solutions relevant to each component is brought forward. The particular emphasis is laid on choice of methods, techniques and computerized tools for pragmatic methodology.

1. INTRODUCTION

Information processing systems extension in an obvious way has prompted the investigation of methodological approaches and their applications. It has resulted from the generalization of increasing experience in the fields of analysis, design and implementation of information processing systems particulary in business and administration organizations. On the other hand the systems developers and users have become conscious of the essential methodological requirements in the light of failures in process of system development. The use of methodology allows gaining a number of advantages like the possibility of precise description and understanding of the investogated universe of discourse (UD) by non-professionals, control of the system design and operation process, coordination of the individual tasks within the interdisciplinary designing team and last but not least the acceleration and shortening of the whole system life cycle with the simultaneous improvement of the system quality. The proposed solutions have been the result, first and foremost, of the research center's efforts aiming at creation of the theoretical basis of so-called information systems science by integration of selected elements of computer science, management theory semantics, ontology , artificial intelligence, psychology and others. The experimental methodological works of industrial organizations have significantly contributed to the empirical side of this field.

As a matter of fact several hundred partial or overall methodological approaches have been proposed up to now. But they have still remained the set of versatile, individual proposals descending from the distinct scientific assumptions and experiences. The succeeding CRIS-1[17], CRIS-2[18] and CRIS-3[19] Conferences (IFIP WG 8.1 Working Conference on Comparative Review of Information Systems Design Methodologies) havent also had a direct effect on consistent, sound, generally accepted methodological foundations of Information Systems Development (ISD).

1/ Gdańsk University, 81-824 Sopot, A.Czerwonej 101, Poland

This diversity has risen from different objective and subjective reasons. Their critical analysis is outlined especially in [4]. In this paper only the difficulty in accurate structuring and definition of data processing tasks in business is stressed. The contradiction of this situation is the possibility of precise problem definition in sciences. The new types and directions in systems development [15] like fourth generation languages, prototyping, knowledge beses and expert systems have been conductive to problems definition's precision and hence the accuracy of the result received.

2. THE CONSTITUENTS OF INFORMATION SYSTEMS DEVELOPMENT METHODOLOGY

The taxonomies of ISDMS have been usually based on the set of well-matched criteria. Their spectrum is either broad, universal [3], [9]) or refers to the specific methodological aspects for example time perspective [11], requirements analysis [5], conceptual modelling ([11], [23]), graphical techniques [7] and computer-aided tools ([1],[12]).

In light of the research perfomed, the following dichotomic classification can be carried out according to the succeeding criteria:

- description of UD behaviour: data - and process-oriented,
- direction of ISD: top-down and bottom - up process,
- degree of formalization [5]: technical and social methodologies.

The viewpoints adopted in the specific methodology are suited the intermediate states of the extreme concepts.

In spite of the diversity among actual methodological approaches there are not many definitions of ISDM. It can be determined as a coherent, logically arranged set of methods and procedures of technical and organizational nature, which enables one to accomplish the system life cycle by system developers. It seems that with the present knowledge such methodology includes the following constituents:

- formalism s , the models of UD description, its statics and dynamics,
- the structure of ISD process in form of sequence of phases, subphases and succeeding tasks,
- the detailed methods and techniques of ISD, their documentation means and graphical notation adequately to the formalism proposed,
- computer - aided tools of ISD,
- quality estimation criteria of project and system and mechanisms of their check-up,
- specification of the essential requirements towards the members of interdisciplinary designing team and training materials,
- the rules of planning and control of system evolution.

3. A COMPARISON OF THE SELECTED METHODOLOGIES

The mentioned above set of potential ISDM constituents has been used for the general presentation of the current state of the art within ISDMs and the estimation of the selected approaches, commercially available. The limited scope of this contribution does not allow accomplishing detailed study in comparison to some prievansly quoted references.

The models of UD, called often conceptual models, should make it possible to describe UD statics and dynamics. The former is being described by such models as [17] entity-attribute-relotionship, predicate logic or database abstractions. The system dynamics is modelled by functions, events, transactions or operations. There are numerous autonomous ways of reality modelling for example in ISAC, SADT, SAK and SA-SD methodologies. They often combine the elements data and process modelling.

The process of ISD in general consists of four phases: analysis, design, implementation and operation. Practically it differs in the succeeding methodologies in structure and terminology according to their specific features. However the system life cycle of individual methodology remains in more or less consequent connection with the general process sequence.

From the pragmatic systems developers' viewpoint the essential role in ISD is played by the set of available methods techniques and computerized tools deciding of the designing style and its effect. Particularly in the last few years the "avalanche" increase of automatic aids for analysis and design as well as automated system generators has taken place. Automatic generation of the adequate graphics and data dictionary|directory offers and significantly accelerates possibilities for construction of creative, alternative proposals of solution . Some commercially available methodologies are presented and compared in tablé 1 in face of the most characteristic methods and tools used.

Table 1

Methods and aids for the selected ISDM

Methodology	Main concepts, methods, techniques	Automatic aids
D2S2-Development of Data Sharing Systems [14]	Entities, atributes, relationships, functions, events, entity diagrams, functional decomposition, data flow diagrams, data/functions matrices, information engineering	BIS/IPSE, DDEW, ER-Designer, Consultant Analyst
IE-Information Engineering or JMA-James Martin Associates [15]	Business models, data and process models, application generators, information center, prototyping, fourth generation languages (4GL)	IEF, IEW, Quick Build Work Bench, TAF
ISAC-Information Systems Work and Analysis of Changes [13]	Change analysis, tables of problems, needs, properties, graphs: A,-I,-C-,D-graphs, data and program structures, process tables, Jackson diagrams	BLUES, GRAPHDOC
LSDM - LBMS Systems Development Method [2]	Data flow diagrams logical data structuring, entity life histories, process outlines, third normal form (3 NF)	AUTO-MAT,MUST, TIP DEFINE IT, SDD
SADT-Structured Analysis and Design Technique [20]	ICOM/input-control-output-mechanism/of data and processes, data dictionary	IMP,PASILA,SAS IDFF 0
SAK -Systems Analysis and Konstruction or PROGRAMATOR [2]	Business diagrams, information requirements graphs, process and dialogue, diagrams,basic information sets, Jackson diagrams,4NF relations	DASAC
SA-SD-Structured Analysis and Structured Design ([6], [8])	Data flow diagrams, data dictionaries, 3NF relations, data immediate-access diagrams, structured English, decision trees, decision tables	AUTOMATE,BIS/IPSE EXcelerator,Must, TIP DEFNE IT,ProMod, Prokit* ANALYST

The most recommended criteria of system estimation are its consistency and completeness. The other elements of methodology are connected with the organization of ISD process Some partial approaches comprise only certain, selected ISDM constituents.

4. CONCLUSIONS

The designers of information processing systems searching for optimal ISDM have still to deal with increasing number and diversity even redundance of methodological approaches. Therefore, it is essential to compare their features and aims. This paper covers the proposal of making such comparative analysis on the grounds of ISDM potential constituents set. The choice of appropriate methodological alternative or specific tool should be based on taking into account a peculiar character of the UD and matching the relevant options.

The growing number of the complete and partial methodological solutions imposes upon analysts and designers the new style of system development. Its synthesis creates the environment or workbench [15] for ISD. It ought to constitute the set of versatile procedures methods, techniques and aids supporting the system developers. They should posses the advanced current knowledge and qualifications in this field to choose the relevant concepts a create original ones for solution of user problems and for meeting his requirements.

REFERENCES

[1] Analyst Workbench, Infotech State of the Art Report, Pergamon, Maidenhead 1987.
[2] Banteman, J.-P.: A Feature Analysis of the LBMS Systems Development Method, in [21], 3-25
[3] Brandt, I.: A Comparative Study of Information Systems Design Methodologies, in [18], 9-36.
[4] Bubenko, J.A.Jr.: Information System Methodologies - A Research View, in [19], 289-318.
[5] Ceri, S.: Requirements Collection and Analysis in Information Systems Design, in [10], 205-217.
[6] De Marco T., Structured Analysis and System Specification, Yourdon Press, New York 1978.
[7] Floyd, C.: A Comparative Evaluation of System Development Methods, in [19], 19-54.
[8] Gane, C.P., T.Sarson: Structured Systems Analysis, Tools and Techniques, Prentice Hall, New York 1979.
[9] Iivari J., P.Kerola: A Sociocybernatic Framework for the Feature Analysis of Information Systems Design Methodologies, in [18], 87-139.
[10] Kugler, H.J. /ed/ Information Processing 86, North-Holland, Amsterdam 1986.
[11] Kung, C.H.: An Analysis of Three Conceptual Models with Time Perspective, in [18], 141-167.
[12] Lokeman, P.C., H.C.Mayr: Information Systems Design, Techniques and Software Support, in [10], 617-634.
[13] Lundeberg, M., G.Goldkuhl, and A.Nilsson: Information Systems Development, A Systematic Approach Prentice Hall, New York 1981.
[14] MacDonald, I.G.: Information Engineering - An Improved, Automatable Methodology for the Design of Data Sharing Systems, in [18], 173-224.
[15] Martin, J.: Fourth Generation Languages, Prentice Hall, New York 1985.
[16] Olive, A.: Analysis of Conceptual and Logical Models in Information Systems Design Methodologies, in [18], 63-85.
[17] Olle, T.W., G.H.Sol, and A.A.Verrijn-Stuart /eds/: Information Systems Design Methodologies: A Comparative Review, North-Holland, Amsterdam 1982.
[18] Olle, T.W., H.G.Sol, and C.J.Tully /eds/: Information Systems Design Methodologies: A Feature Analysis, North-Holland, Amsterdam 1983.

[19] Olle,T.W., H.G.Sol, and A.A.Verrijn-Stuart /eds/: Information Systems Design Methodologies: Improving the Practice, North-Holland,Amsterdam 1986
[20] Ross,D.T.: Application and Extensions of SADT, Computer 2(1985),25-34.
[21] Structured Methods, Infotech State of the Art Report, Pergamon, Maidenhead 1984.
[22] Wigander, K. et al.: Structured Analysis and Design of Information Systems, McGraw Hill, New York 1987.
[23] Wrycza, S.: Conceptual Modelling of the Data Bases of Business Organization Gdańsk University Press, Gdańsk 1985 (in Polish).

MODEL BASED DEFAULT DATA GENERATION FOR NET SYSTEMS

D. Singer[1/]

Introduction

When realizing large databases, the main difficulties arise in the collection of data. In general a great part of data is not available on account of man power capacity for administration, due to the difficulties inherent in the measuring methods or even to the lack of time for gathering the data. Therefore, the goal is intimate to reconstruct missing data based on available ones - in other terms -, to reconstruct the missing data on the basis of implicit informations inherent in the data base.

The reconstruction of data can be made according to:
- functional relationships between the data,
- first-order logic inference,
- statistical considerations,
- fuzzy logic inference.

The reconstructed data are the default data.

The methods of data reconstruction can be used with success for large systems consisting of a few types of elements joined together according to relatively simple relationships. Such systems are primarily electric, water, gas and canal nets. The system parameters are here partly of topological character, partly they are geometrical and material-constants. The functional relationships between the system parameters are determined, beyond the topology, by the constitutive relationships of the single member types (branches) of the net and the conservation laws of energy and mass.

The types of data reconstruction for net systems

From the viewpoint of reconstructibility net data can be classified as follows:

- data which can be reconstructed from the data belonging to a branch incident to one of the two ends of the branch in question. Data obtained in this manner should be named in the following shortly as A-type default data;
- data which can be reconstructed from data of branches incident to both ends of the branch in question: B-type default data;
- data obtained according to statistical considerations concerning the net and its environment: C-type default data.

On Fig. 1, a typical portion of a net is demonstrated. The branch b is incident on the left end 1 with the branches i,j,k, on the right end 2 with the branches p,r,s... . Let us assume that the missing data of branch b should be replaced by default data

1/ Computer and Automation Institute, Hungarian Academy of Sciences, H-1111, Budapest, Kende u. 13/17, Hungary

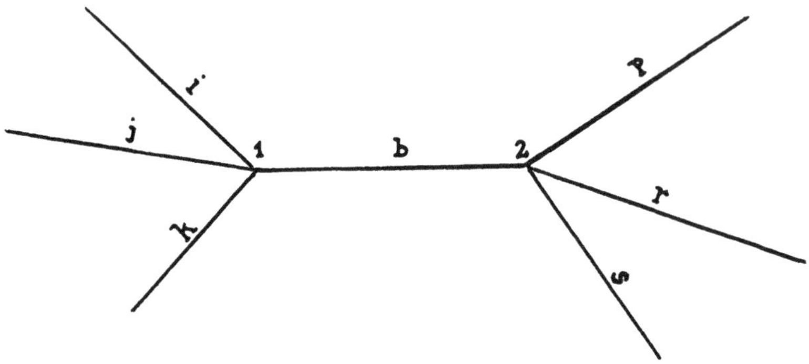

Fig. 1.

derived from the data belonging to the branches i,j,k and p,r,s, respectively. The default value of the parameter belonging to the left end 1 of b equals obviously to the appropriate right-end parameter of branches i,j,k. The default value of the right-end 2 parameter equals to the appropriate left-end parameter of branches p,r,s. Such default data are of type A, e.g., the geodetical height of the ends of a canal section reconstructed from the geodetical heights of sections incident to it, or the end temperatures of a branch of a communal heating net deduced from the incident branch temperatures.

A B-type default value of branch b, e.g., the diameter of a gas pipe, the material or the data of its construction, can be obtained from the corresponding parameters of the branches incident to it on both ends. The reconstruction of the diameter, the material, or date of branch b is based on the hypothesis that if one of the branches i,j,k adjacent in 1 have identical diameters, material or dates with those of one of the branches p,r,s adjacent in 2, one can state with high probability that the same values hold for branch b, too.

Default data of C-type are the reconstructed mud heights in the individual sections of a canal net, A plausible assumption is here that the mud height is in correlation with the slope of the canal section. On the basis of the mud heights, measured in a larger number of canal sections at different time spans one can calculate the correlation coefficient between the slopes of the canal sections and the rate of elevation of the mud and from this the mud heights for not measured canal sections.

A formal approach

The default rules given in the foregoing, can be formulated as follows. For the A-type default data:

$$\text{IF } \exists \text{ FB}(y) \text{ THEN}[\text{BR}(x) = \text{FB}(y)]$$
$$\text{IF } \exists \text{ SB}(x) \text{ THEN}[\text{BR}(y) = \text{SB}(x)] \qquad (1)$$

where BR denotes the branch in consideration, FB the foregoing and SB the subsequent branches, x and y are the values of the investigated left end and right end parameters (we assume that the flow direction is from the left to the right). The default values BR(x) and BR(y) can be found by searching in the database of the net (if it exists), for the records FB and SB containing the values y and x.

In order to determine default data of type B, - for the case of only one foregoing and one subsequent branch -, the rule has the following form:

$$\text{IF}\{[\exists \text{FB}(z),\ \text{SB}(z)] \text{ AND SB}(z) = \text{FB}(z)\}$$
$$\text{THEN } [\text{BR}(z) = \text{SB}(z)] \qquad (2)$$

z denotes the branch parameter.

If there are more foregoing and/or subsequent branches, the rules are slightly more complicated. When there exists only one foregoing branch and two subsequent branches, the rule has the following form:

$$\text{IF}\{[\exists \text{FB}(z),\ \text{SB1}(z),\text{SB2}(z)] \text{ AND } \{[\text{SB1}(z) = \text{FB}(z)]$$
$$\text{OR } [\text{SB2}(z) = \text{FB}(z)]\} \text{ THEN}[\text{BR}(z) = \text{SB}(z)]\} \qquad (3)$$

SB1 and SB2 denote the two succeeding branches.

The case of two foregoing branches and one subsequent branch is also of practical importance. The default rule has in this case the form:

$$\text{IF}\{[\exists \text{FB1}(z),\ \text{FB2}(z),\exists \text{SBz}] \text{ AND } [\text{SB}(z) = \text{FB1}(z)]$$
$$\text{OR}[\text{SB}(z) = \text{FB2})]\} \text{ THEN}[\text{BR}(z) = \text{FB}(z)]\}\} \qquad (4)$$

Based on the above considerations, a FORTRAN program has been elaborated to reconstruct missing data of large municipal canal nets. The program is also capable to prove the consistency of data in the canal data base, comparing the computed default data tothe measured ones (if they exist) [1].

The above rules represent in reality only plausible propositions assigning a set of possible values to the quantified premisse. The rules can, therefore, be interpreted as possibility assignment equations in fuzzy logic sense of the general form [2]:

$$\pi_X(u) = \text{Poss}(X=u) = \mu_F(u)$$

where π_s is the possibility distribution function, $\mu_F(u)$ is the membership function of the fuzzy relation F representing in our case the rules in (1) to (4). Poss{X=u} should be read as the possibility that X may have the value μ. With the discretized membership function $\mu_F(u)$, one can define the characteristic number, the sigma-count

$$\Sigma\ \text{Count}(F) = \sum_{i=1}^{n} u_i$$

which can be interpreted in our case as the measure of reliability of the default value obtained by the above rules.

[1] D. Singer, P. Móritz: DELT - Program for Generation of Default Values and for the Control of Primary Date. Report of the Computer and Automation Institute, HAS, Theme No.: 620258, December 1986, Budapest (in Hungarian).

[2] L.A. Zadeh: The Role of Fuzzy Logic in the Management of Uncertainty in Expert Systems, Fuzzy Sets, 11 (1983) 199-227.

VARIOUS UNITARY BASED ROBUSTNESS STUDIES ON A FURNACE CONTROL SYSTEM

U. Jumar [1)]

1. Introduction

In any model-based controller design procedure the engineer is commonly confronted with some significant model uncertainties, consideration of which is relevant to the design result. Frequently, the plant is not completely known, since the only information is based on measured step responses, either a simplified model is used to deal with a complex decentralized control problem, or some plants cannot be described by a linear model at all.

In the time domain the input-output (I/O-)behaviour of a linear multivariable (MIMO-)system is represented by its impulse response matrix. The error attached to this nonparametrical model may be taken into consideration by tolerance bands for the elements of this matrix in the sense of unknown-but-bounded uncertainties. This corresponds to the engineer's thinking and allows him to deal with various kinds of model errors by a unified approach. In this case, a robustness analysis is aimed at deciding whether the closed-loop systems performance lies in the acceptable range for all possible model uncertainties within the fixed bounds.

The primary concern of this paper is to point out how different aspects of robustness, considered to be essential from a practical point of view, may be treated by the unitary approach of robustness analysis proposed by Lunze [5]. His method starts from a decomposition of the system to be analysed in an approximate model, and an error model which embodies the uncertain model part and is described by means of multivariable upper error bounds. These bounds are represented by the output of a majorizing linear "comparison system". After some simple calculations on the level of comparison systems the method leads to transparent quantitative characterizations of robustness, allowing both the examination of stability and transient performance of the closed-loop system.

Particularly, in this paper robustness analysis is applied to a control system with simple multivariable tuning regulators [9] as well as to a configuration of such controllers in the spirit of Davison's [1] robust decentralized control. As a consequence, additional applications of robustness study arise: evaluating the effect of a decentralized design based on isolated subsystems, checking connective stability, and examining quantitatively the possibility of controller retuning.

1) Technische Universität "Otto von Guericke" Magdeburg
Sektion Automatisierungstechnik und Elektrotechnik
Boleslaw-Bierut-Platz 5   DDR-3040 Magdeburg

2. Principle of Robustness Analysis

For the sake of completeness in this paragraph the results of [5] to evaluate the input-output behaviour of incompletely known systems are briefly summarized. According to figure 1a it is assumed, that the system S is decomposed into two subsystems SS, interconnected through the signal vectors s and z.

Subsystem SS1 is completely known and described by the two convolution equations

$$y = G_{yu} * u + G_{ys} * s, \quad z = G_{zu} * u + G_{zs} * s, \tag{1}$$

where u(t), y(t) are the control input vector and the output vector, respectively, and G denotes impulse response matrices G(t).

Subsystem SS2 has the input vector z and the output vector s, and embodies the incompletely known properties of system S. Instead of an exact description of the multivariable dynamical, possibly non-linear and time varying SS2, a linear MIMO-system

$$r(t) = V_F * w \tag{2}$$

is used. Its impulse response matrix is determined such that for every input z(t) and w(t) = |z(t)| the output r(t) maximizes the absolute value of the output s(t) of SS2:

$$r(t) = V_F * |z| \geq |s(t)| \quad \text{for all } t \geq 0. \tag{3}$$

Then the system (2) is called a comparison system to SS2.

Consider the composite system S, where SS1 is described by (1) and SS2 by (3). Assume that SS1 and the comparison system for SS2 are stable. Then the overall composite system S is stable, if

$$\lambda_p\left\{W_F \, W_{zs}\right\} < 1 \quad (4) \qquad W_F = \int_0^\infty V_F \, dt, \qquad W_{zs} = \int_0^\infty |G_{zs}| \, dt, \quad (4,5)$$

where λ_p denotes the Perron-root and W_F, W_{zs} are the static reinforcements of the comparison systems to SS2 and the G_{zs}-part of SS1 respectively (cf. Lunze [6]).

To estimate the I/O-behaviour of S, the isolated SS1 is used as an approximate model

$$\hat{y} = G_{yu} * u \tag{6}$$

with ^ denoting the approximative character. Assume that (4) has been successfuly tested. Then an upper bound of the approximation error $y - \hat{y}$ is given by

$$|y - \hat{y}| \leq |G_{ys}| * \bar{V} * |G_{zu}| * |u|, \qquad \bar{V} = V_F + V_F * |G_{zs}| * \bar{V}. \tag{7}$$

The general model structure of fig. 1a offers freedom in the choice of the signals s and z, thus enabling the actual decomposition of S to be tailormade with respect to the practical circumstances that lead to the uncertainty of the model [7]. In this paper two special model decompositions are focused which are widely used in literature, too. Let the plant, assumed to be stable, be described by the approximate model

$$\hat{y} = \hat{G} * u. \tag{8}$$

To deal with the uncertainty of this model often one of the structures shown in fig. 1b and 1c is suitable [2]:

$$G = \hat{G}(t) + \Delta G(t), \quad |\Delta G(t)| \le V_F(t) \tag{9}$$

$$G = \hat{G}(t)\ (I + \mathrm{diag}[k_i(t)]), \quad \mathrm{diag}[|k_i(t)|] \le W_F \text{ for all } t \ge 0. \tag{10}$$

By (9) a parallel arrangement of approximate model and error model is described. All errors of the matrix elements from G(t) are assumed to be independent of each other. In case of (10) the model error is in cascade. The errors, interpreted as relative ones, are related to the control inputs. In contrast to (9), here the upper bounds are described by a static MIMO-comparison system with the reinforcement W_F.

3. The Furnace Control System

While preparing a control scheme for industrial tube furnaces the author was occupied with the design on an appropriate pilot plant furnace at the Technical University of Magdeburg. This plant provides an illustrative non-academic example within this paper. Figure 2 shows a simplified scheme of the tube furnace [3].

3.1. The Control Task

The primary task is to regulate the product's output temperature subject to unknown changes in the combustion. At the same time, any asymmetry in heating the two main tubes has to be avoided, thus requiring the compensation of temperature variances in tubes 1 and 2. However, due to processes succeeding the furnace this control should not influence the throughput of product. The oxygen concentration in the exhaust gas is introduced as a controlled variable the stabilization of which is relevant to an efficient combustion. Furthermore, regulation of the pressure in the firebox is desired.

3.2. The Model Situation

The model knowledge was confined to measured step responses. Starting from this point, a nonparametrical approximate model (8) was obtained by visual inspection of the "nominal" step responses

$$\hat{H}(t) = \int_0^t \hat{G}(\tau)\, d\tau, \tag{11}$$

fitting simple maximum third order delay-lag transfer functions, and calculating the corresponding impulse responses [3]. A whole set of H(t)-measurement data offered the opportunity to evaluate upper bounds for the uncertainties of the established model with respect to measurement errors, derivations from the operating point, etc. The way outlined in [2] was followed to obtain tolerance bands for G(t) and to describe these bands by a comparison system according to (9) or (10).

3.3. The Structure of Control

In [3] the authors elaborate the structure of an appropriate control system. Investigating the dynamical behaviour of the furnace it becomes evident that this multivariable plant may be split into a "slow" part of product temperature control and a "faster" one, containing the regulation of combustion. This was considered to be important in choosing a decentralized structure of control.

Figure 2 illustrates the definition of modified I/O-signals. According to §3.1 the temperature difference is directly introduced as controlled variable y_1, the new control input u_1 acts on both tube flows, and input u_3 is identical with the tracking signal of an additional inner control loop. The controller configuration is:

- controller 1: two-variable, PI-type, $[y_1, y_2]$ paired with $[u_1, u_2]$
- controller 2: single loop, I-type, y_3 paired with u_3
- controller 3: single loop, PI-type, y_4 paired with u_4.

4. Robustness Studies

The control system of §3 will provide the background for the selected robustness investigations described in this paragraph.

4.1. Design based on an approximate plant model

For simplicity, this chapter is concerned with the design of controller 1 alone, considering the corresponding part of the plant model. Despite this fact, any special indexing is omitted. Employing the tuning approach [9] a promising PI-type controller is given by

$$u = K_P\, e + K_I \int_0^t e(\tau)\, d\tau, \quad e = y - v, \quad K_P = \beta\, \tilde{K}_P, \quad K_I = \alpha\, \tilde{K}_I \qquad (12)$$

with v, e denoting the command input vector and control error vector, respectively. However, to reduce the number of practical experiments a partial replacement of on-line tuning the parameters $\alpha > 0$ and $\beta \geq 0$ by computer simulations is helpful. In this case, a quantitative robustness analysis becomes necessary, since design and simulation use an approximate model (8) whereas the found controller has to operate on the real plant.

First and foremost, examining the stability of the furnace control system with controller 1 applied, is of interest. We assume that the controller has been tuned by means of a convolution based simulation [2]:

$$y = G_{yv} * v, \quad G_{yv} = -[\hat{G}K_P + \hat{H}K_I] + [\hat{G}K_P + \hat{H}K_I] * G_{yv} \tag{13}$$

$$u = G_{uv} * v, \quad G_{uv} = -[K_I + K_P\delta(t)] + [K_I + k_P\delta(t)] * G_{yv}. \tag{14}$$

Furthermore, we presuppose the unknown-but-bounded uncertainties of (8) having been classified as unstructured in the manner of (9) and fig. 1b. As a result of evaluating these model errors matrix W_F is obtained (cf. [2] for a detailed interpretation).

Hence, the input-output stability of the closed-loop real system can be proved by (4), after SS1 in fig. 1a has been replaced by the controlled approximate plant model. The adequate adaption of the general decomposition shown in fig. 1a yields

$$W_{zs} = \int_0^\infty |G_{uf}|\, dt = \int_0^\infty |G_{uv}|\, dt. \tag{15}$$

Searching for the local extrema of the closed-loop step responses

$$H_{uv}(t) = \int_0^t G_{uv}(\tau)\, d\tau, \tag{16}$$

commonly having been calculated within the simulation (13)-(14), matrix (15) is derived without any additional integration.

4.2. Decentralized controller design

Even when the overall control system is of comparative transparency, the design of decentral controllers is simplified, if only the relevant part of the plant is considered, ignoring all the plant's cross couplings at first. By making use of this way, the two SISO-controllers of the furnace are determined.

Controller agent 2 was chosen to demonstrate the performed robustness analysis, which should confirm, that in addition to a simulation of isolated SISO-control loops the interconnected system fulfils the design requirements, as well. Ignoring the couplings $[u_3 - y_2]$ and $[u_2 - y_3]$ an I-type controller 2 was found to control the plant $[u_3 - y_3]$ satisfactorily.

Since the previously designed controller 1 incorporates an I-part, ensuring that e(t) asymptotically vanishes for constant external signals, it was easy to show with our example, that controller 1 causes a 5 % relative derivation of the $[u_3 - y_3]$ plant's original static gain. Therefore the use of (10) with

$$|k(t)| \leq 0.05 = W_F \tag{17}$$

is obviously a promising attempt to describe the effect of controller 1. Whereas SS1 in fig. 1c now describes the SISO-control loop under study,

SS2 represents the upper bounds (17) for the model error caused by the influence of controller agent 1. If the general case equations (6),(7) are rewritten to cope with the problem of fig. 1c (cf. [2]), the tolerance bands of fig. 3 for the tracking and disturbance closed-loop step responses can be calculated, showing that in spite of design simplification the control is acceptable. Whether controller 1 is connected or not, in any case the transient behaviour of the combustion control loop lies within the given bands.

4.3. Checking connective stability

In a similar way the connective stability of the decentralized control system may be assessed, i.e. the I/O-stability of the overall system when arbitrary disconnections of controller agents occur. With regard to the furnace example a slight extension of the well established results to this problem is necessary, however, for the approximative character of the plant description has to be taken into consideration, and no exact model is at the designers disposal.

Deriving profit of the general structure from fig. 1a, subsystem SS1 is used to describe the decentrally controlled approximate plant model, whereas SS2 represents both, multivariable upper error bounds for the whole set of couplings between the decentral subsystems and for the uncertainties of the approximate MIMO-plant model. This results in an actual system decomposition as illustrated in fig. 4 for the case of two controller agents. As a matter of fact, only the first two decentral subsystems of the furnace are relevant to the connective stability, for the third one is coupled merely in one direction.

According to fig. 4 the matrix W_{zs}, required to check (4), becomes

$$W_{zs} = \text{diag}\left\{ W^1_{uf} \quad W^2_{uf} \right\} \tag{18}$$

with the upper index denoting the number of decentral subsystem 1 and 2. Using the error bounds W^1_F and W^2_F from §4.1 as well as some realistic bounds for the incompletely known couplings $[u_2 - y_3]$, $[u_3 - y_2]$ the static reinforcement matrix W_F of the comparison system for SS2 can be constructed. Finally, a calculation of the Perron-root (4) confirms the connective stability.

4.4. Retuning the found controllers

Once a MIMO-controller has been found during any model based design procedure, a variation of controller parameters within practical implementation of the system or later will provide some difficulties from the theoretical point of view in general. Frequently, nothing can be said about allowed parameter variations nor about the resulting I/O-behaviour of the closed-loop. However, this problem can be tackled easily through the method of robustness analysis utilized in this paper.

Although a more generalized consideration would be possible, we restrict our analysis to the retuning of central tuning regulators as they are explored in [9]. Again, the two-variable PI-controller of the furnace serves as example. Figure 5 demonstrates the assumed situation. The designed and implemented controller of type (12), termed with the upper index "o", is to be retuned by varying the parameters α and β:

$$\alpha^1 = \alpha^o (1 + \Delta\alpha) \quad \text{or} \quad \beta^1 = \beta^o (1 + \Delta\beta). \tag{19}$$

Hence, the question arises, whether such a retuning is allowed with respect to retained closed-loop stability. Moreover, the plant may be incompletely known, as indicated in fig. 5 by an additional error model.

An easy to handle stability test can be deduced in two steps. First, the subsystems SS1-1 and SS1-2 from fig. 5 are combined to SS1, presupposing the stability of the real plant with the original controller applied. Second, the static I/O-behaviour of composite SS1 is described by a new comparison system. Subsequently, stability of this comparison system coupled with SS2 can be examined by (4). In this way the stability condition

$$\lambda_p \left\{ |\Delta\alpha| \tilde{W}^I_{zs} + |\Delta\beta| \tilde{W}^P_{zs} \right\} < 1 \tag{20}$$

$$\begin{aligned} \tilde{W}^I_{zs} &= W^I_{zs} + W_{zf} (I - W_F W_{uf})^{-1} W_F W^I_{us} \\ \tilde{W}^P_{zs} &= W^P_{zs} + W_{zf} (I - W_F W_{uf})^{-1} W_F W^P_{us} \end{aligned} \tag{21}$$

was derived, where all matrices W term static reinforcement matrices which are defined like in (5) and are related to

$$G_{zf} = \hat{G} + [\hat{G}K^o_P + \hat{H}K^o_I] * G_{zf}, \quad H_{zf} = \int_0^t G_{zf}(\tau)\, d\tau \tag{22}$$

$$G^P_{zs} = G_{zf} K^o_P, \qquad G^I_{zs} = H_{zf} K^o_I \tag{23}$$

$$G^P_{us} = K^o_P [I\delta(t) + G^P_{zs}], \qquad G^I_{us} = \int_0^t K^o_I [I\delta(t) + G^I_{zs}(\tau)]\, d\tau. \tag{24}$$

If one of the tuning parameters is assumed to remain unchanged, the allowed variation of the other one is given by

$$|\Delta\alpha| < 1/\lambda_p(\tilde{W}^I_{zs}), \qquad |\Delta\beta| < 1/\lambda_p(\tilde{W}^P_{zs}). \tag{25}$$

Taking into account 10 % relative model errors in the manner of (10), we found $|\Delta\alpha| < 0.79$, $|\Delta\beta| < 0.3$ for controller 1 of our furnace. Mention should be made, that a consecutive repetition of this procedure results in some kind of "automatic" controller design. Starting from a stability guaranteeing initial parameter set, according to (25) α and β can be increased step by step by computing the new closed-loop system and subsequently determining the next possible parameter variation.

5. Conclusions

With the assumption that a nonparametrical impulse response model of the considered pilot plant furnace is available, the paper has demonstrated the application of various unitary based robustness investigations. The method employed requires merely weak computer facilities, for the occuring convolution equations can be solved numerically without any difficulties. Computed tolerance bands for the closed-loop transient behaviour as well as upper bounds for an allowed controller retuning illustrate that quantitative statements on the robustness were derived instead of a stability test alone. Since the concept is based on multivariable comparison systems, the MIMO-character of the assessed control system is preserved even in the description of its uncertain part. Combined with an especially adapted structure of decomposition, this results in a reasonable degree of conservatism of the robustness analysis. It is worthwhile mentioning that, extending the presented simulation based results, a similar robustness study was successfuly accomplished for the industrial crude oil distillation column of [4]. Finally, we note that the method from [5] applied in this paper is also in the spirit as that of other researchers not cited here, e.g. [8].

References

[1] Davison, E.J.: The robust decentralized control of a general servomechanism problem. IEEE Trans. AC 21 (1976) 1, 14-24.

[2] Jumar, U. and U. Korn: The use of quantitative robustness analysis for the design of multivariable tuning regulators. IFAC 10th Congress Preprints, vol. VIII, 251-256, Munich 1987.

[3] Jumar, U. and H. Krüger: Untersuchungen zur Regelungsstruktur eines Industrieofens. Messen Steuern Regeln, Berlin 30 (1987) 5, 199-203.

[4] Korn, U.: A decentralized control system for a distillation column. IFAC 9th Congress Preprints, vol. III, 92-97, Budapest 1984.

[5] Lunze, J.: A unified approach for designing robust linear feedback controllers. IFAC 9th Congress Preprints, vol. IX, 14.2/B-5, Budapest 1984.

[6] Lunze, J.: The design of robust feedback controllers in the time domain. Int. J. Control 39 (1984) 6, 1243-1260.

[7] Lunze, J.: Robust multivariable feedback control. ZKI-Informationen 3/84, Berlin 1984.

[8] Owens, D.H. and A. Chotai: Robust controller design for linear dynamic systems using approximate models. Proc. IEE D-130 (1983) 2, 45-56.

[9] Penttinen, J. and H.N. Koivo: Multivariable tuning regulators for unknown systems. Automatica 16 (1980) 4, 393-398.

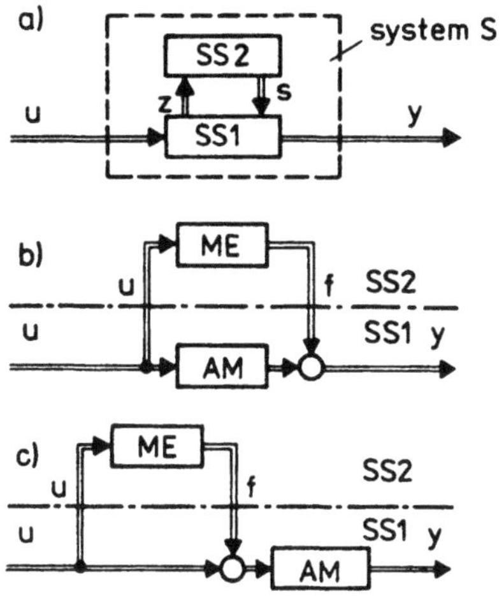

a) general structure of decomposition

b) parallel arrangement of approximate model AM and model error ME

c) model error in cascade

Fig. 1. Model of the incompletely known system

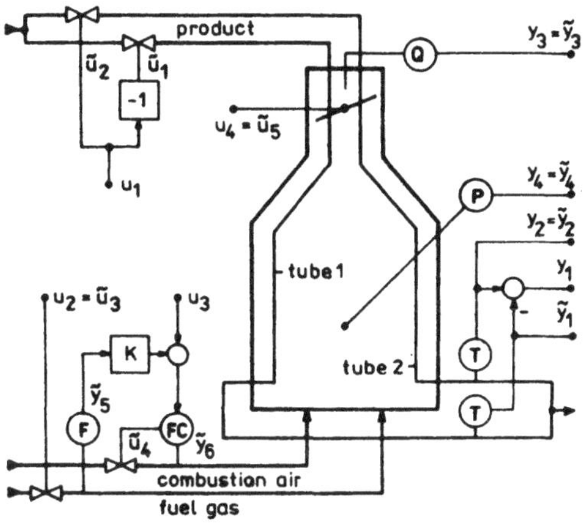

Fig. 2. The pilot plant tube furnace

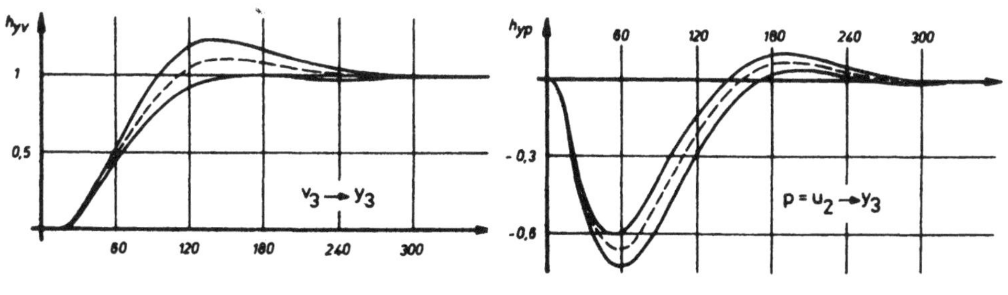

tracking responses

response to an input disturbance

Fig. 3. Evaluation of the combustion control closed-loop step responses

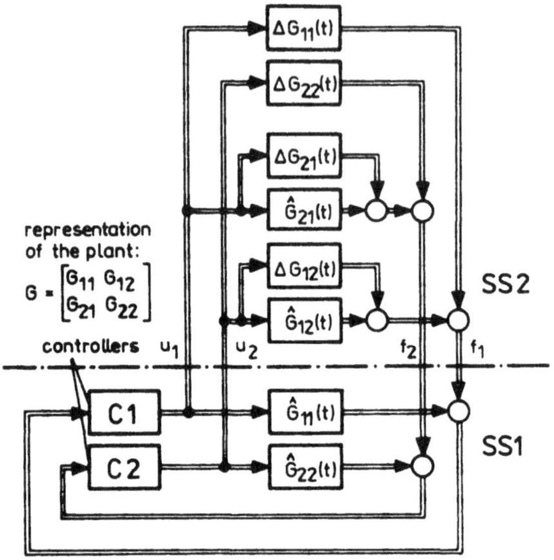

Fig. 4. Decomposed model of the decentralized control system

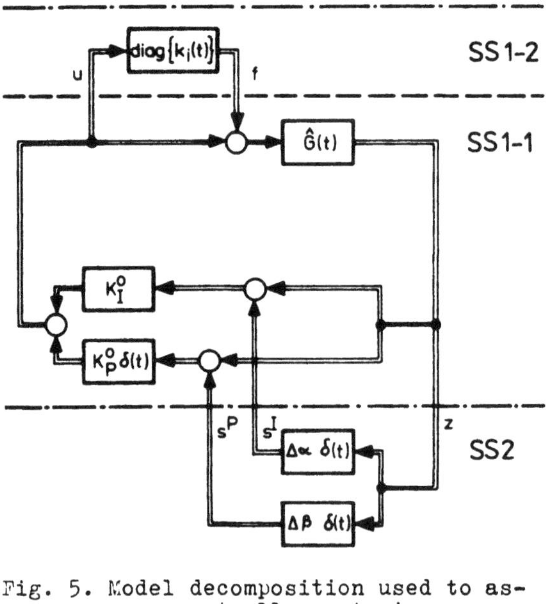

Fig. 5. Model decomposition used to assess controller retuning

OPTIMISATION PROBLEM FOR BY-PASS ROUTES CREATION IN THE CASE OF A TELETRANSMISSION LINE BREAK-DOWN

Andrzej Winnicki, Ph.D.[1], Grzegorz Płoszajski, Ph.D.[2]

1. Introduction

Determination of by-pass routes in teletransmission network with broken trunk-line is investigated. These routes are created from group segments switched in nodes. Technical means used for them are taken away from nonaffected links. Since the grade of service in these links decreases, the right choice of by-pass routes is of prime importance.

The optimisation problem formulated for this purpose is practically unsolvable for real networks. It results from the large scale of the network model as well as from various constraints which must be taken into account. The quality of the transmission, switching principles applied in the teletransmission, capacity of lines, number of filters and the grade of service in nonaffected links are the most important factors [5],[6]. In the paper step by step reduction of the original optimisation problem is presented. Suboptimal solutions one can find for networks of any size.

2. Determination of a failure surrounding

Transmission quality requirements limit the number of group segments used for by-pass routes creation [1]. Hence significant part of possible routes (too long in these sense) must not be taken into account and the problem of optimisation becomes a'priori limited to certain fragment of the network called a failure surrounding. Different algorithms for this purpose were proposed [4]. One of them is presented below.

Topology of the teletransmission network, index of the affected trunk-line, admissible lenght of created routes (expressed by the number of groupe segments – n_{max}) and the list of existing routes are the input data.

Step 1. Determine indexes of affected links;

Step 2. Enclose origin, transit and destination nodes for links obtained in step 1 to the failure surrounding;

Step 3. Enclose all nodes which make possible to create by-pass routes of lenght not exceeding n_{max}

---

1) Institute of Mathematics, Warsaw Technical University, pl. Jednosci Robotniczej 1, 00-665 Warszawa, Poland

2) Institute of Automatic Control, Warsaw Technical University, ul. Nowowiejska 15/19, 00-665 Warszawa, Poland

Step 4. Enclose transmission lines between the nodes appointed in step 2 and step 3.

The effect of the algorithm is illustrated by Fig. 1. Graph of the network is shown on Fig. 1a). The line between nodes A and B used by links AB and AF is assumed to be damaged. Succesive phases of the failure surrounding creation are presented in Fig. 1b), 1c), 1d).

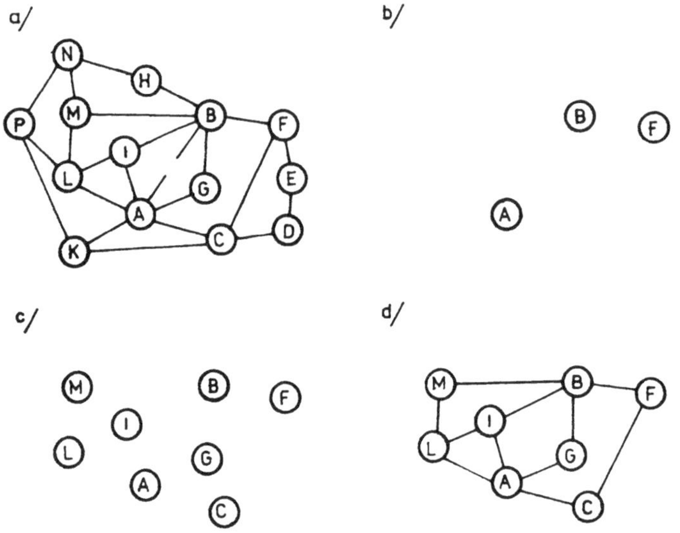

Fig.1. The failure surrounding creation for the link AB: a) graph of the trunk line network, b) execution of step 2 of the algorithm, c) execution of step 3 of the algorithm (n= 3), d) execution of step 4 of the algorithm.

The size of the failure surrounding is determinated by parameter n_{max}. For sufficiently large value n_{max} the failure surrounding is identical with the whole network. In practice, however, only small values of these parameter are used. Thus, the failure surrounding comprises only a part of the whole network. In the next stage of presented method so called *reservation network* within the failure surrounding is determined.

3. Determination of a reservation network

A reservation network is a set of technical means which can be potentially used to form by-pass routes. It consists of group segments and filters forming the proper reserve of the network and taken away from not affected link together with parts of routes which becames free in result of break-down.

Usually there are some group levels and multirouting configuration of links in real network. Therefore different reservation network are obtained, if different routes of links are devoted to potential reduction. The right choice of such network become a problem by itself. Moreoever, reservation network of this type seems to be too much complicated for any optimisation process. Hence, decomposition according to group levels is proposed. In the begining by-pass routing determination is performed for the highest group level and then succesively for lower levels (if necessary). This approach is substantiated by computational reasons (decreases and simplifies optimisation problems) as well as practical requirements (assures decrease of necessary switchings).

The concept of so called *marked nodes* was proposed to assure the right choice of each group levels reservation network. The set of marked nodes consists of nodes prefered as transit nodes for created by-pass routes. According to former experience nodes coincident to those which determine broken links should be enclosed to this set [1].

An algorithm used for reservation network determination consists of the following steps:

Step 1. Determine admissible number of potentially reduced groups for each link in affected area;

Step 2. Enclose elements of proper reserve to the reservation network;

Step 3. Enclose efficient group segments from affected links;

Step 4. From each link consisting of one route enclose group segments and transfer filters of this route with respect to step 2;

Step 5. From each link with several routes enclose group segments and transfer filters in the following succesion:

- from routes which use marked nodes as a transit nodes,
- from other routes (if necessary) with respect to step 1.

For more detailed informations see [2].

Application of this algorithm for determination of reservation network with secondary group segments is illustrated in Fig. 2. Graph of the affected area for link CD is shown in Fig. 2a). Numbers placed at egges of the graph represents capacities of transmission lines. The graph of result reservation network is shown in Fig. 2b).

An optimisation problems formulated within reservation networks of this nature are sufficiently simple to be solved by typical optimisation methods. Numerical experiments are presented in [4].

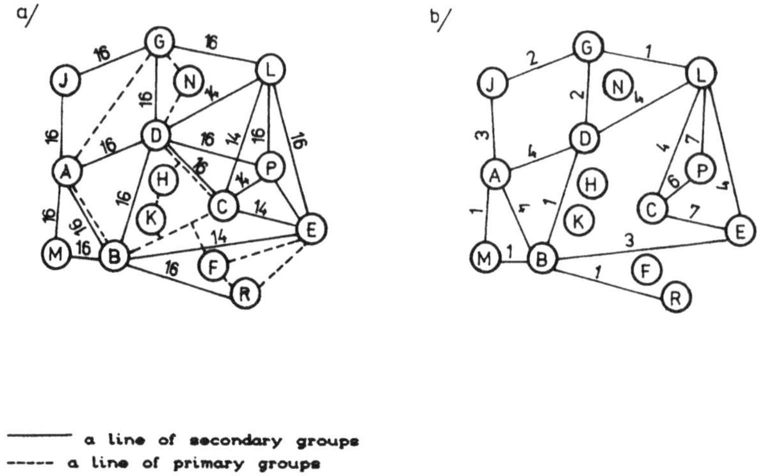

Fig. 2. Determination of the reservation network: a) the graph of the failure surrounding, b) the graph of the reservation of secondary groups.

4. Conclusions

Optimisation problems of discussed nature can be solved by the use of presented method for network of any size. Moreover, decomposition of the computational process makes possible to use the method in conversional manner and in particular to modify the problem according to additional, nontypical requirements. All algorithms can be utilised in real-time control system as the time of computation is relatively short.

5. References

[1] "Opracowanie uproszczonego algorytmu rozmieszczania kanałów rezerwowych w sieci linii teletransmisyjnych", Praca zbiorowa IA PW (1984)

[2] "Wyznaczanie dróg obejściowych przy uszkodzeniach linii teletransmisyjnych z uwzględnieniem ograniczeń komutacyjnych węzłów", Praca zbiorowa, IA PW (1985)

[3] "Opracowanie założeń ekspertmentalnego uruchomienia procedur awaryjnego wyznaczania dróg zastępczych w sieci teletransmisyjnej wraz z odpowiednim oprogramowaniem", Praca zbiorowa, IA PW (1986)

[4] "Badanie metod rozwiązywania zadań przestrajania sieci teletransmisyjnej", Praca zbiorowa, IA PW (1987)

[5] Wydro K.B. "Przeciwdziałanie skutkom awarii linii w sieci telekomunikacyjnej" Przegląd telekomunikacyjny 10-11 (1986)

[6] Wydro K. "Problemy wyznaczania dróg zastępczych zastępczych przy awarii linii w sieci teletransmisyjnej" Przegląd telekomunikacyjny 10-11 (1988)

SUBSTITUTIONAL ROUTING FOR TRUNK LINE BREAK-DOWN IN THE TELETRANSMISSION NETWORK

dr ing. Kornel Wydro [1/]

1. INTRODUCTION

Trunk-line break-down in teletransmission network interrupts some number of the routes of some different links. It makes necessary to restore, at least partially, failed transmission possibilities for each affected link. Usually it can be done by creating suitable substitutional (by pass) routes from the technical means taken away from neighboring links, not being affected by the break-down.

Problem researchings [1] , [2], [3], [4] gave results suitable for practical application. Describing a full problems structure, is interesting from not only telecommunication point of view.

2. SYSTEMS MODEL

The system constitutes a network of the group segments existing in the individual trunk-lines and other transmission devices placed in nodes (exchanges). Each trunk-line is characterised by type and number of groups (sets of transmission channels in suitable frequency bands). Group segments of the separate trunk-lines are in nodes joined by transit filters or connected to the group transfer equipment. Full connections structure depends on distribution of network transmission capacity between separate links accordingly to their service demands.

For service a traffic in the link, there is alloted some number of groups (may be of different groups levels), guided usually by more than one route and using one or more of trunk-lines in each route. Some example is shown on the fig. 1.

Coupling operations on the ends of trunk-lines may be done only on the same groups levels and changes of the group level can be made in the nodes having suitable free group transfer units.

Also, the model consists of very large number of data describing all network elements and their connections. It forces to look after methods allowing to obtain practical solutions using possibly small part of these and/or their synthetic features.

3. METHOD OF SUBSTITUTIONAL ROUTES CONSTRUCTION

Recently elaborated method [3], [4], assumes that substitutional routes (SR) can be coupled as by-passes of the damaged line and/or optional segments of affected links or groups of links.

Trunk-line failure damaging some routes of some links causes necessity of restoring them by means of appropriate SR's with defined grade for given link. At the same, lower boundaries for group structures are imposed - the matter is, that some of damaged groups cannot be restored by equivalent set of groups of lower groups level (nondivisibility condition). However carrying transmission on possibly high group level and possibly short ways are common recomendations. It agrees with practically important requirement of low number of groups switchings in

[1/] Institute of Automatic Control, Warsaw Technical University, Nowowiejska 15/19, 00-665 Warsaw

SR building.

Therefore arise, at a complicated groups structure and distribution of transmitting means accesible for SR building, a task of computation of such a set of SR´s paths, which in a necessairy or better degree with regard to groups structure and possibly highest with regard to recovery conditions reconstructs lost transmission channels. Simultaneusly, commutational conditions in the nodes should be accounted, eg. possibility to join group segments by transit filters and acces to groups of different level.

The first task in the problem solution is to determine a set of technical means called as **reservation network** (RN), which can be used to construct SR. For practical reasons at the beginning should be selected some fragment of network around the failed line. We call this part as **failure surrounding** (FS). Main mathematic problem to be solved in SR construction is to choose from many discrete elements of RN such a set of them, which in possibly largest extend fulfils quantitive transmission requirements, having at the same good operational features (low number of necessairy commutations). The obtained results depends on many decisions which must be taken up during problems solving. First case concerns situations when FS is too little to gain the satisfying SR. In the case, a trial of expanding of FS may be undertaken. It not always will expand transmission possibilities, because there can be limits not exceedable for the failure concerned. In such a case one may admit reduction of required level of restoring for all or selected links, or further diminish not affected links for expanding RN. Even if FS is satisfactory large, choosing of a SR have not yet unique solution. It can be mentioned only that if there are multipathial links, choice of path from which groups and filters are taken away for RN changes the solution.

Another decision must be taken up when the paths of SR are selected in the RN. For the paths of SR are iteratively selected for succesive levels of groups, succesive modification of the RN depends on the decision which of groups was really used in former step of selection.

Definitivly selected RN even does not lead to unique solution, for there can be some number of acceptable solutions differing each other by paths combination and their transmission capability or by numbers of commutation operations. Also from the set of solutions should be selected one solution acceptable for realisation or if there is no such a solution, procedure with modified parameters can be repeated.

The problem can be formulated as an optimisation problem in different forms. For example, with given set of groups segments and transfer filters of different levels, placed in a given configuration in the network (RN), should be constructed a SR enabling to restore failed transmission capability in the desirable degree with lowest number of commutation operations. Fortunately, the problem can be linearised, unless have integer character.

4. PROBLEMS STRUCTURE

As it was shown, many practical conditions influences the possible solutions. An entire complex of the solution steps and conditions is shown on the fig. 2. Different strategies of solving are considerable determined by fact that problem concerns the situations with relatively short time of existence, until the broken line is repaired. In such a case it is not justified to work out and apply too complicated techniques and procedures or admit solutions which needs too many commutational operations as it will prolong time of reaction to break-down. It is desired to have alohoritms relatively quick, giveng results may be suboptimal but satisfying for the considered case of failure.

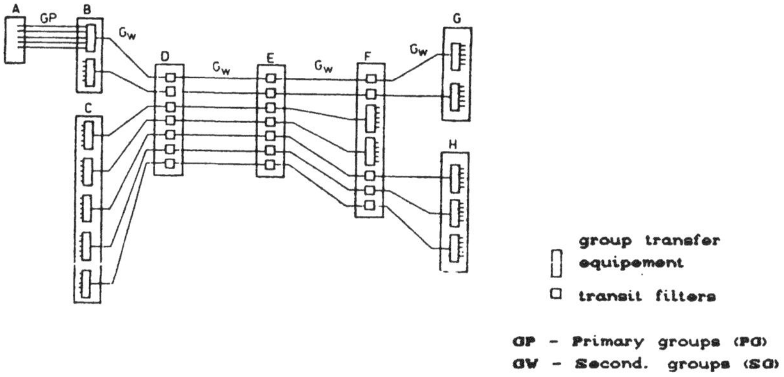

Fig. 1. Example of four links paths: A(BDEF)G - 5 PG (1 SG), B(DEF)G - 5 PG (1 SG), C(DE)F - 10 PG (2 SG), C(DEF)H - 15 PG (3 SG)

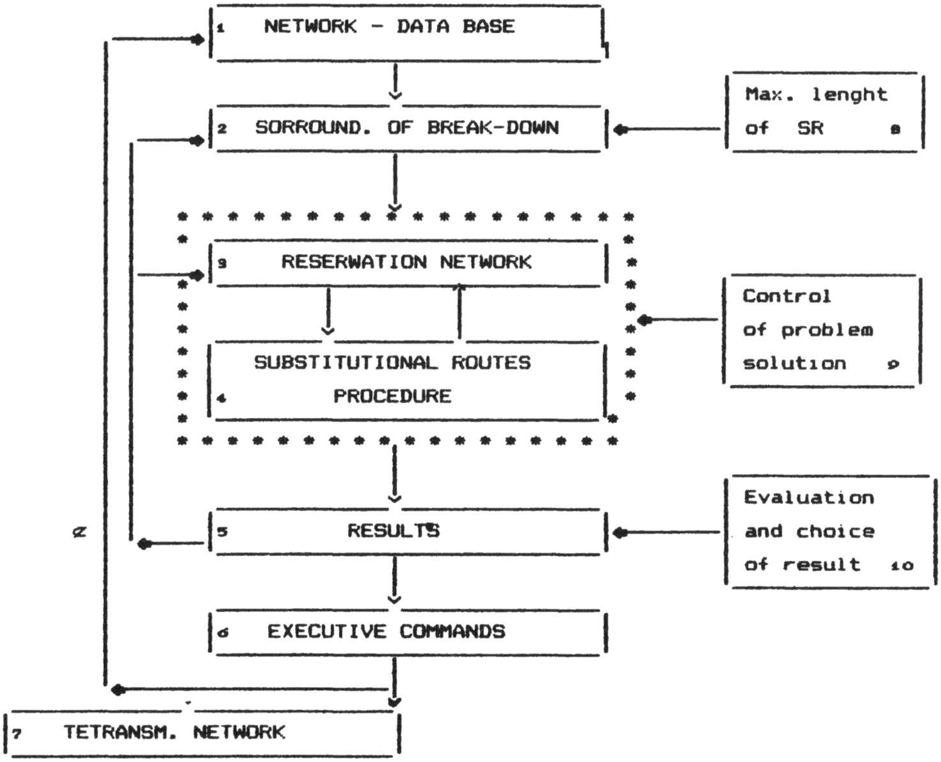

Fig 2. Problems structure

Data base. Assumption that problem will be solved with computers requires to describe all the network in the form of adequate data base. With regard to the big amount of data describing every necessairy characteristics of network elements (estimated number of 10^6 elements) and anticipated variety of tasks to which the data base will be used, should be constructed as universal and convenient as possible. [7].

Failure surrounding. At first it is necessary to detache from the network such a FS, the dimension of which gives fair promise for obtaining satisfactory SR and are not too big (combinatorial problems!) for accesible computational means.

For choosing FS one can use only very general hints. It can be assumed, for example, that paths of subsitutional routes will not exceed given number of sections. Choosing procedure will be in this case as follows. Number of the broken line allows to recognize links routes carried by the line and defining the **areas of break-down**, eg. sets of nodes and lines on the both sides of broken line, carrying affected routes.

First, all the nodes and lines in the areas of break-down are included to FS. Next are there included, together with corresponding nodes, lines directly connecting both areas, lines connecting them by one node (path of two segments), lines conecting by two nodes, and so on, until permissible number of segments is reached. May be usefull to estimate now the transmission capability of FS from the SR point of view and make changes if it is not adequate: eg. one may allows to include paths with higher number of sections. FS defined on the basis of networks graph and informations about lines types, forms the part of network from which will be SR constructed.

Reservation network. The RN is determined of the free groups segments and filters being not occupied by any link, group segments and filters drawn from not affected links, together with this parts of damaged routes which become free after break-down.

For defining of the RN one have to determine quantitively network elements in FS, which can be used for SR [8]. For this purpose it is necessary to recognize in all the trunk-lines of FS links routes carried by them, and decide how many groups from each route can be used for RN. There is some problem when link is multirouted, because choice of routes from which groups to RN will be taken influences the following modifications of RN. It seems, that it is reasonable to take away from rather biggest route but only one in the link. It should lead to concentation of switching operations; in some cases another circumstances may appear.

The case when transmitting posibilities of RN is too low, proper operator's decisions are necessary. Dependently of situation one may decrease the level of restoring for some or all affected links, increase the grades with which not affected links are charged, or apply both measures. (We assumpt there that FS is maximal, as in contrary case it can be enriched without changing of restoring and charging levels). There also should be possible to charge some links in higher grade; it leads to arbitrary decisions based on operators experience or - if it is reasonable - to some more compliated problem of "equal" charging of all angaged links by result of failure.

Determination of substitutional routes. Because a different transmissions systems may occur in separate lines, at first is used a part of RN consisting groups and filters of highest group level existing in the fragment of network being taken into account. Because not all elements of highest group levle are used for SR, the remainders "translated" to so called **apparent lines** of groups of the succesive lower level. (Apparent line joins the nodes in which such a lower groups are accesible). The apparent lines, together with earlier designed groups of lower range become a partial RN of new, (lower) group range and is used to complete SR.

The remainder again is returned to RN with suitable "translation" to succesive group range. Described iterative steps are continued until SR is completed or RN cannot give more of applicable paths. After quantitive solving of the task, real segments of groups in separate lines have to be assigned to specified paths.

SR determination task was formulated as integer linear programming problem. To solve it a computer program fitted to computers of IBM PC type was elaborated. The program works in few steps, i.e. computes the sets of substitutional routes paths for succesive groups levels. Computed results have at begining quantitive character, but reflects the structure of these routes, ie. shows where and haw many technical means is to be used for SR.

If reached result is satisfying, program is finished; when result is not satisfying, there are three possible ways of proceeding. One may go back to the begining of program and repeat computation at changed weights of importance of restoring for different links. The second possibility is come back to RN defining procedure and repeat it with changed parameters of decreasing not affected links and/or restoring of links affected. At last, one can come back to the procedure of FS defining and repeat all the program with some new, greater FS.

Executive commande. The chosen satisfying solution, until now in quantitive form, needs to be presented as a typical verbal report. To generate automatically such a reports, it is necessary to define in advance which groups and in what order in separate links may be taken for SR construction. Then quantitive solution may be assigned with real elements signs and generates an report for execution of switchings. At the same time BD should be temporary modified suitable to temporaty status of elements (disconected, used directly in SR) and configuration of their connections. Even their normal status is to be remembered for easy return to the normal state of network.

* * *

As it was shown, the problem of building SR is quite complicated and needs to solve many partial tasks with considerable difficulty. Sketching the full problems structure is substantially helpfull for correct solving of the problem and such a manner of proceeding leads to results with practical application meanings, e.g. to computer programs and problems solution which can be applied in telecommunication network exploatation.

5. BIBLIOGRAPHY

[1] "Rozmieszczenie kanałów rezerwowych w sieci linii teletransmisyjnych", Praca zbiorowa, IA PW (1983)

[2] "Opracowanie uproszczonego algorytmu rozmieszczania kanałów rezerwowych w sieci linii teletransmisyjnych", Praca zbiorowa, IA PW (1984)

[3] "Wyznaczanie dróg obejściowych przy uszkodzeniach linii teletransmisyjnych z uwzględnieniem ograniczeń komutacyjnych węzłów", Praca zbiorowa, IA PW (1985)

[4] "Opracowanie założeń eksperymentalnego uruchomienia procedur awaryjnego wyznaczania dróg zastępczych w sieci teletransmisyjnej wraz z odpowiednim oprogramowaniem", Praca zbiorowa, IA PW (1986)

[5] Wydro K.B. "Przeciwdziałanie skutkom awarii linii w sieci telekomunikacyjnej", Przegląd telekomunikacyjny 10-11 (1986)

[6] Płoszajski G. "Modele sieci teletransmisyjnej do badania operacji przełączania", Przegląd telekomunikacyjny 10-11 (1986)

[7] Stelmasik H. "Baza danych sieci telekomunikacyjnej", Przegląd telekomunikacyjny 10-11 (1986)

[8] Winnicki A. "Wyznaczanie sieci rezerwacyjnych dla tworzenia dróg zastępczych przy awarii linii teletransmisyjnej", Przegląd telekomunikacyjny 10-11 (1986)

[9] Paczyński J. "Wyznaczanie tras obejściowych przy awarii linii w sieci telekomunikacyjnej" Przegląd telekomunikacyjny 10-11 (1986)

[10] Wydro K. "Problemy wyznaczania dróg zastępczych przy awarii linii w sieci teletransmisyjnej", Przegląd telekomunikacyjny 10-11 (1987).

OBJECT SYSTEM STRUCTURE AND BEHAVIOUR ANALYSIS IN DATABASE DESIGN

Martin Holeňa *)

Object system modelling - a system-analytic part of the database design

During the last decade, conceptual modelling became - much influenced by the ANSI/SPARC and the ISO TC97/SC5/WG3 proposals - a part of the database, resp. information system design process. Just as database implementation design can be divided into the logical design and the physical one, according to the relation to data physical storage characterisation, so we can also divide the conceptual modelling phase into two different sub-phases:

1. Object system mapping. Its subject is determination of those elements and classes of elements of reality (e.g., things, persons, institutions, properties, decisions, processes, relations between things, persons etc.) that could be for some of the future database users so important that they would like to be able to obtain information about them from the database. The term object system (universe of discourse) refers to that part of reality, about which information is to be obtainable from the database; hence, the object system mapping is a type of system analysis of this part of reality.
2. Database information specification. Its subject is specifying the way of identifying and handling the information stored in the database. This sub-phase includes the solving of problems connected with names and identifiers choice, synonyms and homonyms, information deducibility, integrity constraints, consistency conditions, specification of transactions and their pre-and postconditions, synchronization, exception handling etc.

This contribution concentrates on the object system mapping subphase. The basic and most widely used method of object system mapping is the representation of the object system by one or more abstract structures built up by means of several formally distinguishable constructs, each having a specific interpretation in reality. Hereafter, we shall call such abstract structures object system models and the object system mapping by means of them will be called object system modelling. Specificity of interpretation of constructs differentiates object system models from traditional data models.

Object system structure and dynamics models

During the database design, of particular importance is usually the part of the object system that could be termed its dynamic component, i.e. the part formed by changes of things, properties, relationships etc., different events, actions, activities, decisions, evolution, but also by the passage of time. The dynamic component of the object system

*) M.Holeňa, SVT ČSAV, Pod vodárenskou věží 2, 182 07 Praha 8, ČSSR

is important in two aspects:

- Some elements or classes from the object system dynamic component belong to those which the future users are most interested in.
- Many elements from the dynamic component during the database utilization regularly result in certain operations with information and, consequently, also with data (data insertion, deletion, modification and retrieval, transfer of data between main and external storage etc.). On the other hand, for some of the operations with data one or more elements from the object system dynamic component may be found stimulating the operation concerned.

Some of the object system models enable to catch explicitly the fact that an element of the model represents an element or a class from the object system dynamic component - i.e., to catch this fact in such a way that one or more of the model constructs are applied only to represent such elements or classes. We shall call these models <u>object system dynamics models</u>. The other models, in turn, enable this fact to be cought only implicitly, i.e., by means of the semantics of each single element forming the model. In some casses, for instance if we want to specify the operations with information during the database information specification, such a way of representation may be very cumbersome and awkward. These models, for the sake of convention, will be called <u>object system structure models</u>.

<u>An example of the object system structure models - the ER model</u>

So far, object system structure models have been used far more often. Also at our institute, we confined ourselves to a model from this group until recently, namely to a modification of the very popular entity-relationship (ER) model |2, 3, 6|. This model was chosen because of its most common use as an object system model. In this respect, it surpasses both semantically richer models - e.g., SAM, SHM, SDM, or the infological model |1, 6|, and models enabling easier communication with future users - e.g., binary, HIT, or other functional models |5, 6, 7|.

The modified version of the ER model we proposed was formally defined as follows:

An <u>entity-relationship model</u> is a triple $((E_i)_{i \in \hat{p}}, (R_j)_{j \in \hat{q}}, (A_k)_{k \in \hat{r}})$, where $p, q, r \in \mathcal{N}$ &

$$(\forall i \in \hat{p})(\forall j \in \hat{q})(\forall k \in \hat{r})\ E_i, R_j, A_k \text{ are sets } \& \ E_i \cap R_j = \emptyset \ \& \ E_i \cap A_k = \emptyset \ \& \ R_j \cap A_k = \emptyset$$

$$\& \ (\forall j \in \hat{q})(\exists n_j \in \mathcal{N} \setminus \{1\})(\exists p_{j,1}, \ldots, p_{j,n_j} \in \hat{p})\ R_j \subset E_{p_{j,1}} \times \ldots \times E_{p_{j,n_j}} \ \&$$

$$(\forall k \in \hat{r})(\exists V_k \text{ - a set})(\exists D_k \in \{E_1, \ldots, E_p\} \cup \{R_1, \ldots, R_q\})(\forall f \in A_k)\ f \text{ is a}$$

$$\text{mapping } \& \ \mathrm{def}(f) = D_k \ \& \ \mathrm{val}(f) \in 2^{V_k} \ \& \ (\forall d \in D_k)\ \mathrm{card}\ f(d) < \infty .$$

Here $(\forall i \in \hat{p})$ E_i is called an <u>entity type</u>, $(\forall e \in E_i)$ e is called an <u>entity</u>, $(\forall j \in \hat{q})$ R_j is called a <u>relationship type</u>, $(\forall r = (e_i, \ldots, e_{n_j}) \in R_j)$ r is called a <u>relationship</u>, $(\forall k \in \hat{r})$ A_k is called an <u>attribute</u>, $(\forall f \in A_k)$ f is called a <u>constellation</u> of A_k.

Furthermore, we formally defined cardinality of relationships and miltiplicity of attributes, as well as attribute aggregation and decomposition, but the definitions will not be given here due to lack of space. For the same reason, we shall not attempt to describe here the interpretation of the constructs used to build up this version of the ER model.

A brief review of the object system dynamics models

Strating to recognize the importance of the object system dynamics models, we carried out, at our institute, a comprehensive study and comparison of those models belonging to this group that we have found in the literature. Most of them have been proposed directly for the purpose of database or information system design. However, among the models we studied were also three from outside the area of databases and information systems - Petri nets, finite state machines and Markov chains.

A very brief survey of the basic features of the studied models is given in table 1. The limited extent of this contribution does not allow to discuss here the meaning of the individual columns of this table, nor to mention each of the models separately. A more detailed treatment can be found in |4|, where also references to the original presentations of these models are given.

An illustration

We shall now illustrate the difference between object system structure and dynamics models giving a practical comparison of the ER model with the object system dynamics model used in the methodology Remora. As means for comparison we will use convenient graphical representations.

In figure 1, an ER diagram (graphical representation for the ER model)of a little part of the object system of a real database designed at our institute is shown. The same part is depicted by means of a graphical representation for the object system model used in the methodology Remora in figure 2. An explanation of the symbols used in these figures is given in |4|.

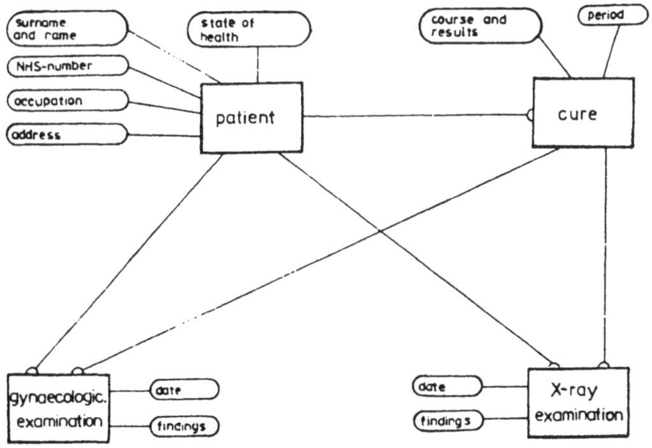

Figure 1

model(s)	included in	constructs for the dynamic component	related constructs	graphical reperesentation	results utilized for
Bubenko´s	conceptual information model, abstract model realm	event, event type	attribute function, time of occurence	nets	entity and event types specifications, information model
entity/ /event + JSD	modelling phase	event, sequence, selection, iteration		behaviour diagrams	network phase, model processes (life history programs), synchronization (messages, state vectors)
PIOCO	I/O model	event, event type, event generalization, event subset, event classification	association, situation	OS (object system) graphs, OA (object system event action) graphs	information model, information process model, interaction model, IS graphs, IT graphs, IP graphs, IA graphs
Remora	conceptual schema design	event, operation, trigger, t-event, t-operation, t-trigger	modify, induce (ascertain), t-modify, t-induce	only for the database information specification	static subschema (c-objects), dynamic subschema (c-operations, c-events), elementary transactions, transaction sequencing graph, design of the ISMS
ISAC	change analysis, activity studies	activity	real flow, message flow	A-graphs (activity graphs)	coordination of information subsystems, precedence and component analysis, information analysis, I-graphs
ABRAHAM	"Istanalyse", "Sollkonzept"	activity	acces path	ABRAHAM symbolism	system design, activity trees, information trees
IPSO + Roussopoulos-Yeh´s	information system analysis, data conceptual analysis, applications analysis, environment and requirements analysis	process, subprocess, synchronization, task, subtask	channel, flow	information flow diagrams, data conceptual schema, process diagrams, task flow diagrams	data logical schema design, module design, external data schemata design, applications structured diagrams, data logical schema, specification of tasks and subtasks
Insyde	phase one	process event, function link, communication link		information flow schema (design schema)	semantic schema (conceptual schema), application events definition
CKM	graphical representation	function, event, trigger	information flow	conceptual requirement graphs, conceptual structure graphs	functional decomposition, conceptual behaviour graphs (actions), acces table generation
Petri nets	first overview net, second overview net, conceptual schema design, views conceptual design	transition (agency)	flow (arc)	Petri nets, behaviour diagrams	high-concurrency net, low- -concurrency net, behaviour descriptions (transactions), views integration, dynamics analysis (operation frequences)
finite state machines	conceptual analysis, detailed analysis	next stage function	life cycle history	ER diagrams, entity life cycle diagrams	data structure and file design, program design, state/ /event matrix hierarchy
homogenous Markov chains	requirements definition	Markov chain, state of a Markov chain		only for the database information specification	decomposition of transactions into processes, estimation of traffic load on the global information structure

Table 1

Further research

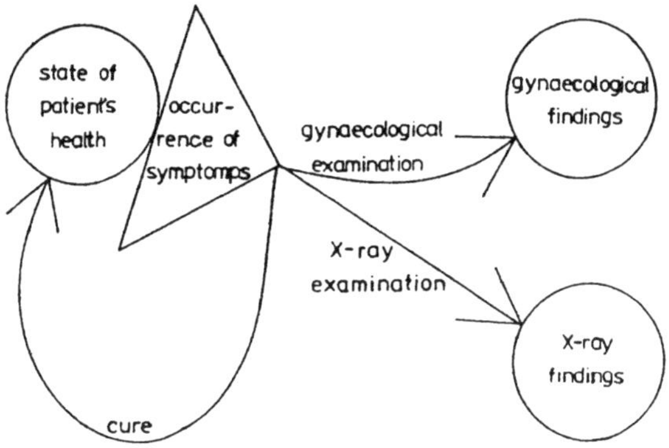

Figure 2

Studying the usability of the above mentioned object system dynamics models we found that for several purposes there were the Markov chains either the only suitable or the most suitable model. We found as well that the exceptional position of this model among the studied object system dynamics models is due to its ability to catch the probabilistic features of many elements of the object system, hence a consequence of the fact that it is a stochastic model. That is why we assume that other stochastic models could be suitable for object system dynamics modelling, too.

We decided to attempt to use stochastic Petri nets for object system modelling. For this purpose, we proposed a new version of stochastic Petri nets, called semiparametric stochastic Petri nets (SPPN), which we connected with the ER model. At present, we are developing a software tool for automatic obtaining of parameters of SPPN. A formal definition of SPPN and of their connection with the ER model can be found in |4|.

References

|1| Brodie, M.L. - Ridjanovic, D.: On the Design and Specification of Database Transactions. In: On Conceptual Modelling. Perspectives from Artificial Intelligence, Databases and Programming Languages. Springer Verlag, Berlin 1984, 277 - 312.

|2| Chen, P.P.S.: The Entity-Relationship Model: Towards a Unified View of Data. ACM Transactions on Database Systems, 1(1976), 1, 9 - 36.

|3| Elmasri, R. - Weeldreyer, J. - Hevner, A.: The category concept: An extension to the entity-relationship model. Data and Knoweldge Engineering, 1(1985), 1, 75 - 116.

|4| Holeňa, M.: Object system dynamics modelling and stochastic models. In: Proceedings of the 11th International Seminar on Database Management Systems (to appear 1988).

|5| Kulkarni, K.G. - Atkinson, M.P.: EFDM: Extended Functional Data Model. The Computer Journal, 29(1986), 1, 38 - 46.

|6| Tsichritzis, D.C. - Lochovsky, F.H.: Data Models. 1. ed. Prentice Hall, Englewood Cliffs 1982:

|7| Zlatuška, J.: HIT Data Model. A Functional Approach to Data Bases. In: Proceedings of the Seventh International Seminar on Database Management Systems. Bulgarian Academy of Sciences, Varna 1986, 21 - 33.

COMPUTER-AIDED ELABORATION OF RESEARCH STRATEGIES - THE EXPERT SYSTEM PHARMEX

Karin Schiele, Manfred Koch, Margrit Nachbar[1)]

1. Introduction

The search for strategies for well directed basic research and applied research is a complex task based on detailed information about different social spheres. Besides, the time factor is a crucial one to the future effect of the cycle Science-Technology-Production-Application. The research strategy to be elaborated has to correspond especially to the following requirements:

1. It has to be derived in such a manner that the scientific and technological solution of the imposed research task meets an essential social need at the moment of its perfection.
2. It has to consider both the international and the national research situation of the respective discipline, i.e. it has to be directed to such research projects which have good chances to be successful in the international competition of research institutions.
3. It has to effect that the obtained scientific and technological solution finds effective conditions in production.
4. It has to be founded on the analysis of all those social frame conditions which are important in connection with the realization and application of the solution.

We try to support the working out of research strategies by help of an expert system. Consequently this expert system has to

- fulfill the mentioned four tasks,
- facilitate the finding out of problem fields which are convenient at the appropriate moment by help of a suitable inference mechanism, enable to carry a distinct dialogue with experts and users,
- contain a comprehensive explanation capability and ability to learn, consider the time factor in a fit manner.

The structure of the expert system PHARMEX designed for strategy-making in R&D on the pharmacological field is shown in Fig. 1. Above all we are confining ourselves particularly to the steroids. The users of the system will be not only the Central Institute of Microbiology and Experimental Therapy but also JENAPHARM. The expert system is leaning upon preliminary works on research strategies in the pharmacological sector which are summarized in [1].

1) Akademie der Wissenschaften der DDR, Institut fuer Theorie, Geschichte und Organisation der Wissenschaften, Prenzlauer Promenade 149-152, Berlin, DDR 1100

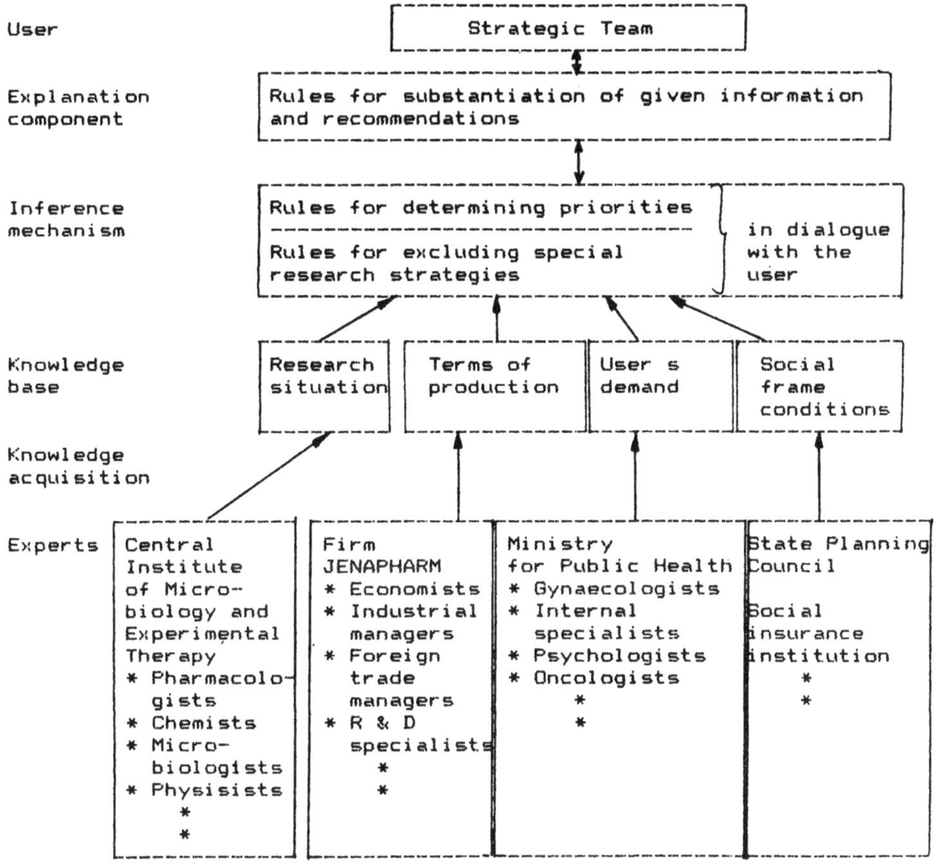

Fig.1. Structure of the expert system PHARMEX for the elaboration of research strategies in the pharmacological sector

2. The knowledge base of PHARMEX

The above named four general tasks to be performed in course of the elaboration of research strategies suggest a division of the knowledge base (kb) of our expert system into four parts. Each of these four parts has a hierarchical structure and is related to the same system of categories, that is the system of steroid agents in his dynamics.[1]

Kb part 1 contains all information needed for a prospective analysis of the demand of drugs on the basis of steroids. Starting-point there are all diseases treated by steroids as well as all other indication fields of steroids (e.g. contraceptive).The main spheres of demand resulting from the indication fields in the GDR as well as in the international scale are found out with the help of special literature, statistical material and expert statements. This happens in a twofold manner: On the one hand the development trend of the quantitative

1) At present about 300 steroids are known in the world.

demand of steroids is assessed, on the other hand a qualitative demand is formulated which represents the requirements referring to qualitative improvements of the agents and drugs. Here we are trying to take into consideration the social assessment of the several indication fields.

Kb part 2 serves to find out such scientific and technological problems being able to become innovations from the view of the national and internatinal research situation. It is based on detailed attainments about the international level of knowledge, the international patent situation, the international situation of competition and the national research situation. The assessment of the international research situation supposes the evaluation of the essential results from several information sources (publications, patents and other) and their systematic registration in three complexes. These are the level of knowledge, the patent situation and the situation of competition. So current research work in the GDR gets the possibility to react flexible upon the development of the international research situation. To elaborate research strategies one can start from the knowledge about the research situation contained in the system in order to formulate scientific and technological problems that might be favourable for the production of drugs in the GDR.

Kb part 3 has to consider not only the national but also the international terms of production. The present terms are the starting-point of both complexes for the purpose of estimating the future terms of production. In order to dertermine the present national production circumstances the experts of JENAPHARM have to put into the knowledge base especially the following economic characteristics

costs of production determined by the output capacity (fixed assets, raw and auxiliary material, wages),
- the share of imports of raw material and the corresponding prices,
- calculation of prices for the output,
- net proceeds representing the surplus product.

The firms in the GDR are interested in raising their net product, the returns and the export because their performances are evaluated according to these categories as well as their investment possibilities are determined by them. Finally the efficiency of their production is considerably affected by the choosen technology. The terms of export depend both on their own patent activity and on patent strategies of competitors in foreign countries.

Kb part 4 includes such information about the social frame conditions (connected with the R&D process of drugs) which is not represented in the other three parts of the kb, i.e. the idle time due to diseases which are treated or might be treated by steroids, and the social insurance costs (sickness pay, costs for drugs, doctors' fees, labor costs and other costs for medical treatment).

3. Inference mechanism, knowledge acquisition, explanation capability and time management

The expert system PHARMEX is projected as a backward-chaining deduction system. Each consultation is beginning with the starting menue which offers the following options:

START	starts a new consultation,
INFO	informs about desired elements of the kb,
PRIOR	determines priorities among different research projects,
EXPERT	tries to acquire knowledge,
HELP	supports handling the system.

The central task of the system is PRIOR, a dialogue-supported mechanism for setting priorities among all conceivable research projects in the corresponding field. Consulting PRIOR the user has to decide at first what type of research strategy he wants to investigate - a short-term imitation strategy, a middle-term using strategy or a long-term offensive strategy. In dependence of this decision the inference engine activates different modules of rules. At first these rules exclude from the further investigation such research projects which are impracticable or ineffective. Then the system makes a decision proposal in dialogue with the user by determining priorities among the remaining research projects. The explanation component of PHARMEX shows the deduction of the given proposal if desired. In the course of the consultation the user may correct or modify some rules as well as he may add new rules to the system. PHARMEX records the taken modifications in its memory in connection with the name of the user. Besides, the user has the possibility to refer all rules which are inexact, doubtful or missing from his point of view to EXPERT, i.e. to require additional expert knowledge about certain connections. EXPERT will ask the corresponding specialist for the stored questions in the next expert consultation. These questions concern i. a. the actualization of time-depending statements whose period of validity is passed and the assessment of future developments.

This time[1)] our working group is dealing with the implementation of a prototype system PHARMEX-PROTO in micro-PROLOG on the personal computer PC 1715. This prototype system is to demonstrate the main ideas of the taken approach and thereby to support the discussion with the potential users of PHARMEX. The next step of the work is the detailed elaboration of the expert system in turbo-PROLOG on the 16-bit computer PC 1834.

Literature

[1] Koch, M., M. Hübner, U. Eberhardt: Forschungsstrategien im Innovationsprozeß auf dem Arzneimittelgebiet, Forschungsbericht ITW-F-87-01, Berlin 1987.

1) Deadline of this paper was march 15, 1988.

A SIMULATION - BASED OPTIMIZATION ALGORITHM FOR OPERATIONS CONTROL IN FMS

Carmen Bobeanu*, Gabriel Neagu* and Florin Filip*

Abstract

This paper is an attempt to model and partially solve the scheduling problem in the environment of modern flexible manufacturing systems. In order to obtain a problem-solving, we propose an efficient algorithm based on simulation.

1. Introduction

An operation scheduling through a FMS can be made over a time horizon when the characteristics of parts and FMS behaviour are known, by using the algorithm to be presented. This has been developed based on an event - oriented simulation model.

The paper is organized as follows. In section 2 the assumptions on FMS organization and operation are formulated while priority classes associated with parts and stations are outlined in section 3. Further details of the developed scheduling algorithm and the significant event list are presented in section 4. Section 5 provides a description of the optimization procedure for the scheduling solution.

Finally, some implementation aspects and development prospects are discussed.

2. FMS Model

The discrete simulation model to be described in the sequel is based on some normal assumptions on FMS operation :

1. FMS comprises a set of stations grouped in various station types (fixture-unfixture stations, loading-unloading stations, machining centers, buffer-stations, washing-stations, verification stations).

2. There is a set of parts launched according with a priority rule. There are more operations, O_i^k, to be performed once the part L_i completed.

3. Each part to be scheduled may or may not have a previously assigned due date, tl_i. The first category of parts belongs to the set

* Research Institute for Computer Technique and Informatics 71316 Bucharest, 1 (ROMANIA)

$\mathcal{L}_1$, the second one belongs to the set $\mathcal{L}_2$ and tl_i, where $L_i \in \mathcal{L}_2$ is set to ∞.

4. If two sequential operations of the part L_i are to be performed on the same station type, these are normally sequentially executed on the same station, except the case when a start of a planned maintenance stop of the station occurs during the execution of the first operation. In this case an attempt is made to assign the part to another station of the same type.

If it is not possible, the parts get on a buffer-station.

5. The processing time values, t_i^k, and the transition time between stations, tt_i^k, for the operation O_i^k are known.

6. If there is no vacant station to perform the next operation the part gets on a buffer-station.

7. If a station becomes vacant, the highest priority part waiting in the common queue of that station type is assigned to it.

8. There are priorities associated with machining centers and buffer-stations.

9. Time intervals for station maintenance or verification may be allocated for each station. If a part is assigned to a station at the beginning of such an interval, the time interval is translated on the instant of accomplishing the current operation for the station.

10. There is an automated guided transportation vehicle for the transition of parts between stations which treats the machining centers as a top priority. If it is vacant, it is removed to the origin-point.

3. Priority Classes

In order to indicate the part to be launched a priority, Pl_i, is calculated, taking into consideration for each part, L_i, the external priority, $P_{e,i}$, and the remaining time, $P_{1,i}$:

$$Pl_i = P_{e,i} + P_{1,i} \tag{1}$$

A composite priority rule-that takes into account for each part, L_i, the external priority, $P_{e,i}$, the remaining time, $P_{1,i}$, the time in queues, $P_{2,i}$, the remaining processing time/current processing time, $P_{3,i}$, the estimated size of the next queue, $P_{4,i}$, and the automated guided transportation vehicle removal time, $P_{5,i}$ - is used to indicate the next part to be assigned to a vacant station :

$$Pc_i = P_{e,i} \cdot \sum_{j=1}^{5} w_j \cdot P_{j,i} \tag{2}$$

where the weights $\{w_1, w_2, w_3, w_4, w_5\}$ form the intervention vector in a two-level hierarchical optimization procedure.

There are priorities associated with working stations that take into consideration the unloading time, $R_{1,j}$, and the automated guided transportation vehicle removal time, $R_{2,j}$:

$$R_j = q_1 R_{1,j} + q_2 R_{2,j} \qquad (3)$$

where q_1 and q_2 are weight parameters.

Priorities associated with buffer-stations, that take into account the automated guided transportation vehicle removal time are also defined.

4. The Scheduling Algorithm

The scheduling algorithm uses a file of future events, FEV.

In the beginning this file stores the stipulated events regarding parts and stations over the scheduling time horizon.

The output of this algorithm is a sequence of significant events, ordered in accordance with the moments of their appearance.

The significant event list is the following :

No	Signifiance
1	New part arrival in FMS
2	Start of an operation
3	Accomplishment of current operation
4	Automated guided transportation vehicle loading
5	Automated guided transportation vehicle unloading
6	Start of an automated guided transportation vehicle transition
7	Automated guided transportation vehicle stop
8	Start of a planned maintenance stop
9	Station set-up after event 8
10	Conclusion of part processing in FMS

5. The Optimal Scheduling

The appropiately chosen values of weights w_j, $j = \overline{1,5}$ are crucial for the efficiency of the schedule. To establish the optimal values of weight parameters asks for defining a performance measure. This measure, heaving an obvious economic meaning, indicates, for each simulated sequence, how high the penalty costs are for each part tardiness or earliness of the completion time obtained by simulation as compared with the due date, for the time each part has been waiting in various queues, for the time each station has been in a idle state and for the removal time of automated guided transportation vehicle.

The optimal values of weight parameters are found-out by using a Gauss-type linear direct search procedure, with each iteration corresponding to a simulation of the evolution of FMS.

6. Conclusions

The above-mentioned algorithm has been developed for Romanian-made INDEPENDENT or CORAL minicomputer using FORTRAN-77 code.

Efforts are being made to take into account the frequency of FMS retooling criterion on solution generation.

References

1. J.C. Emery, "Job Shop Scheduling by Means of Simulation and an Optimum Seeking Search "Proc. of The Third Conf. of Applications of Simulation, Los Angeles, USA (1969), p. 363-371.

2. Jiri Herman, Zdenek Safka, "Dynamic Rescheduling of Operations - a Computer Approach to Production Control", Engineering Costs and Production Economics 6 (1982), p. 131-139.

3. Andrew Kusiak, "Planning of Flexible Manufacturing Systems", Robotica (1985), vol. 3, p. 229-232.

SIMULATION OF MANUFACTURING SYSTEMS

Goals, Methods, Tools and Problems

Peter Lorenz *)
Juri Tolujev #)

Simulation models are built and used in situations where experiments on the real world are not possible or not recommendable. Manufacturing systems are systems of this kind. Experiments are expensive, time consuming and often dangerous. Consequently they have become one of the most often handled and quoted application fields of simulation. The authors, involved in two research projects of Flexible Manufacturing Systems (FMS) Simulation, report about goals, methods, tools, and problems.

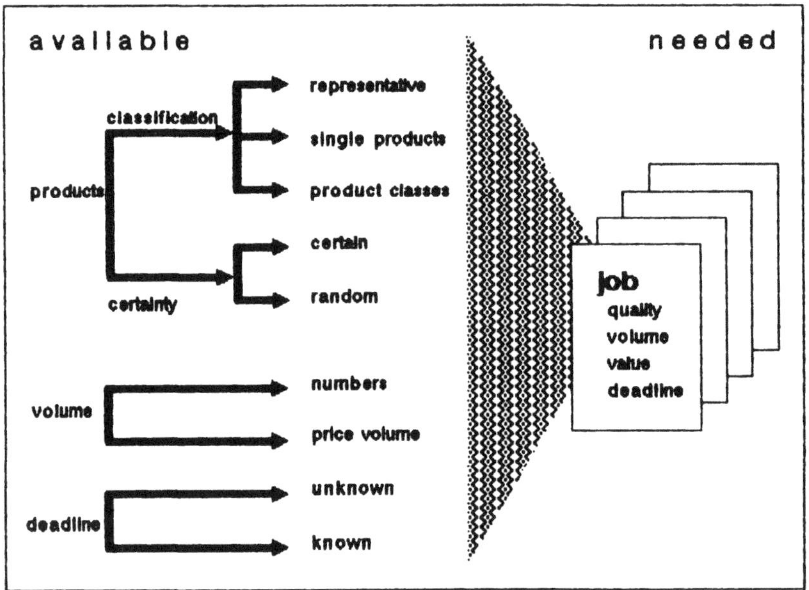

Fig. 1: Availability levels of load data for FMS Simulation

The *goals* of FMS Simulation can be classified in
- decision support for FMS design or reconstruction problems (long term planning) and
- operating support and operator training for FMS control (short term planning).

The differences between these two goal directions are transfered to
- great differences in simulation input and output and
- small differences in simulation models, their data and process structures. [1] reports on the use of SEE WHY (Istel Ltd.) to solve long and short term problems.

---

*) Technische Universität "Otto von Guericke" Magdeburg
Sektion Informatik/PSF 124 Magdeburg DDR 3010

#) Polytechnisches Institut Riga

1. **Basic data transformation and generation**

Input data describing the system under study are usually delivered in a raw mode that is not suitable for system analysis and simulation. Input data must be transformed in a standard-like internal mode. In this mode
- specific keys for production, machines, times etc. are replaced or complemented by internal keys,
- possibly not available job stream data are generated and
- possibly not available initial state data are generated.

Usually in the case of real time problems more and better job stream and state data are available. In the case of long term problems the generation of job stream and state data is a first subject of experiments. Fig.1 shows different levels of input data about the system load and a symbolic transformation in a batch of structural unified jobs or orders. *FMS simulation tools should contain a collection of programs suitable to execute all needed transformation.*

Fig.2 demonstrates two possible ways for gaining daily job streams as a part of the yearly load.

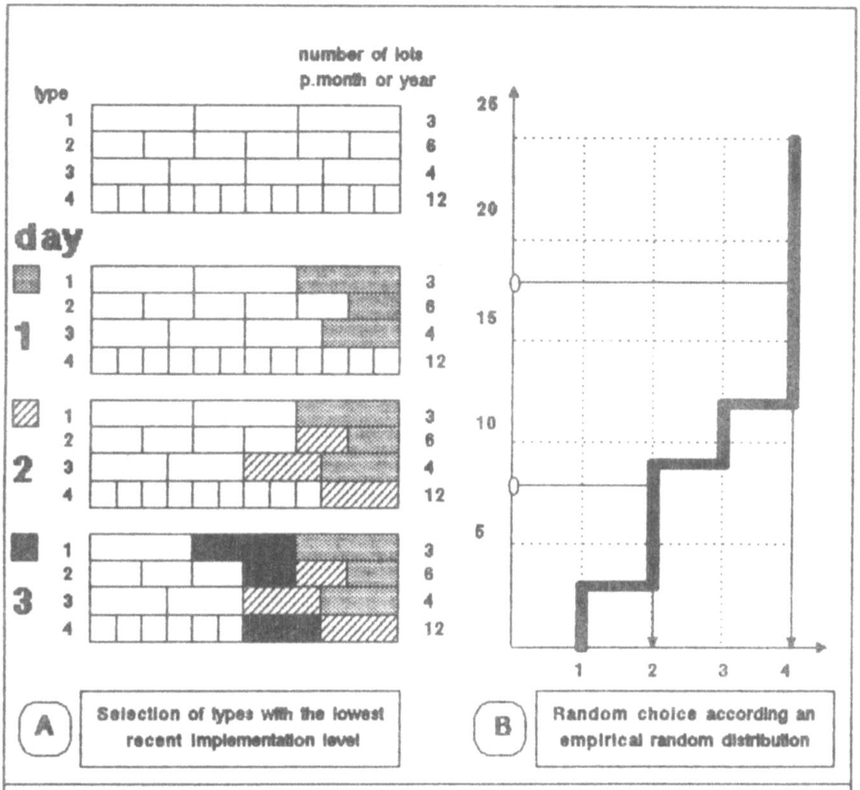

Fig.2: Two ways for generating daily streams from a given monthly of yearly load

Fig.3 provides an overview on FMS input data structure. The arrows represent connections between the different data files. A similar data structure contains [2].

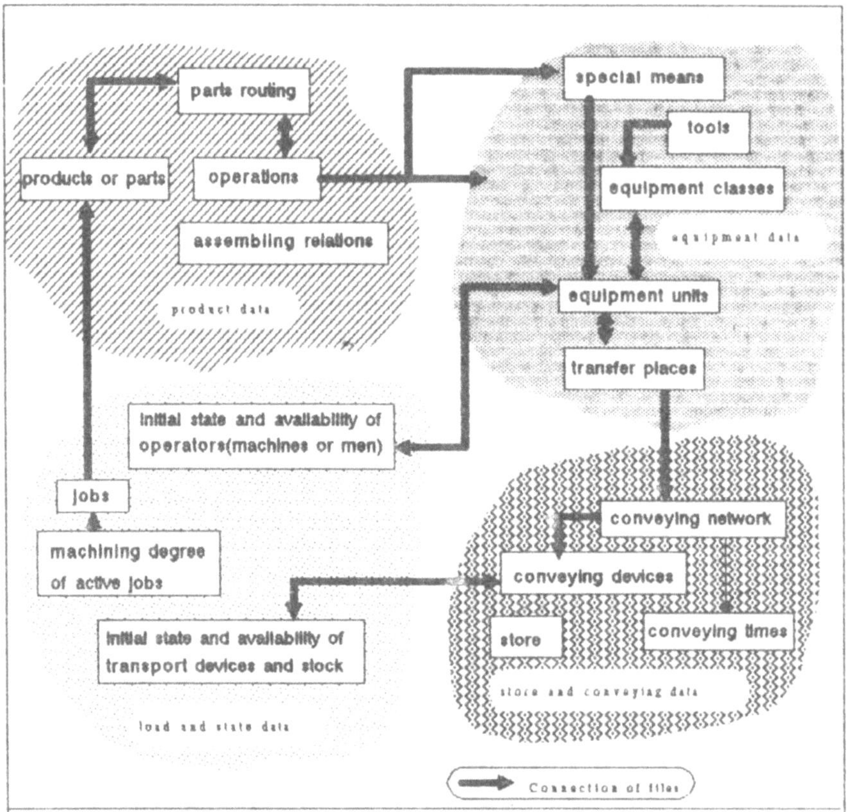

Fig.3: Typical input data structure for complex FMS simulation

2. Load analysis and simulation

An old experience of simulation analysists reads: *Do not simulate, if you can get the needed results by a simplier way!* This experience is fully valid for FMS simulation. The step after input data transformation must be a static load and capacity balance (a "work center loading model" [1]). The simulation analyst loses his credibility, if his complicated models produce results not better then the results of simple balancing. Fig.4 contains an example for a static equipment utilization analysis in a system under study . Only 3 of 9 machine categories have an utilization degree of more then 50%. In these groups 55% of the whole working time is concentrated.

The results of balancing are important for
- the further experimenting strategy and
- the designing of simulation output.

In the present example the six machine categories with an average load less then 50% can be neglected or summarized into one remainder class only.
FMS analysis and simulation tools should contain components for a static balancing of load and capacity.
Disperion analysis for system load and equipment utilization can provide additional insight into the system [2]. Models based on the queueing network theory (e.g. MANUPLAN [1]) offer in many cases an efficient alternative to simulation modelling.

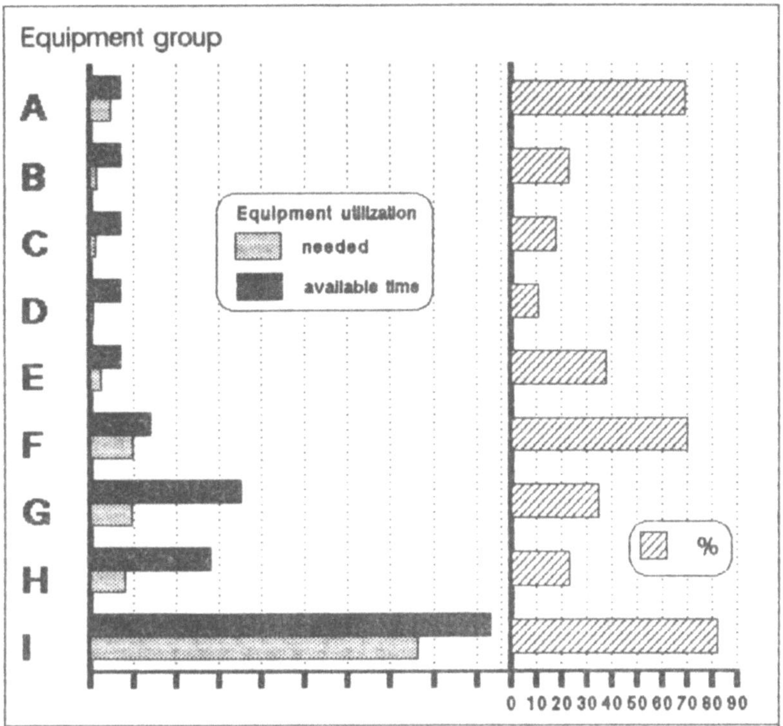

Fig.4: Equipment utilization in a manufacturing system

3. Modular System and model structure

The first column of fig.5 contains a typical model structure of a complex FMS simulation model.
Global Data Management performs data transformation and algorithms for balancing. It computes average values and dispersions. One of the results is the file of equipment groups which contains description of equipment units with identical attributes (e.g. machines which can be exchanged for any working or operating step).
The *Manufacturing Control* performs the job or task management during the jobs stay in the manufacturing process. Usually the externally formed jobs are shared in internal tasks. The Manufacturing Control applies priority or other control rules. It is easy to perform and to model, if the system has a pure work piece machining only. It is complicated, if it has to perform a mixed machining and assembling program.
These and the other submodels are constructed individually or choosen from a tool box accordingly to the *system specification*, made as the first step of the simulation study or task.

4. Precision levels of simulation submodels or modules

An old rule of simulation techniques reads: *Do not make the model finer than necessary!* This rule has a dominant importance for FMS simulations, but it is not true for the system spezification. *The system specification should be as detailed and precise as possible in the actual stage of system design.* The system specification as the working base for the system analyst is his encyclopaedia. On this base he will built raw, medium level and fine models of subsystems as he needs them. The FMS simulation tool box should contain raw and fine submodels of each subsystem, which can be combined without any restriction. Fig. 5

contains examples for combining raw and fine models of subsystems and using results of other subsystems. This technique is used for

- gaining the needed results with minimal effort,
- limiting the volume of the model and
- performing experiments under constant environment conditions.

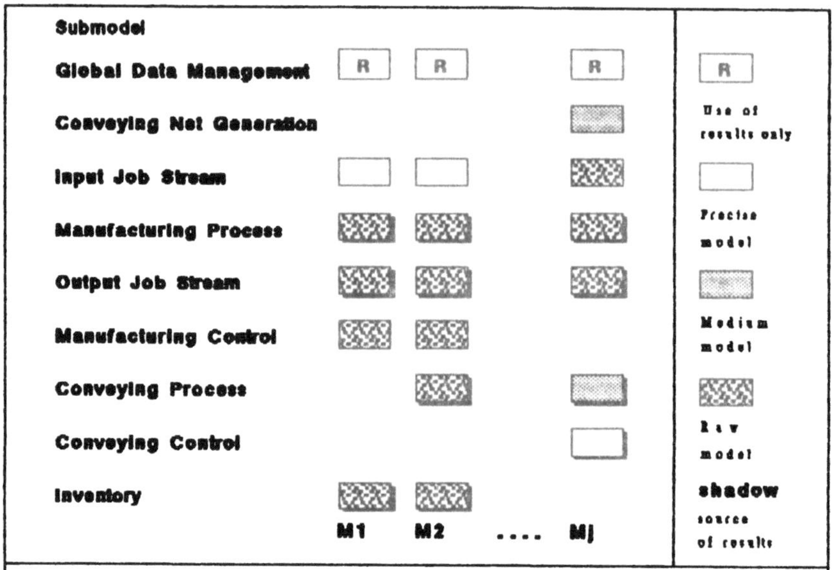

Fig.5: Model creation by combining of submodels with different levels of precision

Model 1 (M1) as the first example of fig.5 is a combination of raw and fine submodels for experiments with different storing policies to investigate its influence on the load and capacity utilization of the equipment (machines). The last example uses results of the Global Data Management for experiments with a precise model of Conveying Control, a medium level model of the Conveying Process and a raw model of the Manufacturing Process.

5. FMS Simulation and Animation

For some years *animation has been a necessary part of FMS simulation*. Animation is used in FMS simulation like in other application fields of simulation

- in the debugging stage of simulation models and programs,
- for simulation results analysis (detection of collisions, unexpected situations or effects),
- for supporting operator training and
- for winning confidence to the model and modelling results.

Animation of FMS simulation is connected with a number of problems and difficulties. In other application fields a look for the real or animated scene gives a reliable impression of the actual situation. In a road traffic or another transporting system you can receive the situation in a few moments. Manufacturing systems are different. You can look in a job shop one or more hours or days and it may be that you have not understood anything. You do not know where the true bottlenecks are, how the job deadlines are kept nor the values of the jobs running times. You do not know whether the job shop is over- or underloaded. Under these conditions animation cannot be limited to representing real movements of parts, conveying devices or men. *Animation of FMS must be implemented on different levels of abstraction.* Over the entity movement animation an animated flow description must be available that describes the manufacturing process on an abstract level. The simulated time on this level runs very quickly. Some hours or a day are represented in some minutes.

6. Tools for simulation and animation

Solid FMS simulation can not be created in a short time. What you need is people and tools. Simulation specialists mostly like one simulation system or language. In the environment of the authors GPSS-like languages and systems have an old tradition. Many students have learned how to use these systems. Some practitioners have experience in this field. *FMS simulation tools should be oriented on a simulation system or language which have many friends in the given environment.*
FMS simulations tools must support a modelling strategy described above. Models with different degrees of refinement must be built and combined. The tools must help the modeller to do this work. The tools should be portable and support the connection with data base management and animation software systems.

R E F E R E N C E S

[1] SOUTHALL, J.T., TWAMLEY, C.J. and R.E. NICHOLLS
Simulation Improves Short Term Planning in Large Batch Manufacturing industry
Proc. of the First Nat. Conference on Production Research
London 1986 p.192-7

[2] YOUNG, A.R. and J. MURRAY
The Development of FMS Operations Procedures Using Simulation
Proc. of the 1st Nat. Conference on Production Research London 1986 p. 198-202

[3] SURI, R. and G. DIEHL
MANUPLAN. A precursor to Simulation for Complex Manufacturing Systems
Proc. Winter Simulation Conference Dec. 1985

[4] LORENZ, P. und E. GOTTSCHALK
Anwendung des PS SIMDIS in der technologischen Betriebsprojektierung
rechentechnik/datenverarbeitung beiheft 3 1978 S.55-59

MODELING AND SIMULATION OF AUTOMATED WELDING MANUFACTURING SYSTEM

W. Krug, F. Blume and F. Erfurth [1)]

1. SYSTEM ANALYSIS MODELING AND SIMULATION

System Analysis and simulation take place in a process acc. to Fig. 1. The starting point is the object to be modelled acc. to a CAD/CAM - problem. It is designated in Fig. 1 as a reality or an original with a given problem. The final result of the process is a usable computer model COMOD which is suitable for carrying out simulation experiments in accordance with the problem to be solved. According to Fig. 1 modeling is regarded as a model development process in clockwise direction and as a model validation process in counter-clockwise direction and computer simulation is a means for assisting the processes /1/,/2/.

To facilitate the understanding of the assisting process acc. to Fig. 1 the following definitions are given:

By reality an original is meant that can be transferred to an existing or hypothetical object or system e.g., in terms of design, technology, manufacturing, economy, information system, etc. that is to be invastigated on a computer system with regard to a definite problem to be solved for instance a computer integrated manufacturing system or CAD/CAM-problem /3/.

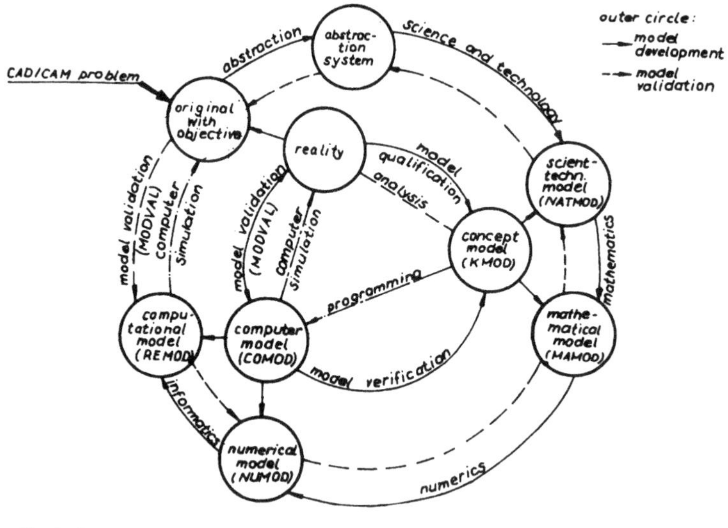

Fig. 1: Process character of modelling and simulation

1) University of Technology Dresden - Section 14 8027 Dresden

In order to anable essential aspects of this reality to be modelled in accordance with a given problem, analysis must be made. This System analysis consists of several components and steps of more precise determination e.g. model qualification of the inner circle acc. Fig. 1 or abstraction of the outer circle, subdivision of the original into partial systems, determination of description possibilities of the essential processes and developments by laws of science and manufacturing process organisation, determination of Bounderies and secondary conditions, parameter studies, etc.!
This gives a concept model KMOD wich by formulating it more precisely as a scientifico-technical model NATMOD or a geometrical and mathematical model or product model MAMOD can be transferred. A computational model consists of a program modul can be run by omplementing the numerical model NUMOD on a micro-computer on the basis of an algorithm. Within the framework of modeling acc. Fig. 1 simulation first serves for model valitation and as a problem-solving method for the analysis and synthesis of modelled processes and systems in computer integrated manufacturing and CAD/CAM.

2. SYSTEM ANALYSIS OF FLEXIBLE AUTOMATION IN WELDING ENGINEERING

Computer aided welding manufacture and planning can be a partial task within a CIM system depending on the production profile or represent its dominating objective. Additionally to an existing welding robot manufacturing cell and welding flow process a flexible welding section can be imagined.
The extended material flow and information flow of such systems aimed at flexible manufacture requires computer aided projecting, production planning and control. For cutting part manufacture solutions of such complex systems have been developed under certain geometrical restrictions - e.g. for rotationally symmetrical workpieces /3/. On the contrary, the development of welding cells towards welding manufacturing systems is still slowly proceeding due to the complexity of welding processes and the variety of structures of welding modules.
Taking into account a complex solution of automated manufacture of welding units aspects of order dispositioning, machine holding time and utilization have to be integrated additionally.

3. MODELING AND SIMULATION OF FLEXIBLE WELDING MANUFACTURING SYSTEMS

For applying a scientific method of technological preparation of flexible welding concepts at first a profound system analysis of the manufacturing process and module assortment will be necessary. The basis for this analysis is the workpiece spectrum to be produced. With the help of welding assembly classifications technical components for the production of welding assemblies can be derived. The results of this classification in the form of requirements concerning design of welding fixtures, devices for feeding and changing welding assemblies, holding tools, but also kinematic aspects, such as the number of required degrees of freedom, preferred loading directions etc. will be included into the material and information flow systems to be established in terms of the welding concept of interest!
For modeling the welding concepts to be investigated a net simulation system based on modified Petri-nets developed by the Berlin Central Institute for Cybernetics and Information Processes was used.
Fig. 2 shows a welding manufacturing cell with automated handling modeled by net components. Acc. to layout variants of system elements (cell in this case) possible movement sequences can be characterized by place and transition relations. The internal cell material flow is coupled to the superior material flow via an input/output pallet (EP, AP). The unit 'cell' consists of welding robot (IR 2), handling robot (IR 1) and a rotating table (DT) as a feeding and changing device of welding assemblies for two welding fixtures. Net -specific inputs for performing simulation experiments are:

- intended part spectrum
- machining and transport times
- treatment run-off

Thus, via a dialogue with the simulation system, which corresponds to the order-independent graph, the generated net is supplied with information on production planning. Then, the actual simulation of manufacturing sequence will proceed in an order-dependent mode.
Additionally to technological parameters, such as number of products produced, utilasation of welding stations, turnaround times, transport intensities, efficiency of stocks and buffers, times of attenuation at manual workplaces and multi-machine attenuation, weak points and bottlenecks can be revealed by means of simulation experiments.
Fig. 3 shows an example of the animation of several system elements (Gantt-diagram). Especially it is possible to acquire influences of disturbance variable (A) in this analysis of behaviour.

During the phase of projecting welding concepts different variants can be investigated and evaluated, and optimization of manufacturing sequences acc.Fig. 3 already existing systems will be enable (B,C), resp..

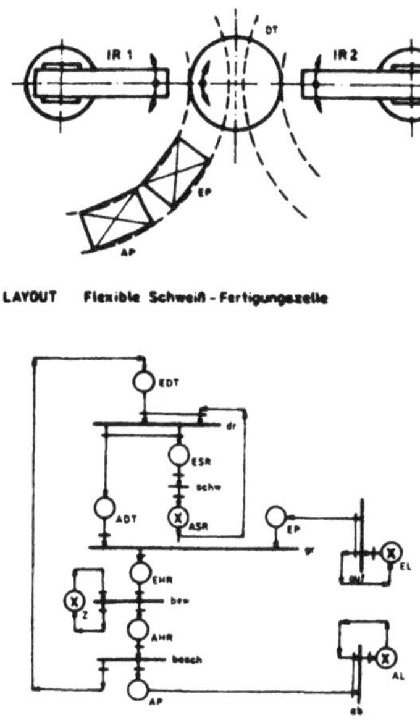

Fig. 2. Layout and net of welding manufacturing cell with automated handling

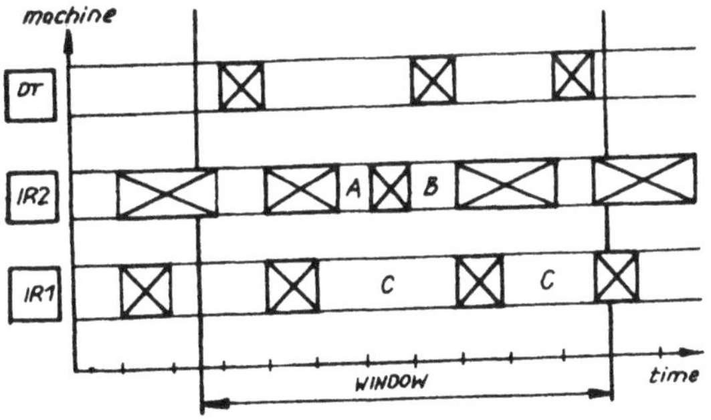

A:failure condition (wire feed system)
B:clearance of welding torch of robot
C:weak point - alternative solution: tool change (grinding operations)

Fig. 3. Utilisation of elements (welding cell)

Literature:

/1/ Oeren, T.I.: Simulation and Model based methodologies and intergrated view. Springer Verlag Berlin, 1984

/2/ Krug, W.: Software Tools for the Model Validation and Simulation. Proceedings of EUROPEAN SIMULATION CONGRESS-ANTWERP, 1986

/3/ Rudolph, Wirth: Gestaltungslösungen integrierter Fertigungen. Verlag Technik Berlin, 1986

ON THE APPLICATION OF ADVISORY SYSTEMS AND SIMULATION PROZESSES FOR THE PLANNING AND CONTROL OF MANUFACTURING SYSTEMS

G.VOIGT[1)]

An efficient flexible automation of manufacturing systems requires the realization of computer-integrated systems. It is recommended to plan and to control manufacturing by aid of cooperating expert and advisory systems realized at a computer network.

Keywords: Manufacturing, flexible automation, CIM- systems, AI, expert systems, advisory systems

1. The problem of planning and control of flexible manufacturing systems

Organization, planning, and control of large- scale manufacturing systems, e.g. of factories, is a very complex task requiring fundamentally new paradigms and new strategies. The factories must meet high demands on flexibility, efficiency, environmental frindliness or at least compatibility with their natural and social environment. Good social and working conditions must be realized in them [1], [2].

The requirements mentioned above can only be efficiently fulfilled by means of computer-integrated systems, including preparation, carrying out and accounting of the production process. The design and management of these systems require thorough analysing and comprehensive modelling of the system, its behaviour, its response to control orders and its interaction with the environment. The factories are hierarchically structured systems, the levels and subsystems of which decide and act responsibly within given limits and are temporarilycapable of working autonomously . Computers or computer networks integrated in this system support the man in terms of man- computer symbiosis and usally work in man- computer dialogue. In [3] we presented such a system. The parameters and relations in it are stochastic or can be only described in a fuzzy way. Knowledge about properties and relations can be only partially formalized and algorithmized and is often vague. Decisions are often based on experience and intuition and are often risky. We can expect a considerable improvement of the situation by profound investigation and modelling of the properties and relations and by integration of the results in expert or advisory systems. Let planning and control of manufacture in a factory serve as an example.

1) Prof. Dr. Gerhard Voigt
University of Engineering at Mittweida
9250 Mittweida GDR

2. Planning of manufacturing programs

To demonstrate the problem and methods of planning manufacturing programs we use a somewhat simplified model compared to reality. The following conditions shall hold. Each final product $i \in \{1,..,I\}$ and its components shall be processed at the work stations $j = 1,..,J$. All the basic times $a_{ji}(s)$ needed at work station j for a product i which shall be delivered in month k^* are to take place in the same month $k \in \{k^*-s+1, k^*-s+2, \ldots, k^*\}$ at the same stage $s \in \{1,\ldots,S\}$ of production, respectively. Periods $k-k^*$ and stags s are assigned one- to- one by

$$\begin{array}{lcl} k-k^* & : & 1-S,\ 2-S,\ \ldots,\ 0 \\ s & : & 1,\ 2,\ \ldots,\ S. \end{array}$$

If a_j are the quantities of products i to be delivered in k^* and $b_j{}^s(k)$ the basic time at work station j then with

$$\underline{A} = [(a_{ji},\ \underline{a}(k^*) = (a_i(k^*),\ \underline{b}^s(k) = (b_j{}^s(k)]$$

$$\underline{Aa} = \begin{pmatrix} ((a_{ji}(1))) \\ ((a_{ji}(2))) \\ \cdot \\ \cdot \\ ((a_{ji}(S))) \end{pmatrix} \underline{a} = \begin{pmatrix} \underline{A}(1) \\ \underline{A}(2) \\ \cdot \\ \cdot \\ \underline{A}(S) \end{pmatrix} \underline{a} = \begin{pmatrix} \underline{b}_j{}^1(k^*-S+1) \\ \underline{b}_j{}^2(k^*-S+2) \\ \cdot \\ \cdot \\ \underline{b}_j{}^S(k^*) \end{pmatrix} \qquad (1)$$

$((a_{ji}(1))) = \underline{A}(1), \forall i,\ j=1,..,J_1;\quad ((a_{ji}(2))) = \underline{A}(2), \forall i,\ j= J_1+1,\ldots,J_2$ etc. holds. If the manufacture schedule is made for the months $k^* = 1,\ldots,12+S$ then

$$\underline{A}\ ((\underline{a}(1)\ \ldots\ \underline{a}(11+S))) = \begin{pmatrix} \underline{b}^1(S-2) & \ldots & \underline{b}^1(11) \\ \cdot & \ldots & \cdot \\ \underline{b}^S(1) & \ldots & \underline{b}^S(11+S) \end{pmatrix} \qquad (2)$$

holds. Matrix $\underline{B}$ with vectors $\underline{b}^s(k)$, $s = 1,\ldots,s$; $k = 1,\ldots,12$ as elements is the matrix of basic times to be spent in year $k= 1,\ldots,12$. If $\underline{B}^F$ is the matrix of the corresponding resources of basic times with elements $bf_j^s(k)$ the following stepwise balances must hold

$$\sum_{\forall j} \sum_{\forall k} (b_j^s(k) - bf_j^s(k)) <= 0 \qquad (3)$$

$$\sum_{\forall k} (b_j^s(k) - bf_j^s(k)) <= 0\ ,\ \forall j \qquad (4)$$

$$|b_j^s(k) - bf_j^s(k)| <= \varepsilon\ ,\ \forall j,\ \forall k \qquad (5)$$

Because stochastic influences are unavoidable the components of the vectors and the elements of the matrices are stochastic quantities in reality [4]. In addition to the balancing conditions economic demands must be met. More detailed considerations are to find in [5].

To fulfil conditions (3) to (5) it is possible to modify the basic times a_{ji}, the quantities a_i and the resources bf_j^s. For this purpose a large quantity of constraints regarding the quantities of products, the delivery times, the disponibility of workers, machines and operations etc. must be taken into consideration. It is necessary to model the relations between basic times, resources, quantities of products, and the manageable technological, organizational, educational, and commercial factors and to use them by establishing balanced manufacture schedules $a_i(k^*)$, $i= 1,\ldots,I$, $k^*= 1,\ldots,11+S$.

Up to now there exists no system of rules or algorithms, which is capable of elaborating a satisfying solution of the planning problem. The relations between time a_{ji} resources $bf_j^s(k)$, and their manageable influence factors can only partially be formalized or to be quantified, be described fuzzily or with verbally expressed rules, and can largely be used by intuition based on experience only. As a rule, the established plans generally fulfil as the balancing conditions, especially condition (5) relatively incorrect. This leads to relatively large queues and to high production costs.

A satisfying solution of the planning problem is only possible by the development and application of expert and advisory systems. For this purpose it is necessary

- to search profoundly in the relations between the manufacturing times, personal and technical resources and their manageable influence factors and the regularities of the turn- around of products and its controllability,
- to collect information about the properties and the actual state of the system,
- to classify the relations, information, criterions, goals, and strategies, and to store them into the knowledge base of the advisory system in suitably structured patterns, and
- to formulate criterions and goals of planning,
- to develop stategies for planning as algorithms, verbally formulated rules, and possibly as vague suggestions,
- to develop an inference system which recommends strategies for planning referred to the situation.

After the selection of a strategy by the user one or more plan variants are established using men- computer dialogue. Important intermediate results and characteristics of the plan are distributed by the computer as data, verbally fomulated information or as graphs. If the results are unsatisfactory the planning process will be repeated with a different strategy. By an interactively working learning system the knowledge base will be completed and the goals and chances of succes related to the situation will be evaluated.

3. The operativ planning and control of manufacture

The monthly allocated manufacturing program is the base of the operative, short- term manufacture schedules and manufacturing control. By elaborating manufacturing programs the stochastic influences are essentially neglected. Operative planning and control must be based on the actual state of manufacture. The deviation between actual and desired values is used for planning and control. The stochastic properties of the manufacturing times, time resources, turn- around times etc. must be considered by planning with aid of simulation of the expected system behaviour. It is appropriate to collect all identical components in series which are to be manufactured in monthly series and to organize planning which is anonymous with respect to customers and orders. In this case an efficient planning and control requires the application of cooperating advisory systems. The problems to solve are analogous to those mentioned in planning manufacturing programs. Because of the stochasticity which can not be neglected in this case the problem becomes very complex and difficult. Investigations into the modes of actions of the system and into the efficiency of control strategies are made in [6], [7]. A program system is realized which is capable of recognizing necessities for decisions by analizing deviations between desired and actual values, and of proposing suitable decisions, and after the decisions are made by the user, to ascertain the consequences. This is a beginning of an advisory system for operative planning and control of manufacture [8].

It is necessary to develop hierarchically structured systems of cooperating advisory and expert systems for an efficient flexible automation of factories. These systems have to be implemented on a computer network. In analizing and modelling the system it is necessary to handle the "hard" part which can be described numerically and by algorithms and the "soft", one which can only be described in a fuzzy, verbal and vague way as a unit. The realizing of flexibly automated factories should be connected with fundamental changes in the goals, products, tchnologies, etc..

References:

[1] Tondl, L.: New paradigmas of design and the function of CAD, ICCADI '87, Karlovy Vary, CSSR, 1987

[2] Voigt, G.: Some remarks on a philosophy and a uniform modelling conception of general flexibly automated manufactoring systems..., ICCADI '87, Karlovy vary, CSSR, 1987

[3] Voigt, G., Lippold, J.: Analisis, modelling and control of complex manufacturing systems, Math.Fschg. 28, Berlin 1085

[4] Voigt, G., Cramer, S.: Diskontinuierliche technologische Prozesse, Akademie- Verlag, Berlin 1986

[5] Foerster, Th.: Probleme und Loesungsvorschlaege der rechnergestuetzten Prozessablaufplanung zur Sicherung der Flexibilitaet, Stabiliteat und Zuverlaessigkeit in der Einzel- und Kleinserienfertigung.., Diss., IH Mittweida 1986

[6] Lippold, J.: Einsatz von Modellen der rechnergestuetzten Prozessablaufplanung u. Fertigungsprozesssteuerung..., Diss. IH Mittw. 1985

[7] Bergmann, F.: Probleme und Loesungsvorschlaege zur Minimierung der Durchlaufzeiten in diskontinuierlichen Fertigungsprozessen..., Diss., IH Mittweida 1988

[8] Eck, K.: Zielsetzung der rechnergestuetzten Steuerung der Produktion im elektrischen Geraetebau..., Diss., IH Mittweida 1987

THE NON-STATIONARY DISTRIBUTION OF NUMBER OF CALLS IN THE M/G/1 PROCESSOR-SHARING QUEUE

S.F. Yashkov [1)]

The M/G/1 queue with egalitarian processor-sharing is studied by the way proposed in [3]. The new results include the solution of the problem of determining the time-dependent queue-length distribution in the form of Laplace transform of a generating function. We also obtain the joint equilibrium distribution of the conditional sojourn time for a call with a length x and the number of calls in the system at its departure.

Keywords: stochastic processes, processor-sharing, random time change, number of calls, sojourn time distribution

1. Introduction.

Queueing models play an important role in analysis of increasingly complicated technical systems. The survey by Cooper gives a good introduction to recent developments in queueing theory [1]. During the last decade considerable attention has been paid to the analysis of various processor-sharing queueing models (see e.g., the survey [4]). One of the basic models among them for investigating the effects of time-sharing is the M/G/1 queue with the discipline of egalitarian processor-sharing (PS). In this model it is assumed that when there are $1 \leq n < \infty$ calls present then all these calls are serviced simultaneously by the single server (processor) with the rate 1/n i.e., they receive 1/n times the rate of service which a solitary call in the processor would receive. Jumps of the service rate occur at the instants of arrivals and departures from the system. Therefore, the rate of service received by a specific call fluctuates with time and, importantly, its sojourn time depends not only on the calls in the processor at its time of arrival there, but also on subsequent arrivals. This makes the PS system intrinsically harder to analyze than, say, the first-come-first-served queue.

The problem of the derivation of the stationary sojourn-time distribution in the M/G/1-PS queue was open for a long time; its solution has been obtained by different methods in [2,3]. This paper deals with new results of an analysis of the M/G/1-PS queue, which are an extensi-

[1)] Institute for Problems of Information Transmission of the USSR Academy of Sciences, 19 Yermolova Street 101447 Moscow GSP-4, U.S.S.R.

on of the results in[3]. In particular, we shall obtain the non-stationary queue-length distribution in this system.

We consider the M/G/1-PS queue where the calls arrive according to a Poisson process with rate λ. The call's lengths (i.e., the required service times with unit rate) are i.i.d. random variables with a general distribution B(x) ($B(0^+)=0$) the first moment of which is $\beta_1<\infty$. We denote $\beta(s)=\int_0^\infty e^{-sx}dB(x)$. Let at time 0 a call with a length which is not less than x arrives in the system. This call finds the system in the state $n,\underline{x}$ where n ($0\leq n<\infty$) is the number of the other calls in the processor at time 0 and $\underline{x}=(x_1,\dots,x_n)$ is the vector of their residual lengths. Let V(x) be the time for which an increment of an attained service time of this call will be equal to x (in other words, V(x) is the sojourn-time of the call with the length x given the state of the queue is $(n;x_1,\dots,x_n)$ at time 0. The number of the calls at time t is described by the process $\{L(t),t\in[0,\infty)\}$ on the standard probability space.

2. Results.

Let $\zeta=\inf\{t: L(t)=0\}$. Introduce an additive functional $X(t)=\int_0^t du/(L(u)\vee 1)$. The inverse of X(t) is defined by $\tau(t)=\inf\{u:X(u)>t\}$ (the process $\{\tau(t)\}$ is a subordinator and, for $\forall$ t, $\tau(t)$ is a stopping time relative to natural σ-fields). Define now the time changed (measurable) process M(t) by $M(t)=L(\tau(t))$, a.s. which is obtained from $\{L(t)\}$ on the first busy period by means of the random time change $\tau(t)$. The new process $\{M(t)\}$ is continuous from the right. Let $\zeta_m=\inf\{t:M(t)=0\}$. We note in passing that $X(\zeta)=\zeta_m$, $\tau(X(t))=t$ and $X(\tau(t))=t$. There is a mutual conformity between the sample paths of the processes $\{L(t)\}$ and $\{M(t)\}$ according to which $\tau(t)=\int_0^t M(u)du$. We note also that the process $\{\tau(t)\}$ is strightly increasing and the sample paths of $\{\tau(t)\}$ are absolutely continuous (a.s.) with the density M(t). Besides for $\forall$ t the random variable $\tau(t)\overset{d}{=}V(t)$ under the initial condition $n,\underline{x}$. We have the

Theorem 1.

$$E\left[e^{-s\tau(t)}\cdot z^{M(t)}1_{\{\zeta_m>t\}}\,\middle|\,(n;x_1,\dots,x_n)\right]=\delta(s,z;t)\prod_{i=1}^{n}\varphi(s,z;x_i,t),\quad \operatorname{Re}s\geq 0,\ |z|\leq 1 \tag{1}$$

where

$$\varphi(s,z;x,t)=\begin{cases}\delta(s,z;t), & x\geq t,\\ \delta(s,z;t)/\delta(s,z;t-x), & x<t,\end{cases}\qquad \delta(s,z;t)=ze^{-t(s+\lambda)}/\psi(s,z;t),$$

$$\psi(s,z;t)=\frac{1}{2\pi i}\int_{c-i\infty}^{c+i\infty}\frac{q+s+\lambda-\lambda z+\lambda z\beta(q+s+\lambda)}{(q+s+\lambda)(q+\lambda\beta(q+s+\lambda))}e^{qt}dq,\quad c\geq 0,\ s\geq 0,\ q>-\lambda\pi(s). \tag{2}$$

Here $1_{\{\cdot\}}$ denotes the indicator function of the event $\{\cdot\}$ and

$$\mathcal{T}(s) = E\left[e^{-s\zeta}\right] = \beta\,(s+\lambda-\lambda\,\mathcal{T}(s)) . \tag{3}$$

Sketch of the proof. The key tool in proof is a study of $\{M(t)\}$ which is interpreted as the number of calls in an $M_n/G/\infty$ queue with arrival rate λn for $n>0$ and λ for $n=0$. Another interpretation of $\{M(t)\}$ consists in the consideration of each call as a particle with the lifetime distribution $B(x)$. Each particle generates another particles (descendants) during its lifetime with rate λ. Then $\{M(t)\}$ is the sum of n+1 independent branching processes, that is, $M(t)= \sum_{i=1}^{n+1} N(x_i,t)$. All these processes are evolved in parallel, each of them is generated by one particle of the length x_i at time 0 and it is terminated at time t if it was not degenerated earlier. The sample paths of $\{N(x,t)\}$ are continuous from the right. The lifetime of the separated branching process $\{N(x,t)\}$ is defined by the stochastic integral $\Phi(x,t)=\int_0^t N(x,u)du$. Let $\varphi(s,z;x,t)= = E[e^{-s\Phi(x,t)} z^{N(x,t)}]$. This transform gives us the joint distribution of the number of particles $N(x,t)$ and the lifetime $\Phi(x,t)$ of one from n+1 i.i.d. branching processes. We note that these branching processes have the same type as the branching processes by Crump-Mode-Jagers.

The unknown function φ is found as the solution of a functional integral equation of the same type as in [3] where were obtained $\varphi(s,x,t)= =\varphi(s,1;x,t)=E[e^{-s\Phi(x,t)}]$ and $E\,[e^{-sV(t)}\mid n,\underline{x}]$. The additional conditions are $\varphi(s,z;0,t)=1$, $\varphi(s,z;x,0)=\delta(s,z;0)=z$. Thus the proof is an extension of arguments of Theorem 4 in [3] to the case when we take into account the number of particles-descendants. We only note that ψ is defined as

$$\psi(s,z;t) = \exp\left(-\lambda\int_0^t\int_0^\infty \varphi(s,z;x,u)\,dB(x)\,du\right). \tag{4}$$

Now we do not require that $\rho=\lambda\beta_1<1$.

Theorem 1 is the starting point to derive the non-stationary distribution of $\{L(t)\}$. We consider $\{L(t)\}$ on the first busy period that begins at time 0 with n calls present (these calls have the lengths $x_1,\dots,x_n$). Thus the state of the system is $n,\underline{x}$ at time 0. We define

$$P_{n,\underline{x}}(s,z) = \int_0^\infty e^{-st} E\left[z^{L(t)} 1_{\{\zeta>t\}} \mid (n;x_1,\dots,x_n)\right] dt .$$

In virtue of Fubini's theorem this equality can be rewritten as

$$P_{n,\underline{x}}(s,z) = E_{n,\underline{x}}\left[\int_0^\zeta e^{-st} z^{L(t)} dt\right]. \tag{5}$$

We made the random time change as it is described above. Then replacing the variable t in (5) with $\tau(t)$ and applying the Fubini's theorem we deduce that

$$P_{n,\underline{x}}(s,z) = z\int_0^\infty \frac{\partial}{\partial z}\left(E_{n,\underline{x}}\left[e^{-s\tau(t)} z^{M(t)} 1_{\{\zeta_m>t\}}\right]\right)dt . \tag{6}$$

Now we do not suppose that the tagged call with a length x arrives in the system at time 0. That is why $E_{n,\underline{x}}[\cdot]$ in (6) is given by (1) without the first factor. Then from (6) and (1) we obtain after removing the condition on $\underline{x}$

$$p_n(s,z) = z\int_0^\infty \frac{\partial}{\partial z}\prod_{i=1}^{n}\left[\int_0^\infty \varphi(s,z;x_i,t)\,dB(x_i)\right]dt .$$

Taking into account (4) this proves

Theorem 2. The distribution of the number of calls on the busy period that begins with n > 0 calls is given by the Laplace transform of their generating function

$$p_n(s,z) = \frac{z}{\lambda^n}\int_0^\infty \frac{\partial}{\partial z}\left[-\frac{\partial}{\partial t}\ln\psi(s,z;t)\right]^n dt , \quad Re\, s \geq 0,\ |z| \leq 1, \quad (7)$$

where ψ is given by (2).

It constitutes a result hitherto unknown for any PS model.

Arguments from the renewal theory allow us to connect (7) with the distribution of the process $\{L(t)\}$ on all half-axis $[0,\infty)$. In particular, if the PS system is empty at initial instant, we can obtain

$$\int_0^\infty e^{-st} E\left[z^{L(t)} \mid L(0)=0\right] dt = (s+\lambda-\lambda\pi(s))^{-1}(1+\lambda p_1(s,z)),$$

where $p_1(s,z)$ is given by (7) and $\pi(s)$ is given by (3).

The Theorem 1 leads also to the joint stationary distribution of the sojourn time of an arbitrary call whose length is x at the arrival time and the number of calls present at its departure (itself included).

Theorem 3. In the statinary situation i.e., when $\rho < 1$

$$E\left[e^{-sV(x)}\cdot z^{L(V(x)^-)}\right] = (1-\rho)\, z\, e^{-x(s+\lambda)} \cdot$$

$$\cdot\left[\psi(s,z;x) - \lambda\int_0^x e^{-u(s+\lambda)}\psi(s,z;x-u)\bar{B}(u)\,du - \lambda z e^{-x(s+\lambda)}\int_x^\infty \bar{B}(u)\,du\right]^{-1},$$

where $\bar{B}(x)=1-B(x)$.

Proof. Similar to the proof of the last part of Theorem 4 in [3].

3. Conclusion.

Theorems 1-3 open the way of efficient numerical studies of equilibrium and time-dependent performance measures for processor-sharing system. The results can be used to derive a number of others, including the conditional and unconditional moments of the marginal distributions of the queue length and the sojourn time.

References

1 Cooper, R.B.: Queueing theory. Handbook of Operations Research and Management Science (D.P.Heyman and M.J.Sobel, eds), Vol.2: Stochastic Models. North-Holland (to appear).

2 Schassberger, R.: A new approach to the M/G/1 processor-sharing queue. Adv. Appl. Probab. 16 (1984) 1, 202-213.

3 Yashkov, S.F.: A derivation of response time distribution for an M/G/1 processor-sharing queue. Probl. Control and Info. Theory 12 (1983) 2, 133-148.

4 Yashkov, S.F.: Processor-sharing queues: some progress in analysis. Queueing Systems 2 (1987) 1, 1-17.

ROBOT MANIPULATOR SYSTEMS: ANALYSIS AND CONTROL

J. A. Tenreiro Machado [1,2] and J. L. Martins de Carvalho [1,2]

1. INTRODUCTION

Manipulator systems are rather complex and highly nonlinear which makes difficult their analysis and control. Classic system theory is well known, however it is inadequate in the presence of strong nonlinear dynamics. Nonlinear controllers produce good results [1] and work has been done e.g. relating the manipulator nonlinear dynamics with frequency response [2-5]. Nevertheless, given the complexity of the problem, systematic methods which permit to draw conclusions about stability, imperfect modelling effects, compensation requirements, etc. are still lacking. In section 2 we start by analysing the variation of the poles and zeros of the descriptive transfer functions of a robot manipulator in order to motivate the development of more robust (and computationally efficient) control algorithms. Based on this analysis a new multirate controller which is an improvement of the well known "computed torque controller" [6] is announced in section 3. Some research in this area was done by Neuman [7,8] showing that better robustness is possible if the basic controller structure is modified. The present study stems from those ideas, and attempts to give a systematic treatment, which results in easy to use standard engineering tools. Finally, in section 4 conclusions are presented.

2. DYNAMIC STUDY OF A TWO DEGREES OF FREEDOM MANIPULATOR

Robot manipulator dynamics is modelled by a set of complex nonlinear matrix differential equations [9,10]; nevertheless, for many aspects of robot system research, the simple two or three d.o.f. manipulator models are sufficient. We highlight this argument, as in the sequel a 2R robot manipulator model is used, without any loss of generality. In fact, the 2R manipulator dynamic model

$$T_1=[(m_1+m_2)r_1^2+m_2r_2^2+2r_1r_2m_2C_2+J_1]\ddot{q}_1+(m_2r_2^2+r_1r_2m_2C_2)\ddot{q}_2-$$
$$-r_1r_2m_2S_2\dot{q}_2^2-2r_1r_2m_2S_2\dot{q}_1\dot{q}_2+m_1gr_1C_1+m_2g(r_1C_1+r_2C_{12})$$
$$T_2=(m_2r_2^2+r_1r_2m_2C_2)\ddot{q}_1+(m_2r_2^2+J_2)\ddot{q}_2+r_1r_2m_2S_2\dot{q}_1^2+m_2gr_2C_{12} \qquad (1)$$

already includes all the configuration dependent inertial, Coriolis/centripetal and gravitational torques that may appear in manipulators with more d.o.f., thus being adequate for our prototype manipulator. This strategy has been used by other researchers [11-13] with the parameter set ($m_2=m_0+m_L$):

$$m_1=0.5Kg; \quad m_0=0.25Kg; \quad m_L=6Kg; \quad r_1=1m; \quad r_2=0.8m; \quad J_1=5Kgm; \quad J_2=5Kgm$$

We shall now linearize (1) by introducing small signal variations around an equilibrium point $\{q_{10},q_{20},T_{10},T_{20}\}$. Substituting these variations in (1) and considering only first order terms, we get after using Laplace transform, the transfer matrix $T(s)=G(s)Q(s)$, where

$$G_{11}(s)=s^2[(m_1+m_2)r_1^2+m_2r_2^2+2r_1r_2m_2C_2+J_1]-[m_1gr_1S_1+m_2g(r_1S_1+r_2S_{12})]$$
$$G_{12}(s)=G_{21}(s)=s^2(m_2r_2^2+r_1r_2m_2C_2)-m_2gr_2S_{12}$$
$$G_{22}(s)=s^2(m_2r_2^2+J_2)-m_2gr_2S_{12} \qquad (2)$$

The corresponding inverse matrix transfer function $G(s)=[H(s)/d(s)]T(s)$ has a characteristic equation $d(s)=G_{11}(s)G_{22}(s)-G_{12}(s)^2$.

From these expressions we can conclude that:

-Coriolis and centripetal terms are of second order influence on the dynamic response, that is in stability terms. Only gravitational and inertial terms are responsible for the transfer function poles and zeros locations.

-Pole and zero variations are a result of the inertial and gravitational dependence on q_{10} and q_{20}. With a manipulator in outer space, where gravity loading is absent or with manipulators with horizontal structure, the system has two poles fixed at the origin of the s-plane, and variable gain due to the inertial terms.

-The roots of d(s) occur in pairs either on the imaginary axis or on the real axis. To each downward link corresponds a complex

1) Faculdade de Engenharia da Universidade do Porto, Dep. Eng. Electrotecnica Computadores, R. dos Bragas, 4099 Porto Codex, PORTUGAL
2) INESC-Instituto de Engenharia e Sistemas de Computadores, R. Jose Falcao 15, 4º, 4000 Porto, PORTUGAL

conjugate pole pair on the Img. axis. To each upward link, corresponds a real pole pair, symmetrical about the origin. Therefore, it is not surprising when some manufacturers and researchers claim better performances for robots with downward links [14,15].

-There is output coupling, due to the nonzero values of $H_{12}(s)$ and $H_{21}(s)$.

Concerning the control of robot manipulators, several considerations can now be drawn. The standard industrial control configuration, having PID controllers in each robot joint, has to compensate the aforementioned nonlinear dynamics. However, the designer is faced with the selection of a maximum of 5n interacting parameters, and no systematic criteria for their selection! Furthermore, a set of satisfactory parameter values for a given region may prove totally unsatisfactory elsewhere as it sometimes occurs with present day industrial robot systems, whose behaviour is "shaky" in some operating regions. Therefore, a different control scheme must be envisaged, enabling an easier and systematic adjustment of the controller parameter values, as well as output decoupling anywhere in the operational space. This problem will be the topic of the next section.

3. ANALYSIS OF THE ROBUSTNESS AND COMPLEXITY OF THE CONTROL ALGORITHM

Because the poles and zeros of the manipulator transfer function change with q_{10} and q_{20} some form of system compensation is required. This compensation must be adaptive, that is, the compensation must adjust itself to the variations of the system dynamics. The computed torque nonlinear controller:

Commanded acceleration $u(t)=\ddot{q}_d+K_1(\dot{q}_d-\dot{q})+K_2(q_d-q)$

Computed torque $T(t)=J_C(q)u(t)+C_C(q,\dot{q})+G_C(q)$

Error $e(t)=q_d(t)-q(t)$

Error equation $\ddot{e}+K_1\dot{e}+K_2e=V(t)$

Disturbance vector $V(t)=J_C(q)^{-1}\{[J(q)-J_C(q)]\ddot{q}+[C(q,\dot{q})-C_C(q,\dot{q})]+[G(q)-G_C(q)]\}$

achieves such compensation, namely for the case of a precise robot modelling, $J_C(q)=J(q)$; $C_C(q,\dot{q})=C(q,\dot{q})$; $G_C(q)=G(q)$ then $V(t)=0$ and the error equation corresponds to a second order, decoupled linear system. It is common knowledge that in general, these conditions are not fulfilled, resulting a complex nonlinear dynamic system; therefore, it is difficult to predict the resulting dynamic behaviour and to gain a feel for the right controller structure. As the precise payload parameter estimation, in real time, still remains at a research stage [16], the problem must be solved through an improvement of the controller algorithm. With exact load estimation the (ideal) system behaves like a linear block with unity transfer function, nevertheless, the (real) computer controlled system will have a non-zero controller computing time, non-zero sampling period and imperfect robot modelling.

These factors justify a more elaborate study of their implications. Rearranging the computed torque equations we get

$$J_C(q)(\ddot{e}+K_1\dot{e}+K_2e)=[J(q)\ddot{q}+C(q,\dot{q})+G(q)]-[J_C(q)\ddot{q}+C_C(q,\dot{q})+G_C(q)] \qquad (3)$$

From equation (3) we conclude:

-The right hand side may be viewed as the model error torque and the left hand side as the required controller correcting torque.

-This controller correcting torque is driven by the feedback loop $J_C(q)$, K_1, K_2.

-The more inaccurate the model the higher the requirements imposed to the feedback loop.

-For digital control systems, the finite sampling frequency, does not allow a continuous torque correction. Therefore, the discontinuous correcting torque may result in a system performance degradation.

These considerations clearly show the dependence of the system dynamic properties upon the correcting feedback loop $J_C(q)$, K_1, K_2. Nevertheless, the system is also required to have good performances in "time independent" situations, that is in steady state. Here, we have $\ddot{q}_d=0$, $\ddot{q}=0$, $\dot{q}_d=0$ and $\dot{q}=0$, resulting the system equation:

$$e_{SS}=K_2^{-1}J_C(q)^{-1}[G(q)-G_C(q)] \qquad (4)$$

As expected, the steady state error e_{SS} is determined by the gravitational model error and inversely proportional to K_2.

Summing up, we have that the controller blocks fulfil different tasks:

-For dynamic aspects (transients, stability etc.) the feedback loop $J_C(q)$, K_1, K_2 is of prime importance

-In steady state, the gravitational model and the constant gain position matrix (K_2) are the main blocks.

Equations (3,4) enable a deeper insight towards a systematic study of an improved computed torque controller, therefore the designer may use standard control techniques such as the use of integral action (i.e. $G_2(s)=I_2+I'_2/s$) so that the position steady state error is eliminated and derivative action (i.e. $G_1(s)=I_1+I'_1 s$) in order to improve stability.

Besides the referred standard improvements, a reduction in the computational time delay, may be obtained through the use of different sampling ratios [17] on the controller loops: the heavily time dependent $J_C(q)$, $G_1(s)$, $G_2(s)$ loop is computed at a higher rate and the less critical loop $C_C(q,\dot{q})+G_C(q)$ is computed at a lower rate.

Therefore, the control computer works in a time sharing way. This aspect is of utmost importance, as it corresponds to a more efficient use of the system resources. A uniform sampling frequency would be a "waste" of computing power on the $C_C(q,\dot{q})+G_C(q)$ loop, which would be more profitable at the $J_C(q)$, $G_1(s)$, $G_2(s)$ loop. Another important observation resides on the computational burden of each block: $G_1(s)$ and $G_2(s)$ are easily computed, $J_C(q)$ and $G_C(q)$ represent a medium load and $C_C(q,\dot{q})$ is highly complex.

In the fast computational loop $J_C(q)$ is the heavier block. Although efficient computacional techniques are available [18,19] an alternative computational method, is the use of memory. Computer memory has already been used for robot manipulator control [20,21]. In the present case, this method may be viewed as another system management action.

Finaly, a third improvement can be implemented by a predictive action [22,23]. As good path tracking is expected, Coriolis/centripetal and gravitational terms can be approximately computed in advance, estimating positions and velocities as the planned ones. Computation between two "slow samples" is made by interpolation. To verify the new controller performances, a straight line test trajectory from point (-1,1) to (1,0) was used. The resulting position tracking error

$$e_P=[(x_d-x)^2+(y_d-y)^2]^{1/2} \quad (5)$$

is shown in Fig. 1 for an overestimation ($\hat{m}_L$=12 Kg) of the payload mass. The modified controller eliminates the steady state error and reduces the position tracking error e_P, namely the maximum tracking is scaled down by a factor of 35.5.

4.CONCLUSION

Based on an s-plane model, insight has been given into the dynamics of manipulators. It was shown that such systems can be open-loop unstable. Present day "standard" control structure of industrial manipulators was discussed and attention was drawn to some of its limitations: traditionally each joint has its own PID controller and position/velocity feedback. Such control structure is simple and may achieve stability, however it doesn't provide neither output decoupling nor satisfactory performance in the entire operational space. Futhermore, the adjustment of their (interacting) parameters is an overwhelming task given the absence of systematic procedures. A controller capable of better performances must have a different structure. Also some kind of compensation needed, particularly with links upwards. The computed torque algorithm is a first step in that direction. Its limitations, namely stability, steady state error and computational burden were identified which motivated the development of several system structural enhancements. The concern for an efficient allocation of the computational resources lead to several modifications in the conventional computed torque controller: multirate sampling according to the influence of each block, possible use of memory, so that computing time is alleviated and predictive action in order to compensate slow computation effects.

As a result, a new multirate and robust controller enabling real time performances with today's microcomputers was obtained.

REFERENCES

[1] E. Freund, Fast Nonlinear Control with Arbitrary Pole-Placement for Industrial Robots and Manipulators, The Int. J. Robotics Research, vol. 1, Spring, 1982, pp. 65-78.
[2] D.F. Golla, S.C. Garg, P.C. Hughes, Linear State Feedback Control of Manipulators, Mech. Mach. Theory, vol. 16, 1981, pp. 93-103.
[3] H. W. Stone and C. P. Neuman, Dynamic Modelling of a Three-Degrees-of-Freedom Robotic Manipulator, IEEE Trans. Syst., Man, Cybern., vol. SMC-14, July/Aug., 1984, pp. 643-654.
[4] M. C. Good, L. M. Sweet and K. L. Strobel, Dynamic Models for Control Systems Design of Integrated Robot and Drive System, ASME J. Dynamic Syst., Meas., Contr., vol. 107, March, 1985, pp. 53-59.
[5] H. Seraji, M. Jamshidi, Y. T. Tim, M. Shahinpoor, Linear Multivariable Control of Two-Link Robots, J. of Robotic Systems, vol. 3, Winter, 1986, pp. 349-365.
[6] J. Y. S. Luh, M. W. Walker and R. P. C. Paul, Resolved-Acceleration Control of Mechanical Manipulators, IEEE Trans. Automat. Contr. vol. 25, June, 1980, pp. 468-474.
[7] V. D. Tourassis and C. P. Neuman, Robust Nonlinear Feedback Control for Robotic Manipulators, IEE Proc., vol. 132, Pt. D, July, 1985, pp. 134-143.

[8] Charles P. Neuman and Vassilios D. Tourassis, Robust Discrete Nonlinear Feedback Control for Robotic Manipulators, J. Robotic Systems, vol. 4, Feb., 1987, pp. 115-143.
[9] R. P. Paul, Robot Manipulators: Mathematics, Programming and Control, The MIT Press, 1981.
[10] M. Brady et al., Robot Motion: Planning and Control, The MIT Press, 1982.
[11] Kar-Keung D. Young, Controller Design for a Manipulator Using Theory of Variable Structure Systems, IEEE Trans. Syst., Man, Cybern., vol. SMC-8, Feb., 1978, pp. 101-109.
[12] Russel G. Morgan and Umit Ozguner, A Decentralized Variable Structure Control Algorithm for Robotic Manipulators, IEEE J. Robot., Automat., vol. RA-1, March, 1985, pp. 57-65.
[13] J. A. Tenreiro Machado and J. L. Martins de Carvalho, A Smooth Variable Structure Control Algorithm for Robot Manipulators, IEE Control 88 Conference, Oxford, U.K., April, 1988.
[14] Ostman I., Holmqvist V. and Allared C.-A, Pendulum Robot, Asea Journal, vol. 58, 3-4, 1986, pp. 12-17.
[15] H. Asada, T. Kanade and I. Takeyama, Control of a Direct-Drive Arm, ASME J. Dynamic Syst., Meas., Contr., vol. 105, Sept., 1983, pp. 136-142.
[16] Christopher G. Atkeson, Chae H. An and John M. Hollerbach, Estimation of Inertial Parameters of Manipulator Loads and Links, The Int. J. Robotics Research, vol. 5, Fall, 1986, pp. 101-119.
[17] Douglas P. Glasson, Development and Application of Multirate Digital Control, IEEE Control Syst. Magazine, Nov., 1983, pp. 2-8.
[18] J. Y. S. Luh, M. W. Walker and R. P. C. Paul, On-Line Computational Scheme for Mechanical Manipulators, ASME J. Dynamic Syst., Meas., Contr., vol. 102, June, 1980, pp. 69-76.
[19] Charles P. Neuman and John J. Murray, Symbolically Efficient Formulations for Computational Robot Dynamics, J. Robotic Systems, vol. 4, Dec., 1987, pp. 743-769.
[20] J. S. Albus, Data Storage in the Cerebellar Model Articulation Controller (CMAC), ASME J. Dynamic Syst., Meas., Contr., vol. 97, Sept., 1975, pp. 228-233.
[21] Marc H. Raibert, Analytical Equations vs. Table Look-Up for Manipulation: A Unifying Concept, Proc. IEEE Conf. Decision Contr., New Orleans, Dec., 1977, pp. 576-579.
[22] Kenichi Yoshimoto and Kunihiko Wakatsuki, Application of the Preview Tracking Control Algorithm to Servoing a Robot Manipulator, Robotics Research, The First Int. Symp., 1984, pp. 883-897.
[23] Carlos D. Vinante, Carlos Bermudez and Francisco Tarre, Predictive Compensation Implemented with a Microprocessor, IEEE Control Syst. Magazine, vol. 6, Feb., 1986, pp. 40-43.

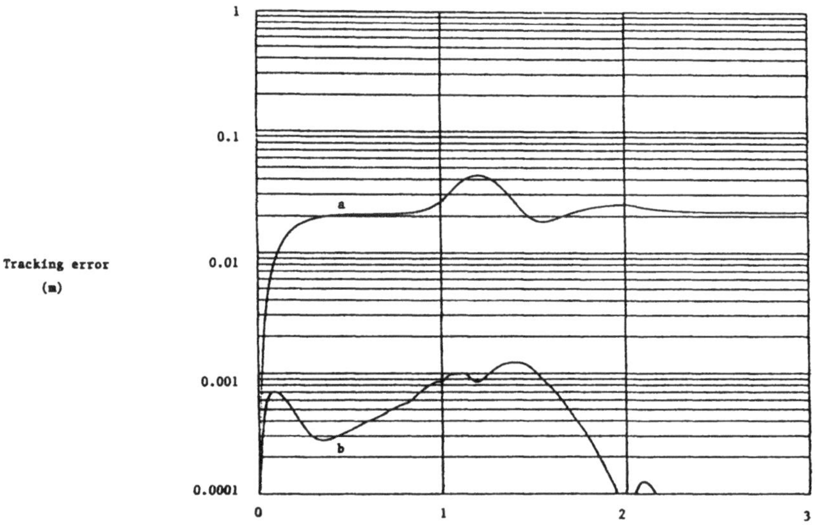

Fig. 1 Tracking error for an overestimation of the payload ($\hat{m}_L$=12 Kg). a-computed torque controller, b-multirate controller. Low sampling rate: 160 msec. High sampling rate: 5 msec. Computational delay on the fast loop: 0.5 msec.

DEDICATED COMPUTER SYSTEM FOR ROBOT MANIPULATORS

J. A. Tenreiro Machado [1)2)] , J. L. Martins de Carvalho [1)2)], Antonio M. C. Costa [2)] and Jose S. Matos [1)2)]

1. INTRODUCTION

In the last decade, robot dynamic computational algorithms have been steadily improved. Lagrangian [1] and Newton-Euler [2,3] recursive numerical algorithms, Horak's mixed calculation [4], Lisp based symbolic derivation of the robot closed form equations [5,6] and recently, costumized computing algorithms [7-9] have been important stages towards more efficient procedures. Nevertheless, the still existing high computational loads, have prevented the industrial application of such algorithms. A new robot computational system which overcomes those limitations is presented. This system is motivated by the following considerations:

-Symbolic formulae pose a high computational burden, due to the large number of arithmetic and transcendental computations.

-These computations, usually performed at a high level compiled computer language are complex, namely, they are compiled to a large number of the microprocessor's machine code instructions.

-These computations must be performed with very high precision; in order to reduce problems associated with finite precision arithmetics, a large word lengh is required.

-There are two situations, as far as precision is concerned: the computer internal representation having very high precision contrasting with the computer interface to the outside world, usually consisting of 8 to 16 bit precision.

A better "management" of the existing hardware resources, should bear in mind the following points:

-The computations should be performed with another algebra simple enough for any processor to perform.

-Computations in this algebra should'nt suffer from finite precision problems, consequently requiring a smaller number of bits.

We conclude on the necessity of a "special" algebra that provides a better management of the existing hardware/software resources. Or to state, alternatively, our position: we need to find a new compiler that generates a more efficient object code in the manipulator hardware/software environment. Such a compiler may be called a dedicated compiler, contrasting with general purpose compilers such as Fortran or C. The Boolean algebra satisfies such requirements. The robot inverse dynamics are converted to an input/output binary table, and afterwords translated to Boolean algebra, through the simplification of the corresponding Boolean truth table. Unfortunately, upon further examination it turned up that this ideal situation is not feasible in practice, as it poses critical obstacles to its computation due to the necessity of implementing a huge Boolean table. Nevertheless, it points to a methodology for a realizable implementation that, with modifications, can achieve a remarkable reduction of the on-line computing time. This realizable (suboptimal) implementation is developed in the next section.

2. A HYBRID COMPUTATIONAL ALGORITHM FOR ROBOT MANIPULATOR DYNAMICS

Noting that the dynamic equations, for a manipulator with n degrees of freddom, are of the form [1,10]

$$T = J(q)\ddot{q} + C(q, \dot{q}) + G(q) \qquad (1)$$

a natural way of achieving a realizable implementation is to split each link torque in its terms. Then each term can be calculated by Boolean algebra, and the final result, i.e. the link torque, by an ordinary arithmetic sum of these terms. Note that for the truth tables required on each Boolean-based term computation, we have the following maximum of input word variables (i.w.v.):

[n (position) + 1 (acceleration)] for the inertial terms [n (position) + 1 (velocity)] for the centripetal terms

[n(position) + 2(velocities)] for the Coriolis terms n(position) for the gravitational terms

where i.w.v. means the appropriate binary-coded representation of each input variable. Therefore, the Boolean computation becomes alleviated as we have a maximum of n+2 i.w.v., contrasting with the requirement of 3n i.w.v. for the fully Boolean-based computation.

One restriction imposed by the coexistence of two different algebras, is that arithmetic summation degrades the overall precision, hence resulting in the need for m extra bits, given by

[1)] Faculdade de Engenharia da Universidade do Porto, Dep. Eng. Electrotecnica Computadores, R. dos Bragas, 4099 Porto Codex, PORTUGAL
[2)] INESC-Instituto de Engenharia e Sistemas de Computadores, R. Jose Falcao 15, 4º, 4000 Porto, PORTUGAL.

$$m = \mathrm{int}[\log_2(p+1)]\ ,\ p = \text{total number of terms} \quad (2)$$

in each term in order to achieve the desired accuracy. Nevertheless, this effect is of small influence since m increases much more slowly than p.

In conclusion, we may say that the off-line "compilation" (i.e. truth table simplification) requirements, either in computing time or in computer memory space, are considerably alleviated, as we pass from one huge Boolean table with 3n i.w.v. to several truth tables, each having a much smaller number of i.w.v., and this without altering significantly the on-line computing time.

3. THE NEW COMPUTATIONAL ALGORITHM IMPLEMENTATION ON A 2R ROBOT MANIPULATOR

To illustrate the implementation of our algorithm, we consider a 2R robot manipulator with the following dynamic model:

$$T_1 = [(m_1+m_2)r_1^2 + m_2 r_2^2 + 2r_1 r_2 m_2 C_2 + J_1]\ddot{q}_1 + (m_2 r_2^2 + r_1 r_2 m_2 C_2)\ddot{q}_2 -$$

$$- r_1 r_2 m_2 \dot{q}_2^2 - 2r_1 r_2 m_2 S_2 \dot{q}_1 \dot{q}_2 + m_1 g r_1 C_1 + m_2 g (r_1 C_1 + r_2 C_{12})$$

$$T_2 = (m_2 r_2^2 + r_1 r_2 m_2 C_2)\ddot{q}_1 + (m_2 r_2^2 + J_2)\ddot{q}_2 + r_1 r_2 m_2 S_2 q_1^2 + m_2 g r_2 C_{12} \quad (3)$$

In our study we use the same data as Young [11], Russel et al. [12] and Tenreiro et al. [13], namely:

$$m_1 = 0.5\ \text{Kg};\quad m_2 = 6.25\ \text{Kg};\quad r_1 = 1\ \text{m};\quad r_2 = 0.8\ \text{m};\quad J_1 = 5\ \text{Kgm};\quad J_2 = 5\ \text{Kgm} \quad (4)$$

and assume the amplitude of each link variable within the ranges:

$$-\pi\,\text{rad} \le q_i \le \pi\,\text{rad};\quad -1\,\text{rad/s} \le \dot{q}_i \le 1\,\text{rad/s};\quad -1\,\text{rad/s}^2 \le \ddot{q}_i \le 1\,\text{rad/s}^2;\quad i = 1,2 \quad (5)$$

For these ranges and for the load (m_2, J_2) referred in (4) we have: -152.5 Nm $\le T_1 \le 152.5$ Nm and -69.1 Nm $\le T_2 \le 69.1$ Nm; nevertheless, motivated by the natural assumption that the torque computation shall be a block of a larger control structure, the feedback loop may demand higher torques, and therefore we assume

$$-200\ \text{Nm} \le T_1 \le 200\ \text{Nm};\ -100\ \text{Nm} \le T_2 \le 100\ \text{Nm} \quad (6)$$

From these considerations it results that equations (5) and (6) give the quantization ranges for the input and output word variables respectively.

The truth table simplification may be alleviated by the use of some general rules. These rules stem from considerations like:

a) The quantization ranges (6) must be the same for all terms of each equation. As usual, amplitude range of each term only covers part of the quantization interval, and the remainder becomes "unused".

b) The accuracy compensating extra number of bits m, as represented in equation (2), varies in discrete steps, and much more slowly than p.

c) Bearing the physical (mathematical) robot manipulator requirements (specifications) in mind [14], it would appear that the Gray code is the more suitable and this, indeed, was confirmed by experimentation.

d) For even arithmetic functions of a single input variable, it was observed that the most significant bit of the corresponding input word variable could be dropped out if the Gray code was used.

e) On terms which have several input variables, the required final precision implies a similar accuracy on each input word variable.

The application of these rules to T_2 gives the results depicted in Fig. 1. The on-line computing time can be further optimized by the use of Binary Decision Diagrams (B.D.D.) [15,16]. Figure 2 shows an histogram of the required computational time for the term $T_{24} = r_1 r_2 m_2 S_2 \dot{q}_1^2$, comparing the conventional arithmetic method with the new algorithm. Both codes were written in Turbo Pascal V4.0, and running on a 8086 8 MHz machine under MSDOS V3.2. Notice the remarkable improvement achieved with the new algorithm.

4. PARALLEL COMPUTATION

We have shown that the proposed hybrid computational algorithm offers considerable performance improvement over purely arithmetic alternatives, even in a monoprocessor sequential computing environement. In this section we show that the algorithm also leads naturally to simple parallel architectures allowing unlimited speed-up of the on-line computations, without accuracy restrictions. In fact, the operations involved in the on-line computations required by our algorithm, allow simple distribution of the computational load amongst several processors. Indeed, this can be achieved without any of the complex scheduling problems, common to arithmetic manipulator inverse dynamics parallel computing structures [17-19]. Parallel architectures like the one depicted in Fig. 3, are a natural consequence of the calculation decoupling allowed by our algorithm. The improvement in the on-line computing time is proportional to the number of processors, and it is not subject to any theoretical limit. The importance of this fact must be emphasized since the other manipulator inverse dynamics parallel computing structures achieved a limited improvement, in the sense that the resulting speed-up is not proportional to the number of processors, and, is in fact, restricted to a maximum of n (a condition that is achieved only if some errors are allowed [20]). Finally, it should be noted that the B.D.D. algorithms are bit oriented instead of word oriented. Consequently, general purpose microprocessors with 8, 16 or 32 data buses, and large instructions sets (unused in this algorithm), that require several clock cycles for each of the machine code instructions, are of little use in improving the overall computing performance. Much more promising is the use of single bit, reduced instruction set and special purpose microprocessors. The design of a dedicated processor based on B.D.D., is being investigated.

5. CONCLUSIONS

A new computational algorithm for robot manipulator inverse dynamics was presented. This algorithm is very efficient because it takes full advantage of both hardware and software capabilities of the robot manipulator system. Another important consequence of the proposed calculation method is the natural appearance of simple, yet powerful, parallel computing structures.

REFERENCES

[1] Richard P. Paul, Robot Manipulators: Mathematics, Programming and Control, Cambridge, MA: Mass. Inst. Tech., 1981.

[2] J. Y. S. Luh, M. W. Walker, R. P. C. Paul, On-Line Computational Scheme for Mechanical Manipulators, ASME J. Dynamic Syst., Meas., Contr., vol. 102, June, 1980, pp. 69-76.

[3] John M. Hollerbach, A Recursive Lagrangian Formulation of Manipulator Dynamics and a Comparative Study of Dynamics Formulation Complexity, IEEE Trans. Syst., Man, Cybern., vol. SMC-10, Nov. 1980, pp. 730-736.

[4] D. T. Horak, A Simplified Modeling and Computational Scheme for Manipulator Dynamics, ASME J. Dynamic Syst., Meas., Contr., vol. 106, Dec. 1984, pp. 350-353.

[5] M. C. Leu, N. Hemati, Automated Symbolic Derivation of Dynamic Equations of Motion for Robotic Manipulators, ASME J. Dynamic Syst. Meas. Contr., vol. 108, Sept. 1986, pp. 172-179.

[6] J. Koplic, M. C. Leu, Computer Generation of Robot Dynamic Equations and the Related Issues, J. of Robotic Systems, vol. 3, n. 3, Fall 1986, pp. 301-319.

[7] Charles P. Neuman and John J. Murray, The Complete Dynamic Model and Costumized Algorithms of the Puma Robot, IEEE Trans. Syst., Man, Cybern., vol. SMC-17, July/Aug. 1987, pp. 635-644.

[8] Charles P. Neuman and John J. Murray, Customized Computational Robot Dynamics, J. of Robotic Systems, vol. 4, n. 4, Aug. 1987, pp. 503-526.

[9] Charles P. Neuman and John J. Murray, Symbolically Efficient Formulations for Computational Robot Dynamics, J. of Robotic Systems, vol. 4, n. 6, Dec. 1987, pp. 743-769.

[10] Michael Brady, John M. Hollerbach, Timothy L. Johnson, Tomas Lozano-Perez, Mattew T. Mason, Robot Motion: Planning and Control, Cambridge, MA: Mass. Inst. Tech., 1982.

[11] Kar-Keung Young, Controller Design for a Manipulator Using Theory of Variable Structure Systems, IEEE Trans. Syst., Man, Cybern., vol. SMC-8, Fev. 1978, pp. 101-109.

[12] Russel G. Morgan, Umit Ozguner, A Decentralized Variable Structure Control Algorithm for Robotic Manipulators, IEEE J. Robotics and Automation, vol. RA-1, March 1985, pp. 57-65.

[13] J. A. Tenreiro Machado and J. L. Martins de Carvalho, A Smooth Variable Structure Control Algorithm for Robot Manipulators, IEE Control'88 Conference, Oxford, U.K., April, 1988.

[14] J. A. Tenreiro Machado, Antonio M. C. Costa and J. L. Martins de Carvalho, Robot Manipulator Dynamics - Towards Better Computational Algorithms, INESC Internal Report, 1987.

[15] Jose S. Matos, The Binary Decision Diagram: A Tool for Logic Design and Implementation, PhD Dissertation, Syracuse University, Syracuse, N. Y., 1983.

[16] Jose S. Matos and John V. Oldfield, Binary Decision Diagrams: From Abstract Representations to Physical Implementations, 20th IEEE/ACM Design Automation Conference, Miami Beach, Florida, 1983.

[17] Chang-Huan Liu, Yen-Ming Chen, Multi-Microprocessor-Based Cartesian-Space Control Techniques for a Mechanical Manipulator, IEEE J. Robotics and Automation, vol RA-2, June 1986, pp. 110-115.

[18] J. Y. S. Luh, C. S. Lin, Scheduling of Parallel Computation for a Computer-Controlled Mechanical Manipulator, IEEE Trans. Syst., Man, Cybern., vol. SMC-12, March/April 1982, pp. 214-234.

[19] T. Watanabe, M. Kametani, K. Kawata, K. Tetsuya, Improvement in the Computing Time of Robot Manipulators Using a Multimicroprocessor, ASME J. Dynamic Syst. Meas. Contr., vol. 108, Sept. 1986, pp. 190-197.

[20] Eli K. Binder, James H. Herzog, Distributed Computer Architecture and Fast Parallel Algorithms in Real-Time Robot Control, IEEE Trans. Syst., Man, Cybern., vol. SMC-16, July/Aug. 1986, pp. 543-549.

r=8	r=8 , m=2 => $V_{INITIAL}$=10 bits						PRECISION
T_2	T_{21}^*	T_{22}^*	T_{23}	T_{24}	T_{25}^*	T_{26}^*	TERM
8	6	6	7	6	8	8	V_{FINAL}
40	11	11	7	11	16	16	TOTAL NUMBER OF INPUT BITS
$0.76\ 10^{11}$	1474	1474	92	1638	64226	64226	USED STATES
$0.34\ 10^{11}$	574	574	36	410	1310	1310	UNUSED STATES
100	6.25	6.25	12.5	6.25	25	25	QUANTIZATION AMPLITUDE RANGE
69.1	4.5	4.5	9	9	24.5	24.5	AMPLITUDE RANGE OF EACH TERM

Fig. 1 Source code chart for the truth table implementation. $T_2=(T_{21}^*+T_{22}^*)+T_{23}+T_{24}+(T_{25}^*+T_{26}^*)$
$T_{21}^*=T_{22}^*=(4+5C_2)\ddot{q}_1/2$, $T_{23}=9\ddot{q}_2$, $T_{24}=5S_2\dot{q}_1^2$, $T_{25}^*=T_{26}^*=49C_{12}/2$

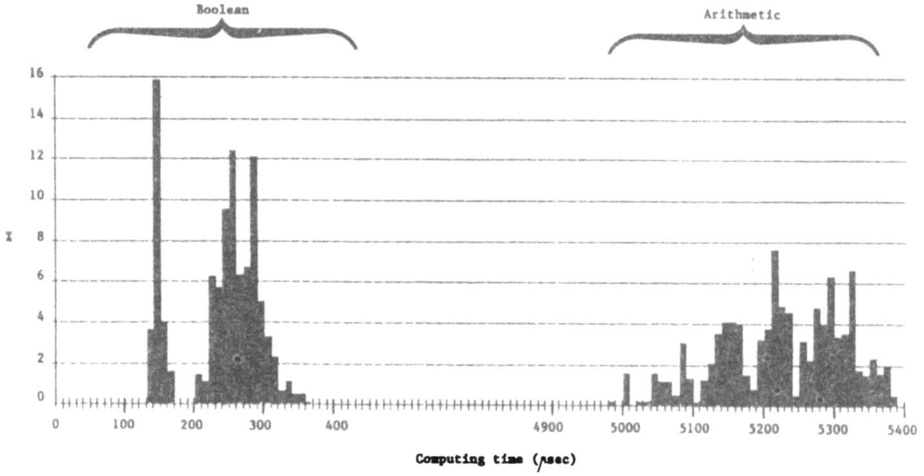

Fig. 2 Histogram of the computational time required by the term $T_{24}=r_1r_2m_2\dot{q}_1^2$.

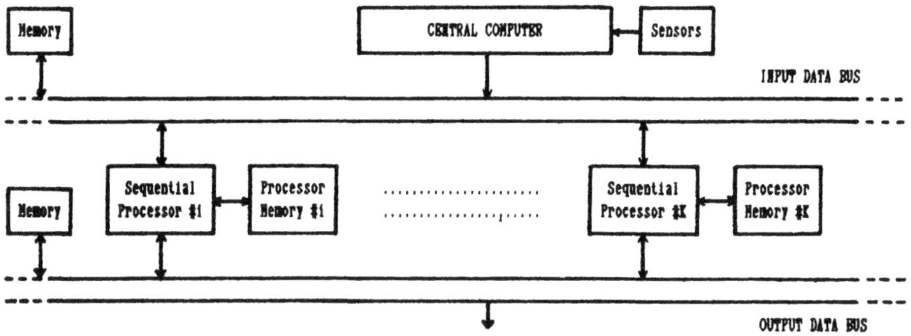

Fig. 3 A possible (decoupled) sequential/parallel computer structure, suggested by the new algorithm.

PERFORMANCE APPROXIMATION OF ROBOTIC SYSTEMS USING RENDEZVOUS NETWORKS

J.W. Miernik [1], M.B. Zaremba [2]

1. Introduction

One of the challenging topics in robotics is the performance evaluation of robot control software. In complex multisensory robotic systems the control software comprises a large number of tasks executed concurrently on different processors, with real-time constraints [2]. The tasks communicate in synchronous manner in order to prevent unexpected large delays, buffer overflows, etc. [1].

The two important aspects of the control software, i.e. stochastic nature of concurrently running tasks and synchronous intertask communications can be modelled with stochastic rendezvous networks (SRVN), proposed by Woodside [5]. An application of the SRVN model to analysis of robot control software - however, with no contention due to the hardware resources - was reported in [6]. Since then, a research problem related to the influence of restricted number of processors has been resolved in [3].

The objective of this paper is to describe the rendezvous modelling of supervisory vision-based control of an industrial robot for the purpose of estimating software performance in terms of software throughput.

2. Rendezvous Network Modelling

The software system is modelled as an acyclic send/receive graph of interconnected tasks. The independent requests generated with known average frequencies and the destination tasks are random. The task execution times are also random. It means that they just may vary in function of received messages. In the present paper it is assumed that execution times are exponentially distributed. The randomness is associated with the receiving task, i.e. server behaviour is independent of customer identity (see [4] for caller-specific service).

The throughput of the task system is evaluated as the throughput of independent reference tasks at the top of the graph. Reference tasks simulate applications running at maximum speed, using subgraphs of common tasks. Basic analytical expressions used for reference task r throughput f_r approximation is the following:

$$f_r = 1/\sum_p x_{rp} \tag{1}$$

where: $$X_{rp} = S_{rp} + \sum_j Y_{rjp} W_{rj} + \sum_j Y_{rjp}\, Wfp_r \tag{2}$$

with averages:

X_{rp} - total task r occupation in phase r,

S_{rp} - service time of task r in phase p,

Y_{rjp} - number of requests from task r in phase p to task j,

W_{rj} - task r delay (waiting + RV service) of task j

Wfp_r - processor delay of task r

In more detail:

---

1) Department of Systems and Computer Engineering, Carleton University, Ottawa, Ontario K1S 5B6, Canada.

2) Département d'informatique, Université du Québec à Hull, Hull, Québec, J8X 3X7, Canada.

$$W_{rj} = \text{Waiting}_{rj}\ 1 + \text{Waiting}_{rj}\ 2 \qquad (3)$$

$$Wfp_r = Wfp_r\ 1 + Wfp_r\ 2 \qquad (4)$$

where numbers 1 and 2 denote mean residual service time and waiting in the queue respectively, for task service and for processor execution. The type 1 waiting is calculated on the basis of the second moment of service time.

The same expressions are used for evaluation of task occupation on each lower level of the task communication graph. Effectively, the algorithm calculates the lowest level task waiting times and builds up occupation time moving up to reference tasks.

3. Modelled Control Software

The software analyzed in the paper is designed for the control of vision-based robotic systems. The functional diagram of the tasks is shown in Fig. 1.

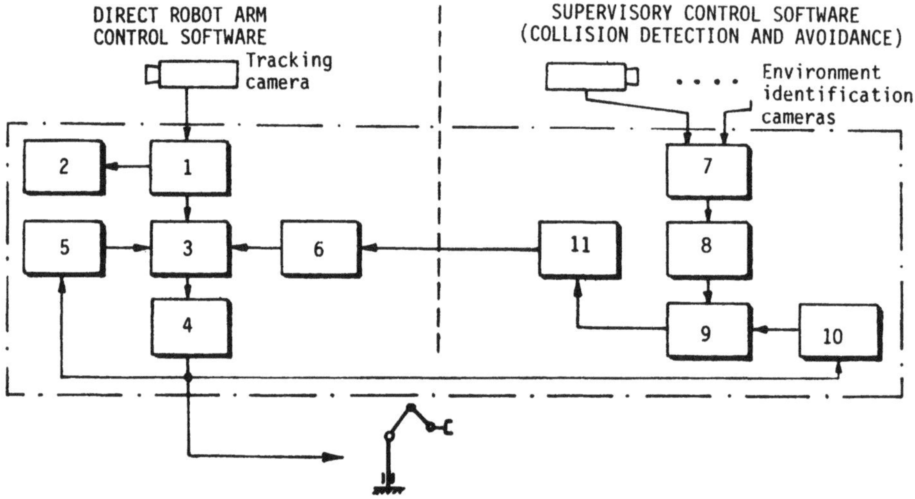

Fig. 1. Functional diagram of the robot control software.

The modules operating on lower level process data from the tracking camera and deal mainly with path control [7]:

1 - calculation of the vision system solution,
2 - generation of graphics display,
3 - calculation of desired robot end effector location,
4 - prediction and generation of control signals,
5 - robot dynamics modelling
6 - calculation of the stand-off between the robot and the manipulated part.

The modules operating on the supervisory level detect collision situations and, when required, generate trajectory correction signals:

7 - image processing,
8 - rudimentary collision detection,
9 - generation of a detailed environment model,
10 - inverse kinematic solution,
11 - obstacle avoidance strategy.

4. Application of the SRVN Model

The purpose of the application is to compare the accuracy of approximation against simulation results in the case of maximum possible concurrency, i.e. when each task is executing on a separate processor. The software to be modelled does not use rendezvous intertask communication explicitly. However, the analysis of the software in Fig. 1 allows the equivalent SRVN representation as in Fig. 2.

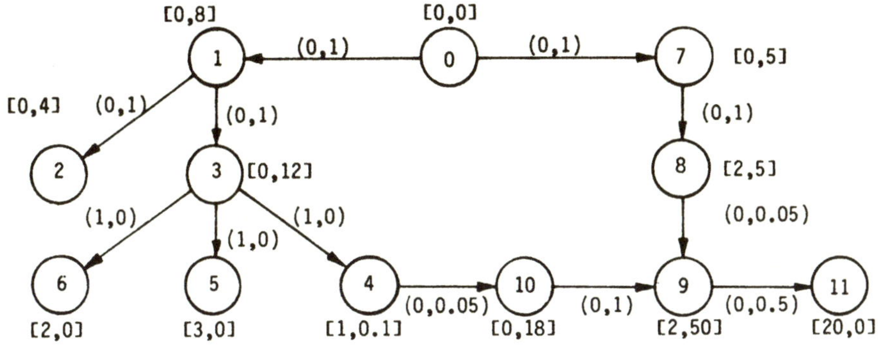

Fig. 2. Send-receive graph of the robot control software.

The symbol (k,l) indicates probability k of the generation of rendezvous request during caller's rendezvous phase of service, and probability l - during its post-rendezvous phase. There are also shown the tasks execution times for the both phases. Reference task 0 is the task driving the entire application and represents camera pictures arriving for processing.

Along with the development of control theory for robotics, more sophisticated and more computationally heavy control schemes are available. Therefore, the designer has to carefully weigh the gains in the performances indices offered by more complex control algorithm versus the losses induced by the increased time delay in the control loop.

In our case, the tasks critical from the standpoint of this analysis are task 3 and task 9. Figure 3 illustrates the software performance in terms of the execution time T. Given maximum execution time (e.g. 29.64 ms for the Cincinnati Milacron T3 with DEPC - Direct External Path Control - communication link), the designer can easily find the maximum execution times of the tasks under consideration. This, in turn, will make it possible to select adequate control algorithms.

In addition to the time considerations, figure 3 compares the results obtained by the SRNV networks method with those simulated in SIMSCRIPT II.5 v. 1.2.

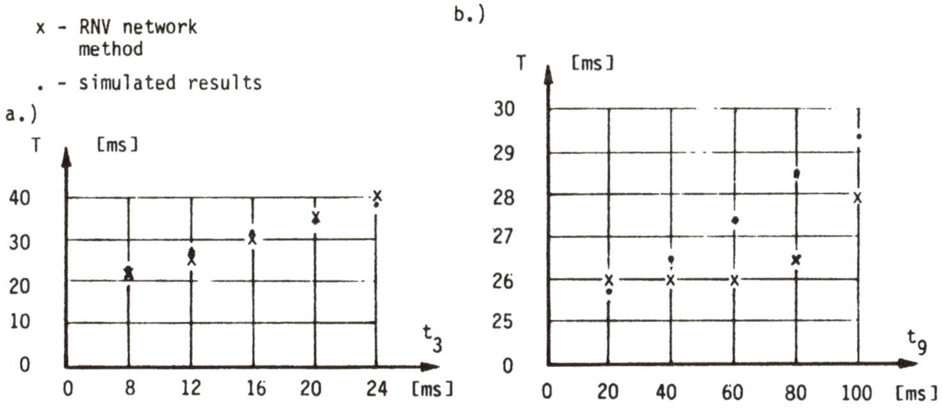

Fig. 3. Software execution time in function of: a.) task 3, b.) task 9.

The maximum error was 4.0% for Fig. 3a. (22.24 ms - RNV network method; 23.16 ms - simulation), and 7.5% for Fig. 3b. (26.40 ms - RNV network method; 28.55 ms - simulation).

5. Conclusions

The application of SRVN modelling for robotic software performance evaluation reveals good accuracy, as compared with simulation with empirical data. The SRVN iterative modelling has the great advantage of being able to be up to hundreds of times faster than the simulation.

Partitioning of real software into concurrently executable tasks is a key factor for throughput increase. The partitions can be quickly evaluated with SRVN, making it a useful performance analysis tool in the design process.

Further research problems are related to:
- stochastic/deterministic characteristics of RV calls and service times,
- construction and analysis of task networks from the standpoint of logical task interactions (task or subgraphs concurrency, exclusions, and optimization),
- implications of robot programming in object-oriented languages on uncertain task execution times.

6. Literature

[1] Gauthier, D., Freedman, P., Carannis, G., and A.S. Malowany: Interprocess Communication for Distributed Robotics. IEEE J. of Robotics and Automation. RA-3 (1987) 6, 493-504.

[2] Kasahara, H., and S. Narita: Parallel Processing of Robot-Arm Control Computation on a Multiprocessor System. IEEE J. of Robotics and Automation. RA-1 (1985) 2, 104-113.

[3] Miernik, J.W., Woodside, C.M., Neilson, J.E., and D.C. Petriu: Throughput Calculation for Stochastic Rendezvous Networks with Processor Contention. Report SCE-87-4, Carleton Univ., Ottawa 1987.

[4] Miernik, J.W., Woodside, C.M., Neilson, J.E., and D.C. Petriu: Throughput of Stochastic Rendezvous Networks with Caller-Specific Service and Processor Contention. Proc. IEEE Infocom '88, New Orleans 1988.

[5] Woodside, C.M.: Throughput Calculation for Basic Stochastic Rendezvous Networks. Report SCE-86-9. Carleton Univ., Ottawa 1986. Under review for Performance Evaluation.

[6] Zaremba, M.B., and J.W. Miernik: Throughput Analysis of a Robot Control Software Using Stochastic Rendezvous Networks. Proc. ISMM Int. Symp. on Software and Hardware Applic. of Microcomputers. Fort Collins 1987, 26-30.

[7] Zaremba, M.B.: Vision-Based Robot Guidance System. Microcomputer Applications. 7 (1988) 1.

SOLUTION OF PARAMETER OPTIMIZATION AND CONTROL PROBLEMS IN THERMAL SYSTEMS BY MEANS OF A LOCAL ADAPTIVE FILTER

Yu.M.Matsevity and A.V.Moultanovsky[1)]

An attempt is made to develop a single procedure for solving parameter optimization and control problems in thermal systems. An adaptive iterative filter algorithm is considered, suited to parameter identification with the use of special-purpose computers. The problem of controlling the granulation of copper smelting slag in drum-type crystallizers was solved, and the results of the solution are presented.
Keywords: slag granulation, identification, optimization, control, adaptive filter, mathematical model.

Continuing prior research of the present authors [1,2] , an attempt is made to develop a single procedure for solving parameter identification and control problems for thermal systems and the processes taking place in such systems.

When dealing with real thermal systems, the probabilistic approach to identification and control problems seems to be the most expedient one, since the characteristics of both the thermal process and observation are random.

Generally, the dynamics of a thermal system is described by the nonlinear equation

$$\vec{X}_{K+1} = f(\vec{X}_K, \vec{U}_K) + \vec{W}_K , \tag{1}$$

where $\vec{X}$ and $\vec{U}$ are vectors of state and control, respectively. "Observations", in their turn, can be described by

$$\tilde{\vec{Y}}_{K+1} = h(\vec{X}_{K+1}) + \vec{V}_{K+1} , \tag{2}$$

where $\tilde{\vec{Y}}$ is the vector of measurements, and the noncorrelated disturbances $\vec{W}$ and $\vec{V}$ are white Gaussian sequences which affect the object in the additive way. The system behaviour is determined by the action of the control signals $\vec{U}_K$ applied at discrete moments of time. The control action is selected through solution of a stochastic optimum control problem or statistical parameter evaluation. To develop a single algorithm for solving these problems we need to discuss their common and distinctive features. In the optimum control problem, we seek a feasible, physically realizable piecewise continuous function $\vec{U}_K$ that should satisfy certain constraints $(\vec{\beta}_K)$ and minimize (maximize) the selected quality criterion.

The other problem involves identification of a set of parameters (boundary conditions, thermal and physical characteristics, etc.) with a concomitant determination of the state vector (temperature field). The sought-for set of parameters characterizes the system behaviour

---

1) Ukr.SSR Academy of Sciences, Institute for Problems in Machinery 2/10 D.Pozharski street, Kharkov, 310046 USSR

from the point of view of formulation and objective of the investigations. If the objective is control of a thermal process, then the sought-for quantity in both problems is a control action (for example, boundary conditions).

A stochastic optimum control problem must be formulated in such a way that all the constraints $\vec{\beta}$ were included in the quality criterion [3]. Therefore, if a control problem is so stated that $\vec{\beta}$ is included in the observation vector $\tilde{\vec{Y}}$ (which implies automatic fulfilment of the condition that the constraints be included in the quality criterion), then a general procedure can be developed for identification and control problems within the framework of the probabilistic approach. As often occurs in thermal system engineering, temperature $T(\tau)$ at some point of the object being engineered is the only "observed" and limiting factor. For example, in the control of nonferrous metal slag granulation it is the slag solidification temperature which plays the part. The value is constant in the case, however the process is characterized by a varying location of the crystallization front with time. For this case, the search for a nonterminal control [3] is automatically transformed into a parameter identification problem in a thermal system. The latter problem is underlied by exterior, or geometrical inverse heat conduction problems (IHCP) in which the unknown heat transfer parameters are determined or the knowledge of the object geometry (in the above example, location of the crystallization front) is improved proceeding from scarce and uncertain data on the temperature field of the system. In case the thermal characteristics of the object material which are necessary for constructing a mathematical model are not known to a sufficient accuracy, then the search of a desired control action should be preceded by a combined IHCP aimed at simultaneous identification of all the unknown conditions of unambiguity.

To implement the procedure for solving identification problems, we suggest an iterative dynamic filter [4] with a regularizing halt of the iteration process at each time step. The process is halted when the summary error is matched with the rms error in the initial measured data. In its original form, this filter has a number of positive characteristics, for example, high accuracy of solution, selection of the number of iterations as a regularizing parameter, and the possibility of using in the computation procedure $\vec{Z}=0$ as an equation for unknown parameters $\vec{Z}$. However, the filter has substantial disadvantages too. First, transition matrices of the system need to be known exactly, while the algorithm allows obtaining only estimates of these, though refined during the iteration process. Second, it requires large memory capacities and high computation speeds, since the algorithm operates with matrices and vectors whose dimension is determined by the total number of the grid model nodes and the number of parameters to be found.

To implement solution procedures for identification and control problems, we suggest here an adaptive filter based on the iterative filter algorithm [4]. It does not require linearization of the initial model equation (1) and hence permits to exclude the extra errors arising from the linearization. Considering the required storage capacity, the algorithm suggested is more efficient, since, instead of the extended vector of state [1], it provides estimates of the stochastic vector of parameters. This allows to do without constructing transition matrices (which necessity is one of the main shortcomings of the filter), thus bringing the iterative filter into the category of adaptive ones. The thermal parameter estimates obtained with the help of a similar filtration algorithm prove to be unbiased and are characterized by a minimum variance. The low dimensionality of the filter (conforming to the number of the sought-for parameters) and the practical absence of rounding-off error accumulation (the method works within the framework of stochastic modelling algorithms) enable the use, in the case of necessity, of short word computers.

The so called artificial filtration technique can be applied to improve the quality of the estimates obtained and to substantially refine the computation algorithm. It consists in introducing extra known model parameters into the vector of measurement $\tilde{\vec{Y}}$ ("rigid" parameters). All the unknown ("soft") parameters will be included in the term $h_1(\vec{z})$ of the right-hand part of the measurement equation, with $\vec{z}$ denoting a vector formed by all the parameters to be identified and/or corrected. Indeed, the quality functional [1] that should be minimized by the estimates obtained with the help of the local adaptive filter with artificial filtration will be written as

$$\mathcal{J}[\vec{z}_K, \tilde{\vec{Y}}_K, j] = \| h_1(\hat{\vec{z}}^{(j)}_{K/K}) - \tilde{\vec{Y}}_K \|^2_{R^{-1}_K} + \frac{1}{j} \| \vec{z}_K - \hat{\vec{z}}^{(j/j-1)}_{K/K} \|^2_{[P^{(i)}_{K-1/K-1}]^{-1}}. \qquad (3)$$

Here, j and i are the numbers of iterations at the (K+1)-th and K-th time steps, respectively; R_K and $P_{K-1/K-1}$ are covariance matrices of the measurement errors and estimate errors. Then the value $\hat{\vec{z}}^{(j)}_{K/K}$ to minimize the functional (3) is determined as the best estimate of $\vec{z}_K$, i.e. the unbiased estimate, with the expectation $M[\vec{z}_K - \hat{\vec{z}}^{(j)}_{K/K}] = 0$, and with a minimum variance. To be more exact, the estimate can be made practically unbiased to within the standard error of the statistical linearization method applied to the initial mathematical model. But it is just another advantage of the suggested method that it requires no linearization of the initial mathematical model. Therefore, first, the estimate can be assumed unbiased and, second, additional linearization-induced inadequacies of the mathematical model to the process under investigation are excluded. Still another important advantage of the method is its stability against eventual measurement anomalies.

The set of equations which determine the computation algorithm of the adaptive filter is as follows:

$$\hat{\tilde{z}}_{K/K}^{(j)} = \hat{\tilde{z}}_{K/K}^{(j-1)} + K_K^{(j)}\left[\tilde{\tilde{Y}}_K - \hat{H}_K^{(j)}\hat{\tilde{z}}_{K/K}^{(j-1)}\right]; \tag{4}$$

$$K_K^{(j)} = P_{K-1/K-1}\left[\hat{H}_K^{(j)}\right]^T\left\{\hat{H}_K^{(j)}P_{K-1/K-1}\left[\hat{H}_K^{(j)}\right]^T + R_K\right\}^{-1}; \tag{5}$$

$$P_{K-1/K-1} = \left[I - K_{K-2}^{(i)}\hat{H}_{K-2}^{(i)}\right]P_{K-2/K-2}\left[I - K_{K-2}^{(i)}\hat{H}_{K-2}^{(i)}\right]^T + K_{K-2}^{(i)}R_{K-2}\left[K_{K-2}^{(i)}\right]^T, \tag{6}$$

where I is the unit matrix, and K is the weight matrix. Construction of the artificial measurement matrix $\hat{H}_K$ corresponding to the operator h in the equation (2) is the crucial point of the algorithm (4)-(6). The matrix terms are partial derivatives of the "measured" parameters with respect to the estimated (identified, corrected) ones,

$$\hat{H}_K^{(j)} = \left\{\partial\tilde{\tilde{Y}} / \partial\hat{\tilde{z}}\right\}_K^{(j)}. \tag{7}$$

Here, we shall not discuss in any detail the principles of computation according to equation (7). It will be enough to say that construction of $\hat{H}^{(j)}$ requires several cycles of solving the initial heat conduction equations with boundary conditions.

The paper presents the results obtained for a number of identification and control problems concerning thermal processes in units of power and metallurgical equipment. In particular, a solution was sought to the problem of controlling the granulation process of copper smelting slag in drum-type crystallizers. The principal controlling parameter was the rate of the drum rotation or the contact time of the working surface with the slag. The magnitude of the rate rigidly determined the thickness of the slag skin that solidified during crystallization, i.e. control of the thermal process of slag formation was achieved.

REFERENCES

[1] Matsevity Yu. and Moultanovsky A.: Simulation of Thermal Processes and Identification of Heat Transfer Parameters. Syst. Anal. Model. Simul.1987, vol.4, No5, pp.363-377 (in English).

[2] Matsevity Yu.; Moultanovsky A. and Recada M.: Identification of Thermal Diffusivity in Metal Hydride Systems. In: Systems Analysis and Simulation, 1985, vol.II, Application Akademie-Verlag, Berlin 1985, pp.334-337 (in English).

[3] Kazakov I.E.: Statistical theory of control systems in state space. - Moscow: Nauka publishers, 1975.-432p. (in Russian).

[4] Matsevity Yu.M. and Moultanovsky A.V.:Identification in heat conduction problems. Kiev: Naukova dumka publishers, 1982.-240p. (in Russian).

OPTIMIZATION OF HEAT DELIVERY TO APARTMENT BUILDINGS USING AN IMPROVED BUILDING MODEL

M. VAJTA - J.E. RIJNSDORP - S.F. van der MEULEN - J.F.C. VERBERNE

(Applied Physics Department, Measurement and System Technique Group,
University of Twente
P.O. Box 217., 7500 AE Enschede, The Netherlands)

INTRODUCTION

In Utrecht, the company PEGUS operates some 100 large apartment buildings (about 10000 apartments). The thermal power (primary energy) is provided by the Combined Heat and Power Station (CHP) in Utrecht. There is a deep believe that by lowering the installation water temperature during the night decrease the heating cost can be reduced. In fact this leads to a huge morning peak in power consumption. Because of the higher cost of power during the peak hours the total heating cost can even be **higher** than keeping the installation temperature constant during the whole night. The aim of our research was to elaborate an optimal control policy avoiding the morning power peak (so reducing energy and costs) meanwhile satisfying some temperature requirements concerning the apartments.

PROCESS MODEL

Our process model consists of 3 major parts: a delivery station (D.S.), radiators and piping in the building, and a dynamic room model.

a./ *delivery station* (D.S.) : Fig.1. shows the simpflified scheme of the delivery station. ϕ_a is the input water flow-rate (with temperature T_a). The mixing valve is controlled by a local Landis & Gyr controller. The insta lation water flow-rate ϕ_m with temperature T_m is kept almost constant by on-trolling the pump. So the relation between these variables is:

$$T_m(t) = \frac{T_a}{\phi_m} \phi_a(t) + [1 - \frac{\phi_a(t)}{\phi_m}] T_{ret}(t) \qquad (1)$$

b./ *radiators and piping*: The installation water is circulating along the pipe lines and takes the thermal power from the D.S. to the radiators in the rooms. The pipelines are described by a transportation dead-time depending on the water flow-rate. We assumed a first order transfer function between the radiator power (P_{rad}) and the mixing temperature (T_m) and a first order with dead-time one between the return temperature (T_{ret}) and T_m.

c./ *room model*: Because of the similarity of the apartments in the building we elaborated only one room model and multiplied by the number of apartments. Our room model is based on the power balance (see Fig.2.):

$$T_1 \frac{dT_r}{dt} = P_{rad} - Q_{wi} - Q_s \qquad (2)$$

where T_1 represents the total heat capacity of the air and internal mass, Q_{wi} is the heat loss through the windows and ventilation, Q_s is the overall heat losses through the walls. In our room model we paid special attention to the dynamic effect of the walls. In literature many building models are available based upon the lumped parameter description of the walls [1]. This approach leads to an inaccurate high-frequency behaviour. The wall is a typical distributed parameter system where the distributed nature can not be neglected. The frequency response of a wall goes to a - 10 [dB/dec] asymptotic value in amplitude and - 45 [°] in phase in contrary with a simple

---

* The author is on leave from the Technical University of Budapest, Chair of Automation, Budapest, HUNGARY.

lumped wall model which goes to - 20 [dB/dec] and - 90 [°], respectively. To improve the high-frequency dynamics of the room model the walls were taken into account by their distributed parameter transfer function $G_1(s,0)$ and $H_1(s,0)$ (where the second argument stands for inside surface location). In the Appendix we give the relations for a single layer wall. In many cases the walls have more than 1 layers (insulation, air gaps etc.) so we have developed a general method to determine the frequency-response of a multilayer wall [2]. The wall model is linear and the inside wall surface temperature depends on the room temperature and ambient temperature as follows:

$$T_s(s) = G_1(s,0)\ T_r(s) + H_1(s,0)\ T_{amb}(s) \tag{3}$$

The heat losses through the walls are proportional to the temperature difference between the room and the inside wall surface. The distributed transfer functions are rather complicate because they contain complex hyperbolic functions with argument $\sqrt{s}$. To overcome this difficulty the authors elaborated a technique to approximate $G_1(s,0)$ and $H_1(s,0)$ by a low-order lumped model keeping the high-frequency properties alike. The approximation technique is based on locating the first pole of the DTF. In a 4 decade frequency range it gives a good result by a 3-rd or 4-th order approximation. The building model may contain different kind of walls all described by their distributed transfer functions. Real-time measurements from an apartment building in Utrecht showed good matching with our process model. Fig.3. shows the structure of our control loop including the current L&G valve controller, our PI controller for the water flow-rate and the optimal controller for the consumed power.

CURRENT CONTROL POLICY

The current control policy is to keep the installation temperature (T_m) at a prescribed value by the L & G controller. At midnight the reference value is lowered and kept at that value till 6^{30} in the morning. Fig.4a. and 4b. shows the consumed power, the installation temperature and the return temperature. After closing the valve the installation temperature soon becomes equal to the return temperature (the small steps in the temperature curves indicate the time the water needs to circulate in the building). Between midnight and about 1^{30} the control valve is completely closed, there is no heat delivery. To keep the installation temperature constant the heating process is restarted at t_{s1} (about 1^{30}). The huge morning power peak (about 690 kW) occurs when the reference value of T_m is changed back from the night value to the day value.

OPTIMAL CONTROL POLICY

Using a dynamic optimization technique we have showed that the morning power peak is mainly due to the current control policy [3]. The consumed power can be expressed as follows:

$$P_a = c\ \phi_a\ (T_a - T_{ret}) \qquad \text{where} \qquad T_{ret} = f(\phi_a) \tag{4}$$

The morning power peak is due to the fact that opening the valve leads to a step like change in the water flow-rate ϕ_a and in T_m (see Eq.1). If we want to avoid the morning power peak we have to control the valve in a way which counterbalance this effect. The best we can do is to "schedule" the delivered power (P_m) according to different circumstances. The optimal controller keeps the consumed power according to a prescibed schedule which sets the reference value of the flow-rate. The flow-rate is controlled by a PI controller setting the reference value for the L&G valve controller. The effect of the outside temperature is taken into account by changing the reference temperature T_{mref}. Fig.4a. shows our real-time experiment on 29-th of March, 1988 in an apartment building in Utrecht. The heating was restarted at t_{s2} (4^{00}) and the consumed power was kept at 200 [kW] till the morning. Fig.4c. shows the corresponding installation- and return temperature curves. During the day we dont't keep the consumed power at a prescribed value so the controller changes back to the L&G controller (keeping T_m at constant). Comparing the consumed power in the 2 different control policies we can see the amount of energy saved by using the optimal controller.

If the input water flow-rate is limited because of some reason we can also take that fact into account. Of course in this case we have to start reheating the building earlier and it will last longer. Besides our simulation results the first real-time experiments confirmed our optimal heat delivery strategy. Now we have been working on a more sophisticated control algorithm involving an adaptive load predictor.

ACKNOWLEDGEMENTS: This work was done while M. Vajta held a Research Fellowship at the Applied Physics Department, University of Twente. The research was supported by VESTIN. The authors would like to thank John Schelhaas for his contribution in setting up and evaluating the real-time measurements.

REFERENCES:

[1] ACHTERBOSCH, G. G. J.-de JONG, P. P. G.-KRIS-SPIT, C. E.-van der MEULEN, S. F.-VERBERNE, J. F. C.: The Development of a Convenient Thermal Dynamic Building Model, Energy and Buildings, 8, (1985), pp. 183-196.

[2] VAJTA, M.: A Matrix Method to Determine the Frequency Response of a Multilayer Wall, Report No. 40121979, U. of Twente, TN-MST, 1987,

[3] VAJTA, M.: Dynamic Optimization of Heat Delivery to Apartment Buildings in District Heating Systems, Report No. 40121981, U. of Twente, TN-MST, 1987.

APPENDIX

We have determined the distributed transfer functions of the wall by solving the Kirchhoff-Fourier heat conduction equation assuming convection on both sides (θ_s is the inside surface temperature):

$$G_1(s,0) = \frac{\Delta\theta_s(s)}{\Delta\theta_r(s)} = \frac{\lambda q\alpha_o \cosh(qd) + \alpha_o\alpha_1 \sinh(qd)}{\lambda q(\alpha_o+\alpha_1)\cosh(qd) + (\alpha_o\alpha_1+\lambda^2 q^2)\sinh(qd)} ; \quad (5)$$

$$H_1(s,0) = \frac{\Delta\theta_s(s)}{\Delta\theta_{amb}(s)} = \frac{\lambda q\alpha_1}{\lambda q(\alpha_o+\alpha_1)\cosh(qd) + (\alpha_o\alpha_1+\lambda^2 q^2)\sinh(qd)} ; \quad (6)$$

where $q = \sqrt{s/a_t}$; and $s = j\omega$ (Laplace operator), a_t is the heat diffusivity, α_o and α_1 are the heat transfer coefficients, λ is the heat conductivity of the wall and d is the thickness of the wall.

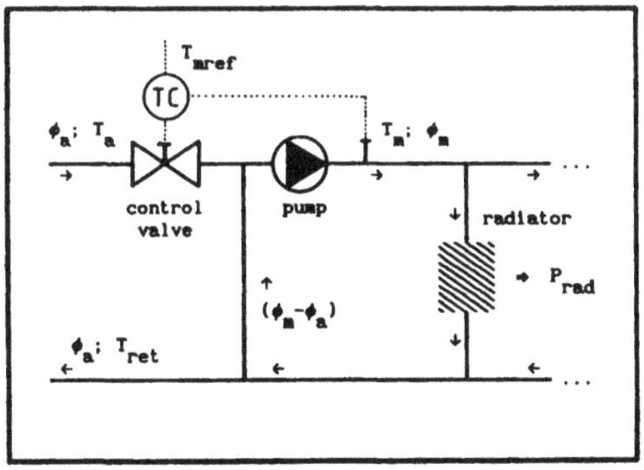

Fig.1. Simplified scheme of the delivery station (D.S.) and its connection to the radiators

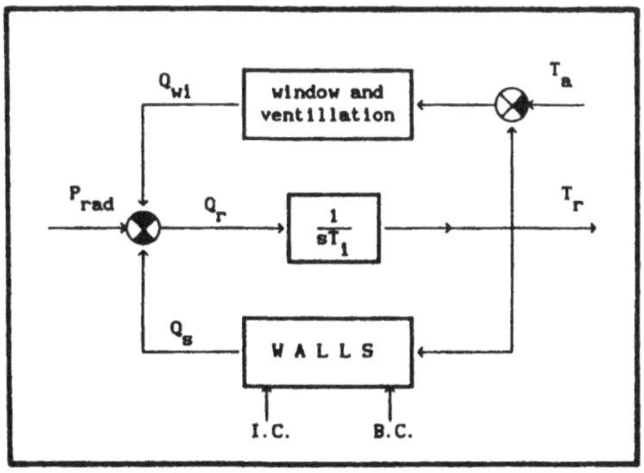

Fig.2. Block scheme of a room model (the walls are described in a distributed way)

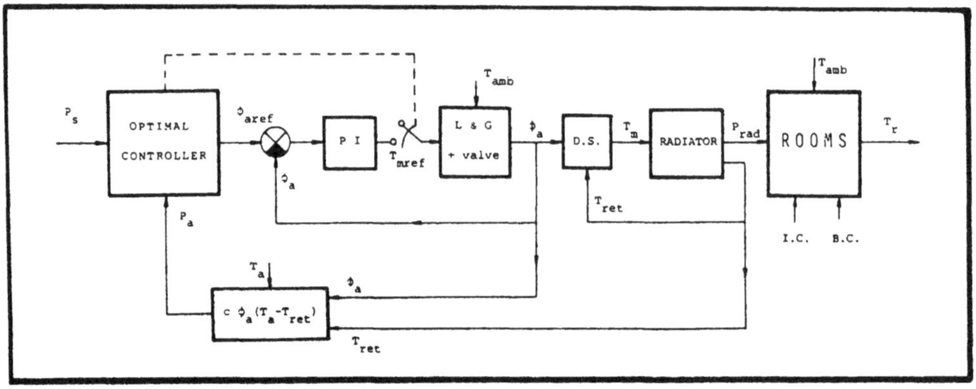

Fig.3. Block scheme of the control loop (L&G denotes the T_m controller)

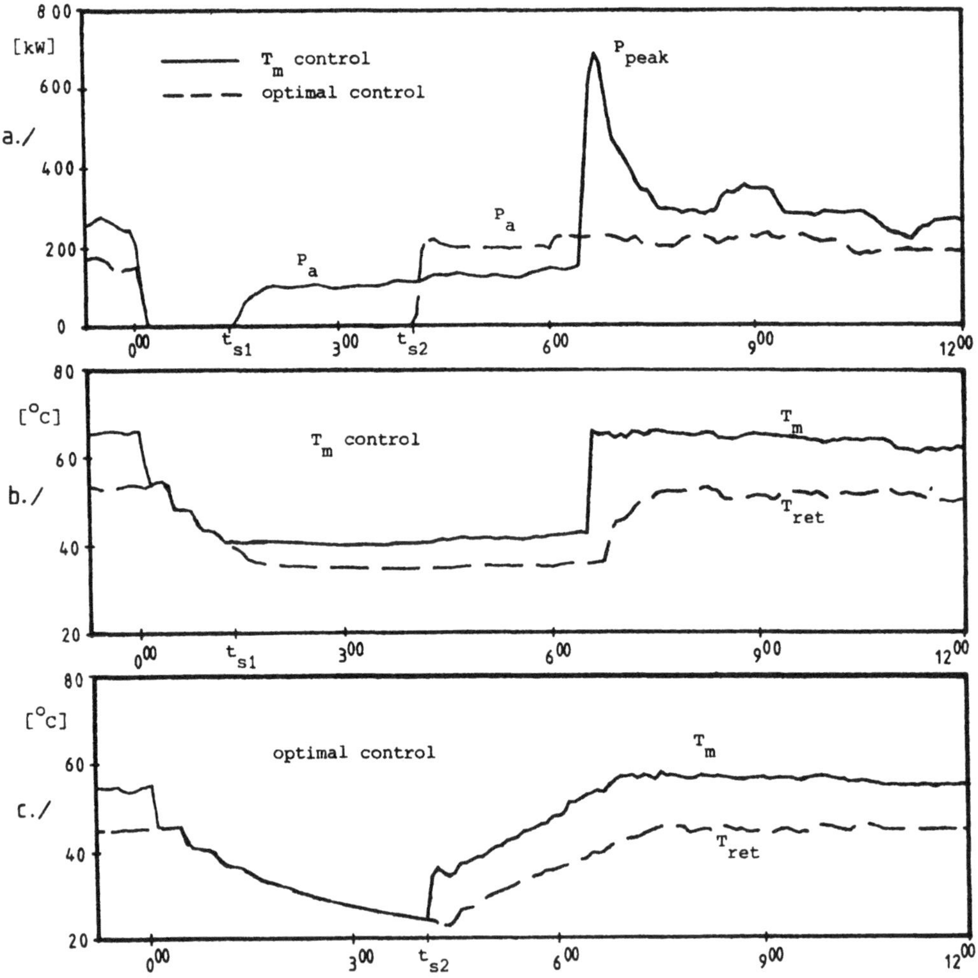

Fig.4. Comparison of the 2 different control policy (real-time)
a./ consumed power,
b./ mixing- and return water temperature (T_m control),
c./ mixing- and return water temperature (optimal control),

PERFORMANCE MESUREMENT AND MATHEMATICAL MODEL IDENTIFICATION OF THE HYDRAULIC SERVO CONTROL SYSTEM OF A VACUUM ELECTRIC-ARC FURNACE

Zhou Shichang, professor
Cao Xinming, professor
Gao Chengrui, assoc. professor
Yang Yanyong, assoc. professor

Department of Mechanical Engineering
Northeast University of Technology
Shenyang, Liaoning, P. R. China

ABSTRACT

The static and dynamic performances of the hydraulic servo control system of the electrode in a VHD vacuum electric-arc furnace have been measured. There is a large dry friction area in the performance of cylinder, which is provided by the vacuum seal on the electrode. Through theoretical analysis, the block diagram of the system is obtained. The dry friction can be represented by a negative feedback of cylinder velocity and is proportional to the unit velocity vector of cylinder. For the identification of the mathematical model of the system, a pulse signal is applied into the electrical panel and the responses of the spool displacement of the servo valve and cylinder velocity are recorded under open-loop condition. By discretization of the response curves, the Laplace transforms of valve displacement and cylinder velocity can be calculated. Then the transfer function of the servo valve and the cylinder can be derived by curve fitting. In the procedure of derivation, the dry friction is treated as the unit velocity vector feedback.

NOMENCLATURE

A_p -- effective area of cylinder
B_p -- damping coefficient of cylinder
C_i -- internal leakage coefficient of cylinder
e -- input voltage to amplifier
$E(s)$ -- Laplace transform of input voltage to amplifier
$I_p(\omega)$ -- imaginary part of $V_p(J\omega)$
$I_v(\omega)$ -- imaginary part of $X_v(J\omega)$
K_c -- flow-pressure coefficient of servo valve
K_q -- flow gain of servo valve
M -- mass of moving part of cylinder
N -- number of steps of discretization
P_1 -- cylinder pressure in the case of upward movement
P_2 -- cylinder pressure in the case of downward movement
P_c -- pressure in cylinder
$P_c(s)$ -- Laplace transform of cylinder pressure
P_s -- pressure of fluid source
$R_p(\omega)$ -- real part of $V_p(J\omega)$
$R_v(\omega)$ -- real part of $X_v(J\omega)$
T -- dry friction force applied on cylinder
v_p -- cylinder velocity
$V_p(s)$ -- Laplace transform of v_p
V_o -- volume of cylinder chamber
x_v -- main spool displacement of servo valve
$X_v(s)$ -- Laplace transform of x_v
β_e -- bulk modulus of fluid medium
δ_p -- damping ratio of valve controlled cylinder
δ_v -- damping ratio of servo valve
ω_p -- natural frequency of valve controlled cylinder
θ_1 -- phase angle of velocity vector of cylinder velocity
ω_v -- natural frequency of servo valve

INTRODUCTION

Hydraulic servo control is widely used to control the motion of the electrode in the electric-arc furnace. But there are few papers discussing the dynamic performance of its servo control system. In order to determine the mathematical model of the hydraulic servo control system of electrode in a VHD vacuum electric-arc furnace, experiments were carried out on the hydraulic system of the furnace. It was found that there was substantial dry friction associated with the cylinder. This caused the existance of deadband nonlinearity in the performance of the system. Therefore, the method of identification commonly used in linear system can not be used to identify this servo hydraulic control system. Based on the theoretical analysis, the characteristic of dry friction is explored and the block diagram of the whole system is drawn. After taking certain steps in the procedure of identification, the system with deadband can be identified by the method commonly used in the identification of a linear system.

OPERATING PRINCIPLE OF THE SYSTEM

Fig.1 shows the scheme of the hydraulic servo control system of the electrodes. Three electrodes have the same control system. Fig.1 only shows the control system for one electrode. The principle of operation is as following: The detected signals of arc current and voltage are fed to control panel for comparison with the desired values and the errors are amplified and input into the servo valve to control the cylinder motion, which drives the electrode up and down. For fire resistance purposes, emulsion is used as the medium in the main system and oil in the pilot stage.

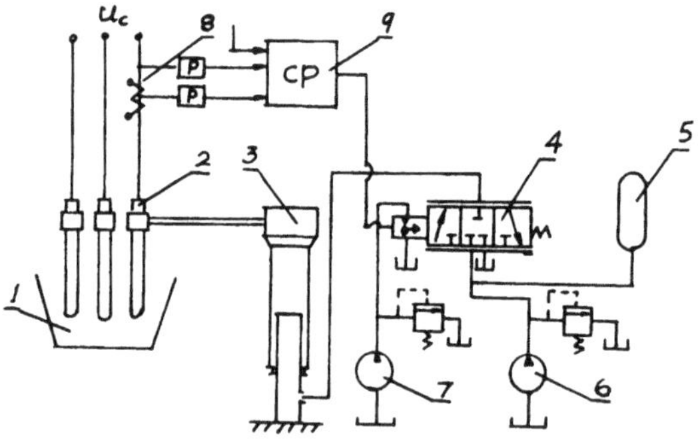

1—steel ladle
2—electrode
3—piston ram
4—servo valve
5—accumulator
6—hydraulic pump of emulsion
7—hydraulic pump of oil
8—current and voltage transformer
9—electrical control panel

Fig.1 Scheme of the hydraulic servo control system of electrode

DETERMINATION OF DRY FRICTION

Due to the existance of a vacuum seal on the electrode, there is evidence of significant dry friction in the cylinder motion. The dry friction is determined by the following method: When the cylinder moves upward with constant low velocity,

$$p_1 A_p = G + T \tag{1}$$

When cylinder moves downward,

$$p_2 A_p + T = G \tag{2}$$

$$T = (p_1 - p_2) A_p/2 \tag{3}$$

Measuring the pressures p_1 and p_2 in the case of moving upward and downward, respectively, the dry friction force T can be calculated by Eq. (3). Due to the velocity is low, the viscous friction can be neglected. The dry friction measured is T = 8134 N.

BLOCK DIAGRAM OF SYSTEM

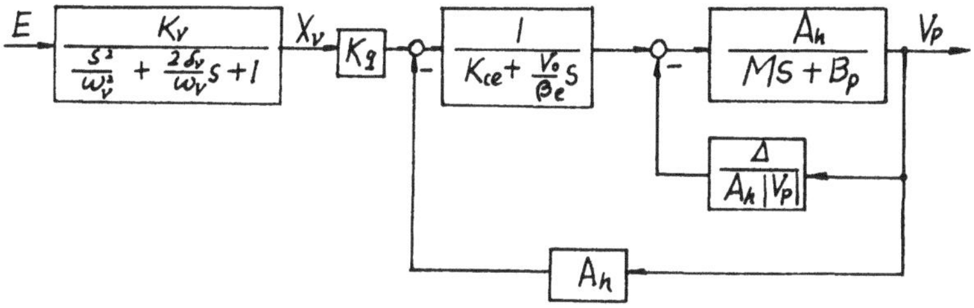

Fig.2 Block diagram of system

The transfer function of servo valve can be expressed as [2]

$$G_v(s) = \frac{X_v(s)}{E(s)} = \frac{K_v}{\frac{s^2}{\omega_v^2} + \frac{2\delta_v}{\omega_v}s + 1} \tag{4}$$

The flow equation of servo valve is [3]

$$Q_L = K_q x_v - K_c p_c \tag{5}$$

The equation of continuity in cylinder is

$$Q_L - C_i(p_s - p_c) = A_p v_p + \frac{V_0}{\beta_e}\frac{dp}{dt} \tag{6}$$

The dynamic equation of cylinder is

$$p_c A_p = M\frac{dv_p}{dt} + B_p v_p + T\frac{v_p}{|v_p|} \tag{7}$$

Taking the Laplace transforms of Eqs.(5),(6),(7) and simplifying them, we have

$$P_c(s)A_p = MsV_p(s) + B_p V_p(s) + T\frac{V_p(s)}{|V_p(s)|} \tag{8}$$

$$K_q X_v(s) - A_p V_p(s) = \left(\frac{V_0}{\beta_e}s + K_{ce}\right) P_c(s) \tag{9}$$

Where $K_{ce} = K_c + C_i$ (10)

Base on Eqs.(4),(8) and (9) the block diagram of the system can be drawn as Fig.2.

Note $B_p K_{ce}/A_p^2 \ll 1$, the mathematical model of valve controlled cylinder can be derived from Fig.2 as

$$V_p(s) = \frac{\frac{K_q}{A_p}X_v(s) - \frac{K_{ce}}{A_p^2}\left(1 + \frac{V_0}{\beta_e K_{ce}}s\right)T\frac{V_p(s)}{|V_p(s)|}}{\frac{s^2}{\omega_p^2} + \frac{2\delta_p}{\omega_p}s + 1} \tag{11}$$

Where

$$\omega_p = \sqrt{\frac{A_p^2 \beta_e}{M\, V_0}} \tag{12}$$

$$\delta_p = \frac{K_{ce}}{2A_p}\sqrt{\frac{\beta_e M}{V_t}} \tag{13}$$

Let

$$G_p(J\omega) = \frac{1}{\frac{(J\omega)^2}{\omega_p^2} + \frac{2\delta_p}{\omega_p}(J\omega) + 1}$$

Eq.(11) can be expressed as

$$G_p(J\omega) = \frac{V_p(J\omega)}{\frac{K_q}{A_p}X_v(J\omega) - H(\omega)Te^{j(\theta_1+\theta_2)}} \tag{14}$$

Where

$$e^{j\theta_1} = \frac{V_p}{|V_p|} \tag{15}$$

$$H(\omega) = \frac{K_{ce}}{A_p^2}\sqrt{1 + \frac{V_t^2\omega^2}{K_{ce}^2\beta_e^2}} \tag{16}$$

$$\theta_2 = \tan^{-1}\frac{V_t\omega}{K_{ce}\beta_e} \tag{17}$$

MEASURING OF THE STATIC PERFORMANCES OF SERVO VALVE AND CYLINDER

A triangular wave signal of low frequency was fed into the amplifier of the servo valve, the main spool displacement of the servo valve and cylinder velocity were measured, the static performances of the servo valve and the valve controlled cylinder can thus be obtained as Fig.3 and Fig.4, respectively. In Fig.4, the cylinder velocity is converted into flow rate by multiplying by the cylinder area. It can be seen from Fig.4 that the deadband of cylinder is significant.

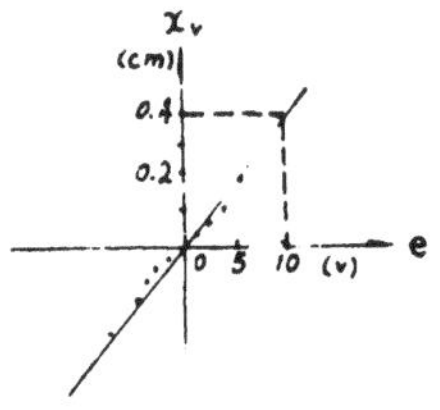

Fig.3 Static performance of servo valve

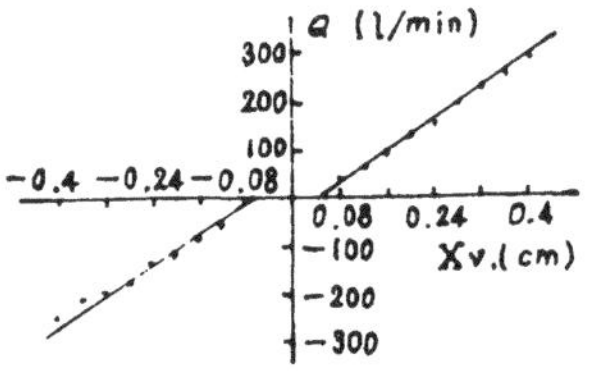

Fig.4 Static performance of valve controlled cylinder

IDENTIFICATION OF THE SYSTEM PERFORMANCE

The identification was carried out under open-loop conditions. The scheme of test is shown as Fig.5.

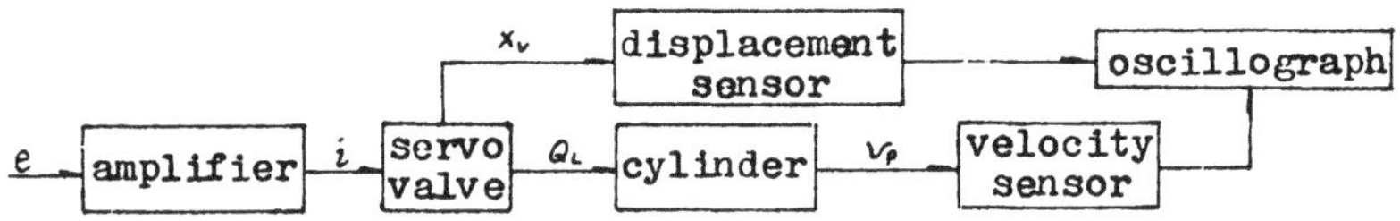

Fig.5 Scheme of test

With a pulse signal input of amplitude 8 V and duration 0.05 s into the amplifier of the servo valve, the responses of main spool displacement of servo valve and cylinder velocity were recorded as in Fig.6 and Fig.7.

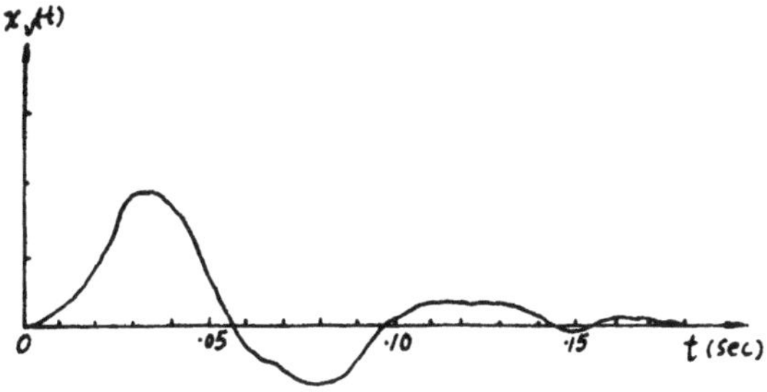

Fig.6 Response of main spool displacement

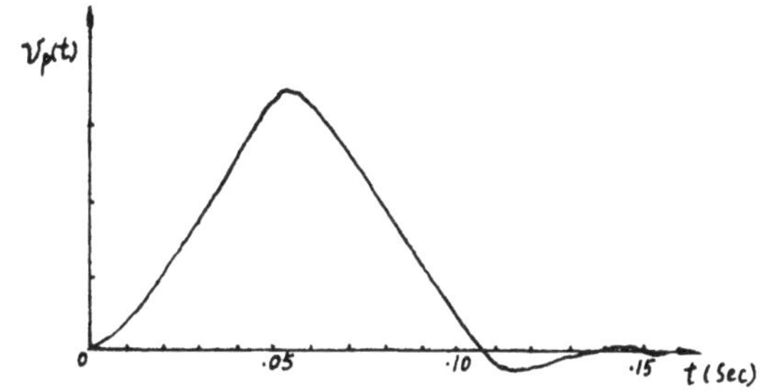

Fig.7 Response of cylinder velocity

By discretization of the curves in Fig.6 and Fig.7, the Laplace transforms of the spool displacement and cylinder velocity can be derived as

$$X_v(J\omega) = \left[\sum_{i=1}^{N} x_v(i\Delta t)\cos(i\omega\Delta t) - J\sum_{i=1}^{N} x_v(i\Delta t)\sin(i\omega\Delta t)\right]\Delta t$$
$$= R_v(\omega) + JI_v(\omega) \tag{18}$$

$$V_p(J\omega) = \left[\sum_{i=1}^{N} v_p(i\Delta t)\cos(i\omega\Delta t) - J\sum_{i=1}^{N} v_p(i\Delta t)\sin(i\omega\Delta t)\right]\Delta t$$
$$= R_p(\omega) + JI_p(\omega) \tag{19}$$

According to Eq.(15), θ_1 is the phase angle of $V_p(J\omega)$

$$\theta_1 = \tan^{-1}\frac{I_p(J\omega)}{R_p(J\omega)} \tag{20}$$

The transfer function of servo valve were obtained from the calculated results of Eq.(18) by curve fitting [5] as

$$G_v(J\omega) = \frac{4\times10^{-4}}{\frac{(J\omega)^2}{55^2} + \frac{2\times0.47}{55}(J\omega) + 1} \tag{21}$$

Where the gain 4×10^{-4} m/v is measured from the static performance of valve in Fig.3.

The following quantities were measured or tested by experiments:

$A_p = 4.153\times10^{-2}\ m^2$, $V_0 = 3.158\times10^{-2}\ m^3$, $K_{ce} = 2.214\times10^{-9}\ m^5/s\cdot N$, $K_q = 1.42\ m^2/s$, $\beta_e = 9.8\times10^8\ N/m$, $M = 4772$ Kg.

Input Eqs.(16),(17),(19),(20),(21) and the above measured quantities into Eq.(14), assume a series of ω, the real part $R_p(\omega)$ and imaginary part $I_p(\omega)$ of $G_p(J\omega)$ with respect to different ω value can be obtained. By curve fitting [5], the expression of $G_p(J\omega)$ can be derived as

$$G_p(J\omega) = \frac{1}{\frac{(J\omega)^2}{110^2} + \frac{2\times0.37}{110}(J\omega) + 1} \qquad (22)$$

The response of the main spool displacement to a 8 v step input was calculated with Eq.(21) by digital simulation. The calculated results are plotted as the dash line in Fig.8. Whilst the solid line represents the experimental result of the response to 8 v input of the valve. The calculated result is basically agree with the experimental result.

According to Eq.(12) and Eq.(13), the natural frequency and damping ratio of cylinder can be obtained as

$$\omega_p = \sqrt{\frac{9.8\times10^8\,(4.153\times10^{-2})^2}{3.158\times10^{-2}\times4772}} = 105.9 \text{ rad/s}$$

$$\delta_p = \frac{2.214\times10^{-9}}{2\times4.153\times10^{-2}}\sqrt{\frac{9.8\times10^8\times4772}{3.158\times10^{-2}}} = 0.324$$

These calculated results are agree with the identified results in Eq. (22), where $\omega_p = 110$ rad/s $\delta_p = 0.37$.

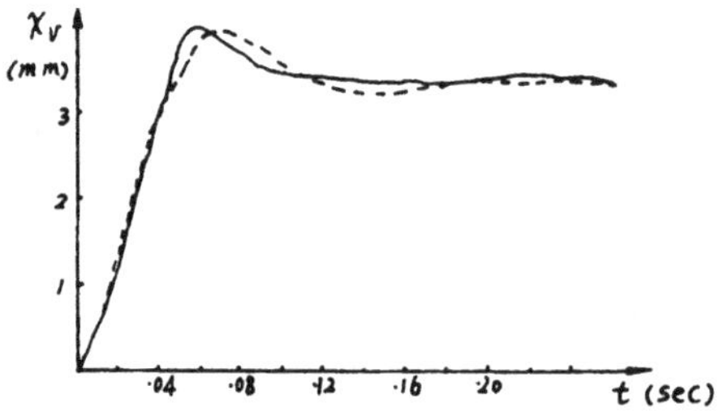

Fig.8 Comparison of calculated and experimental results of the step response of servo valve

Input Eqs(21),(22) into Eq.(11), the expression of the mathematical model of the whole system will be

$$V_p(s) = \frac{\dfrac{1.368\times10^{-2}}{\frac{s^2}{55^2} + \frac{2\times0.47}{55}s + 1}E(s) - \left[1 + 1.46\times10^{-2}s\right] 8134\frac{V_p}{|V_p|}}{\frac{s^2}{110^2} + \frac{2\times0.37}{110}s + 1}$$

CONCLUSION

1. The hydraulic servo control system of electrode of a VHD vacuum

electric-arc furnace can be represented by the block diagram in Fig.2

2. Due to the existance of dry friction induced by the vacuum seal on the electrode, there is a significant deadband on the cylinder performance. The dry friction can be represented by the feedback of a unit velocity vector.

3. Under the circumstances of the existance of deadband nonlinearity, by the treatment stated in this paper, the identification of a nonlinear system with a deadband can be carried out by the method commonly used in linear system.

REFERENCES

1. Zhou Shichang, Cao Xinming "Mathematical Model and Numerical Simulation of the Electrode Hydraulic Servo Control System of a Vacuum Electric-arc Furnace", Proceedings of 1987 American Control Conference, Minneapolis, Minnesota, June 10-12, 1987

2. H. E. Merritt "Hydraulic Control System" John Wiley & Sons Inc. New York, 1967

3. D. McCloy, H. R. Martin "The Control of Fluid Power" Longman Group Limited, 1973, London

4. Katsuhiko Ogata "Modern Control Engineering" Prentice-Hall Inc., Englewood Cliffs N. J., 1970

5. W. D. T. Davies "System Identification for Self-adaptive Control" John Wiley, 1970

IDENTIFICATION OF THE MATHEMATICAL MODEL OF A SERVO VALVE BY MEANS OF A SQUARE WAVE SIGNAL

Zhou Shichang Cao Xinming Gao Chenrui Yang Yanyong
Mechanical Engineering Department
Northeast University of Technology
Shenyang, Liaoning, P. R. China

ABSTRACT

Paper describes the experimental results of identification of a two-stage servo valve with moving coil force motor by means of a square wave signal and the method of mathematical treatment. The transfer function of the valve identified is obtained as

$$G(s) = \frac{0.04767}{\frac{s^2}{59.4^2} + \frac{2 \times 0.47}{59.4}s + 1}$$

The step response of this transfer function was computed by digital simulation and was compared with the experimental results of the valve. Two lines lie within 5 % of one another.

NOMENCLATURES

E -- height of the square wave signal (V)
K -- total gain of servo valve (cm/V)
K_c -- spring constant (N/m)
K_i -- gain of amplifier (A/V)
K_v -- gain of the spool part of servo valve (cm/cm)
K_o -- gain of the moving coil part (cm/A)
m_c -- mass of moving coil (Kg)
t_p -- time of occuring peak value (s)
Δt -- sampling step (s)
U_g -- input voltage of amplifier (V)
x_v -- displacement of spool (cm)
$x(it)$ -- value of discrete input signal (V)
$X(j\omega)$ -- Laplace transform of input signal (V.s)
$y(i\Delta t)$ -- value of discrete output signal (cm)
$Y(j\omega)$ -- Laplace transform of output signal (cm.s)
ω -- frequency (rad/s)
ω_f -- natural frequency of the first order component of spool part (rad/s)
ω_h -- natural frequency of the second order component of spool part(rad/s)
ω_n -- natural frequency of servo valve calculated from step response(rad/s)
ω_i -- natural frequency of servo valve identified from square wave response (rad/s)
ω_c -- natural frequency of servo valve calculated from spring constant and mass of moving coil (rad/s)
τ -- width of square wave (s)
σ -- overshoot
ζ_c -- damping ratio of moving coil part
ζ_h -- damping ratio of spool part
ζ_i -- damping ratio of servo valve identified by square wave response
ζ_n -- damping ratio of servo valve calculated from step response

INTRODUCTION

Previous papers have dealt with the identification of the mathematical model of a servo valve [1,2]. The methods usually used are sinusoidal signal response and pulse signal response. The identification by sinusoidal signal response needs many test points. When using the method of identification by pulse signal response, it is sometimes difficult to generate ideal pulse signal. The square wave signal has been used to identify a moving coil type servo valve and obtained promising results.

COMPOSITION OF THE VALVE AND ITS BLOCK DIAGRAM

The valve identified is a moving coil type servo valve of type TRm7/001 of EMG. Fig.1 illustrates the composition of the servo valve.

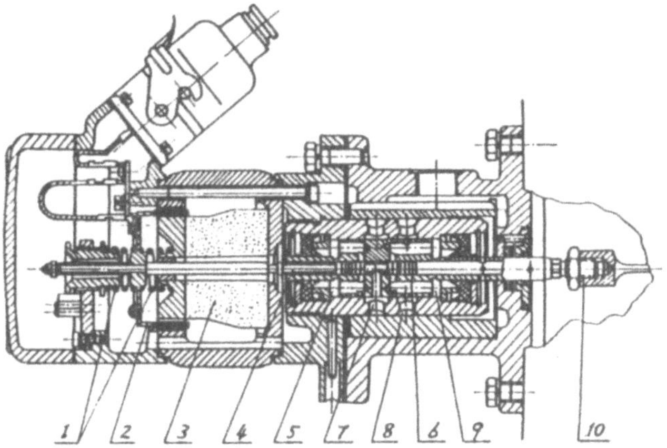

1--centering spring; 2--moving coil; 3--permanent magnet; 4--pilot valve; 5--upper control chamber; 6--spool valve; 7--valve port to oil tank; 8--valve port to oil source; 9--lower control chamber 10--joint to main spool

Fig.1 Composition of the servo valve TRm7/001

When the input current signal applies on the moving coil 2, the moving coil will carry the pilot valve 4 to move, it will induce the pressure difference between two control chambers 5 and 9 and then drive the spool valve 6. The spool valve 6 will push the main spool through the joint 10. The displacement of the main spool was detected by a sensor.

If the input voltage U_g of the servo valve amplifier is used as the input signal and the displacement x_v of the spool valve (equal to the displacement of main spool) is used as the output. The block diagram of the servo valve can be illustrated by Fig.2 [3]. According to the sizes of valve components, we have calculated that ω_h = 2150.5 rad/s; ω_f = 0.3 rad/s; K_v = 216.5 cm/cm. It seems that ω_h is very large and ω_f is very small. Therefore, $\left(\frac{s^2}{\omega_h^2} + \frac{2\xi_h}{\omega_h}s + 1\right)$ can be treated as 1 and ω_f can be neglected. The transfer function of the small closed loop will be

$$\frac{K_v/s}{1 + K_v/s} = \frac{1}{1 + s/K_v}$$

Because K_v is large, s/K_v can be neglected and the transfer function of the small closed loop becomes 1. Therefore the block diagram of the servo valve can be simplified as Fig.3.

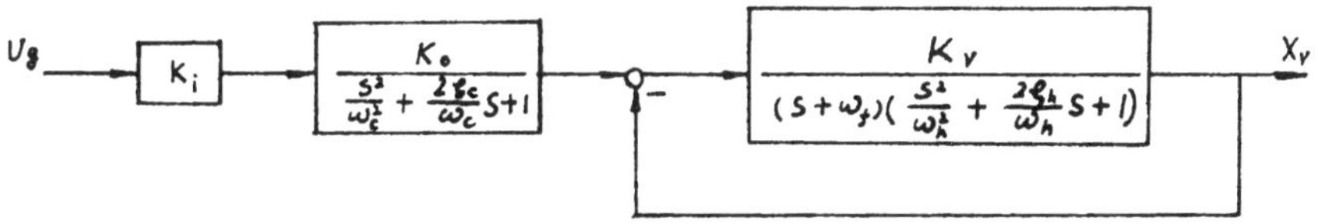

Fig.2 Block diagram of servo valve

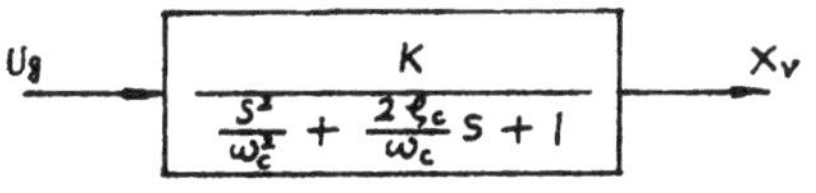

Fig.3 Simplified block diagram of servo valve

The transfer function of the valve will be

$$G(s) = \frac{X_v(s)}{U_g(s)} = \frac{K}{\frac{s^2}{\omega_c^2} + \frac{2\zeta_c}{\omega_c}s + 1} \quad (1)$$

Where $K = K_i K_o$

IDENTIFICATION PROCEDURES

If the input signal in the time domian $x(t)$ is square wave, its frequency spectrum will be [4]

$$X(j\omega) = \frac{2E\sin(\frac{\omega\tau}{2})}{\omega} e^{-j\frac{\tau}{2}} \quad (2)$$

We now treat the output signal in time domian $y(t)$ by discretization. Thus the Laplace transform of $y(t)$ can be calculated as

$$Y(j\omega) = \left[\sum_{i=0}^{N} y(i\Delta t)\cos(i\omega\Delta t) - j\sum_{i=0}^{N} y(i\Delta t)\sin(i\omega\Delta t)\right]\Delta t \quad (3)$$

Therefore, the transfer function will be

$$G(j\omega) = \frac{Y(j\omega)}{X(j\omega)} = \frac{\left[\sum_{i=0}^{N} y(i\Delta t)\cos(i\omega\Delta t) - j\sum_{i=0}^{N} y(i\Delta t)\sin(i\omega\Delta t)\right]\Delta t}{\frac{2E\sin(\omega\tau/2)}{\omega} e^{-j\frac{\tau}{2}}}$$

Separating the real and imaginary parts, the above equation will be

$$G(j\omega) = \frac{\left\{\left[\sum_{i=0}^{N} y(i\Delta t)\cos(i\omega\Delta t)\right]\cos\frac{\tau}{2} + \left[\sum_{i=0}^{N} y(i\Delta t)\sin(i\omega\Delta t)\right]\sin\frac{\tau}{2}\right\}\Delta t\omega}{2E\sin(\omega\tau/2)}$$

$$-j\frac{\left\{\left[\sum_{i=0}^{N} y(i\Delta t)\sin(i\omega\Delta t)\right]\cos\frac{\tau}{2} - \left[\sum_{i=0}^{N} y(i\Delta t)\cos(i\omega\Delta t)\right]\sin\frac{\tau}{2}\right\}\Delta t\omega}{2E\sin(\frac{\omega\tau}{2})} \quad (4)$$

Then, the real part $P(\omega)$ and imaginary part $Q(\omega)$ of $G(j\omega)$ corresponging to different ω values can be calculated by Eq.(4). According to the method of Levy [5], the expression for $G(j\omega)$ can be derived by curve fitting.
In our experiment, the signal generator of type XFD-8 generates a square wave signal U_g of 10 v height and 0.01 s duration. This square wave signal is the input of the servo valve amplifier. The output signal of spool displacement $x_v(t)$ is measured by a sensor and recorded by an oscilloscope. The recorded output signal is shown in Fig.4.

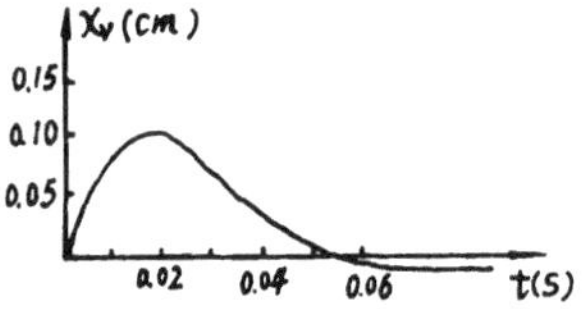

Fig.4 Output signal response

By discretization of the curve of Fig.4 with the step size of 0.005 s , measuring the height of x_v on every step point, we can have a set of $y(i\Delta t)$ Setting a lot of ω values, the frequency response of servo valve corresponding to different ω values can be calculated by Eq.4. Therefore , the transfer function G(s) can be obtained by curve fitting as

$$G(s) = \frac{0.04767}{\frac{s^2}{59.4^2} + \frac{2\times0.47}{59.4}s + 1} \quad (5)$$

From Eq.5, we have that the natural frequency and damping ratio of the servo valve $\omega_i = 59.4$ rad/s, $\zeta_i = 0.47$. For checking the accuracy of identified result Eq.5, we input a step signal into the servo valve amplifier and recorded the response of the main valve displacement shown as the solid line in Fig.5.

We also calculated the step response of the mathematical model (Eq.5) by digital simulation shown as the dotted line in Fig.5. The two lines

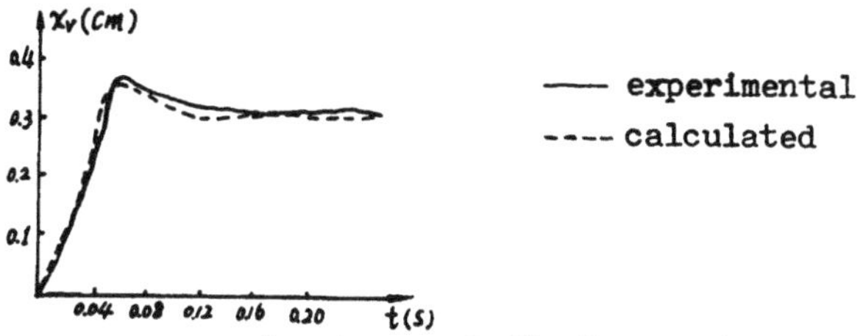

Fig.5 Step response of main spool displacement

appear to lie within 5 % of one another.

According to the experimental curve of step response, we can measure the overshoot as

$$\sigma = \frac{x_{v\,max} - x_v(\infty)}{x(\infty)} = 20.3\ \%$$

and the peak time as $t_p = 0.06$ s . According to the relationship between the parameters of time domain and frequency domain[6], we can calculate ζ_n and ω_n from the values of σ and t_p as $\zeta_n = 0.456$ and $\omega_n = 58.8$ rad/s. The spring constant and the mass of moving coil were measured as K_c = 1837 N/m, m_c = 0.5 Kg. Therefore, we can theoretically calculate the natural frequency ω_c as

$$\omega_c = \sqrt{\frac{K_c}{m_c}} = \sqrt{\frac{1837}{0.5}} = 60.7 \text{ rad/s}$$

Now, we have obtained 3 sets of natural frequency and damping ratio from 3 different ways. They are listed in the following Table

Table Natural frequency and damping ratio

method	natural frequency(rad/s)	damping ratio
square wave response	$\omega_i = 59.4$	$\zeta_i = 0.47$
step response	$\omega_n = 58.8$	$\zeta_n = 0.456$
theoretical calculation	$\omega_c = 60.7$	

The results of the 3 different methods differ only by 3 % and shows that the method of identification by means of a square wave response gives a reliable result.

CONCLUSIONS

1. The square wave response used to identify the mathematical model of the servo valve is reliable.
2. Possibly that this method may be used to identify other hydraulic system.

REFERENCES

1. Tian Jishong, Liu Lianshang " Mathematical Model Identification of Electro-hydraulic Servo Valves " Hydr. and Pneu.,1983,4,p.15(Chinese)
2. Wu Hang " Test and Analysis of Moving Coil Type Electro-hydraulic Servo Valve ", Paper on 1st Conference of Metallurgical Hydraulics(Chinese)
3. H. E. Merritt " Hydraulic Control Systems " John Wiley & Sons,Inc. 1967
4. Mathematics Division of Nanjing Engineering Institute " Integration Transform " P. E. Press 1978 (Chinese)
5. E. C. Levy " Complex Curve Fitting " IRE Trans. on Automatic Control, May,1959
6. J. J. D'Azzo, C. H. Houpis " Linear Control System Analysis and Design" McGraw-Hill Book Co. 1981

ANALOG SIMULATION OF TRANSIENT BEHAVIOR OF CONVERTERS WITH THE APPLICATION OF DYNAMIC THYRISTOR MODELS

Jan Piłaciński[1)]

INTRODUCTION

Simulation of transients in converters of electrical energy is useful when designing and examining systems operating at a high frequency, especially those with internal commutation. Transients can be simulated by different methods - dynamic models of semi-conductor valves can be used for this purpose. In analog modelling, parametric models are especially promising. A synthsis of the model can be made by the branch method, adapted to analog technology by J. Deskur [2]. The method, based on topological analysis, permits an easy arrangement of an analog model of any branched electric circuit with diodes and thyristors.

PRINCIPLES OF BRANCH MODELLING

A full description of the electric circuit is contained in Kirchoff's laws and the characteristics of its particular elements. Kirchoff's laws are topological in character. They are conveniently formulated on the basis of an oriented graph, for which a matrix of cycles $\mathbf{B} = [\mathbf{B}_T \; \mathbf{1}_\mu]$ and a matrix of cutsets $\mathbf{D} = [\mathbf{1}_\rho \; \mathbf{D}_L]$ are determined [1]. Using the relation between the matrix of cutsets for the chords and the matrix of cycles for the tree $\mathbf{B}_T = - \mathbf{D}_L^t$, we get a description of the electric circuit model by means of one matrix only, e.g. the matrix of cutsets $\mathbf{D}_L$ which pertains to the chords. A structure of a branch model corresponding to the above notation is shown in Fig. 1. It was applied to analog modelling following the development of unified models of elementary branches present in electric circuits, including branches containing diodes and thyristors [2].

1) Department of Industrial Electrical Engineering, Technical University, Poznań

Models of tree branches have a causality of the i ➧ u type; models of cords have a causality of the u ➧ i type. Considering the amplification value for high frequencies, the elementary models can be arranged into a series [2]: E-C-G-G_o-modulated for i ➧ u, and then elements R_o-R-L-I-modulated for u ➧ i. Models of semiconductor devices belong to the groups of "peculiar" elements G_oR_o.

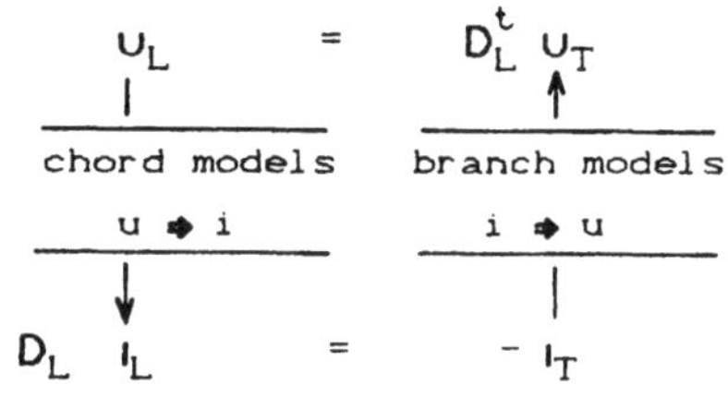

Fig. 1. A general structure of a branched electric circuit

THYRISTOR MODELS

In view of a limited number of the operational elements of an analog device, it is recommended that the simplest thyristor models be used. In order to simulate transients of a converter of electrical current, it is necessary that models of the devices reflect their dynamic properties, at least for the forward direct off-state voltage. A number of thyristor dynamic models, derived from Linvill's physical model of a semiconductor, are discussed in [3]. The models are different with regard to complexity and i ➧ u causality. In the synthesis of the circuit model it appeared necessary to define the graph tree in such a way that it contain devices used for dynamic simulation. The other devices, commutted naturally, could be simulated as static. Fig. 2 a, b shows analog diagrams for the two cases of a dynamic thyristor model. They were formed by combining respective circuits of static models [2] with a circuit which represents the recombination of carriers in the device (Fig. 2c). The thyristor model does not take into account the turn-on process and the junctions were adopted to be ideal. Despite considerable simplifications, the model can be applied to the simulation of the thyristor turn-off process, thus representing the recovery time of the blocking properties of the reverse voltage and parameters Q_{rr}, t_{rr}, I_{RM} (the Γ-type model), as well as time t_q after the time needed to recover the blocking properties of the positive voltage (the ΓΠ-type model). In [4] a method by which parameters of the model are determined on the basis of experimental investigations of physical devices is given.

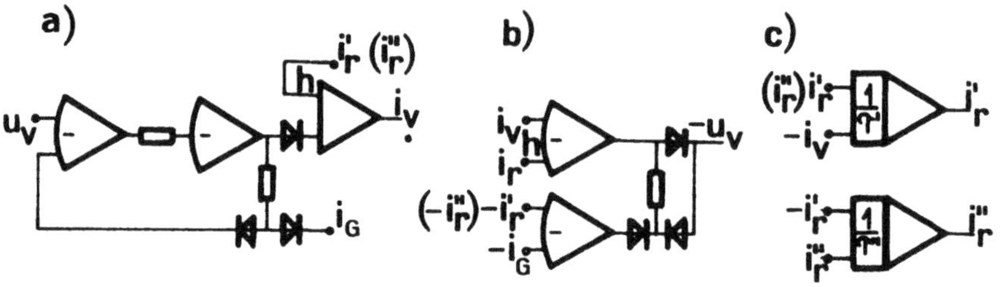

Fig. 2. Analog diagrams of the thyristor model: a - for the u ⇒ i causality, b - for the i ⇒ u causality, c - a dynamics representation circuit (Γ and ΓΠ).

SIMULATION OF TRANSIENTS IN THE CONVERTER

A simple impulse converter circuit with one thyristor has been selected to simulate transients (Fig. 3a). In order to test the operation of the thyristor model under both conditions of causality, two different graph trees were determined (Fig. 3b, c).

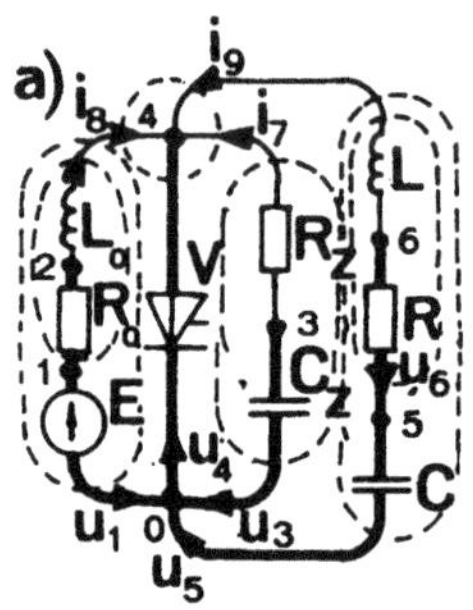

	i_7	i_8	i_9
1		1	
2		1	
3	1		
4	-1	-1	-1
5			1
6			1

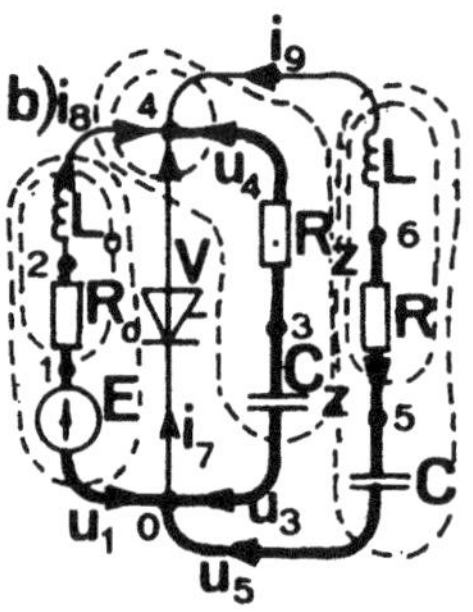

	i_7	i_8	i_9
1		1	
2		1	
3	-1	-1	-1
4	-1	-1	-1
5			1
6			1

Fig. 3. Graphs of an impulse converter.

For each case matrix D_L was determined; directly on its basis an operational diagram of an analog model of the coverter was drawn.

Curves of the current and voltage in the thyristor and condenser, shown in Fig. 4, illustrate the operation of the model. Fig. 4b shows an unsuccessful thyristor commutation. Because the reverse voltage time was too short, the thyristor was not turned on.

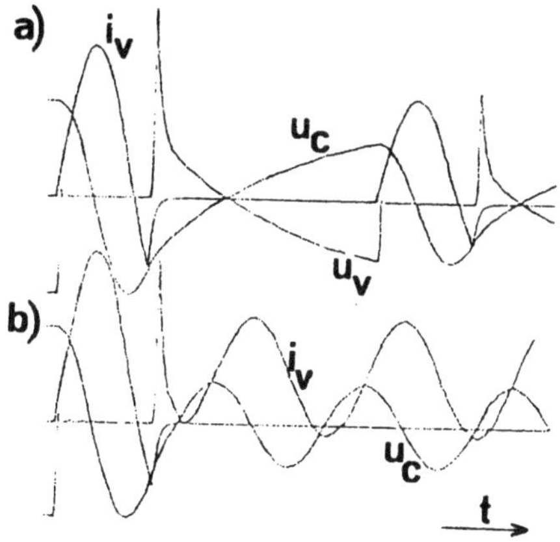

Fig. 4. Curves of thyristor current i and voltage u and condenser voltage: a - operation of the circuit with a higher frequency, b - unsuccessful thyristor commutation.

CONCLUSION

An analog modelling method allows for a simulation of transients in a converter owing to the application of simple thyristor dynamic models. The advantage of the model should be seen in the ease with which an analog model is synthesized. Knowledge of the graph tree makes a traditional description of the circuit with differential equations redundant.

REFERENCES

[1] Chua L.O., Lin P.-M.: Computer-aided analysis of electronic circuits, New Jersey, 1975.975.

[2] Deskur J., A. Szaflarski, (eds): Modelowanie analogowe tyrystorowych układów napędowych. PWN, Warszawa 1986.

[3] Piłaciński J.: Modelowanie półprzewodnikowych zaworów mocy w stanach przejściowych, (diss.). Politechnika Poznańska, Poznań 1983.

[4] Piłaciński J., R. Muszyński: Identification of model parameters using experimental design theory, AMSE Press, Vol. 3, No 4 (1985), 53-64.

DESIGN AND SIMULATION OF ADVANCED CONTROL FOR A PULP DIGESTER HOUSE

N. HVALA, S. STRMČNIK and J. ČERNETIČ [1)]

ABSTRACT: It is reported about the development of advanced control algorithms which are to be implemented in a computer control system for a pulp digester house. The goal of these algorithms is to control the production rate, to improve the quality of pulp and to reduce steam consumption. The control functions considered in the paper include the following: batch sequencing, digester heating control, modification of cooking temperature, prediction of pulp quality during the cooking phase, etc.

1. INTRODUCTION

Since more than a decade, the pulp industry has been recognized as an important area of process control applications. A pulp mill consists of many different processing units, of which the most are interesting for the design of an individual control system [1,5]. Recently, the top level of process management can be also automated to achieve effective coordination of interconnected plant parts and better production control of the whole mill. Up to date, simulation and modelling methods have been widely applied in the pulp industry when the whole mill control has been considered [2]. The aim of this control is to set the production rate of interdependent plant parts and to keep the tank levels as near as possible to the desired values. Models which are used for these purposes are generally very simple, mostly steady-state, based on material and energy balance calculations. On the contrary, simulation is rarely used for the design of a control system for a particular processing unit. This paper reveals how the problem of the design of control algorithms for computer automation of such a typical unit - a pulp cooking process - is being solved by simulation and modelling techniques. Using procedural models of the pulp digester house, various strategies for production, quality and energy control have been designed and tested by simulation. The results presented here were obtained during the development of a computer control system for a particular batch pulp cooking process.

2. CHARACTERISTICS OF THE PROCESS

In the pulping process considered, a batch on a particular digester consists of a fixed and ordered sequence of various operations: charging the digester with wood chips and chemicals, heating-up the digester contents to the cooking temperature (performed in two steps), cooking at a specified cooking temperature, reliefing (high pressure in the digester) and blowing, i.e. emptying the digester contents into the storage reservoir (blowtank). After blowing, a new batch cycle begins with charging.

A typical pulp digester house consists of several digesters which are operating independently each from the other. In our case up to seven digesters can be in operation at the same time. Because of limited energy

[1)] "Jožef Stefan" Institute, Yugoslavia, 61000 Ljubljana, Jamova 39 consumption and other operational constraints, the time schedule of individual operation on each digester must be planned and evenly sequenced in relation to all other digesters. Operational constraints are:

- simultaneous wood charging on several digesters is impossible because a single conveyer system for wood chips is available;
- simultaneous blowing of several digesters into the same blowtank is impossible;
- simultaneous reliefing of more than three digesters should be avoided;
- the level of pulp in blowtanks must be kept between specified values in order not to disturb the continuous processes of washing and bleaching.

3. CONTROL FUNCTIONS

The main aims of the cooking-plant control system are:

- to obtain a regular quality of pulp with small deviations from the desired quality,
- to maintain a specified production rate (usually as high as possible),
- to minimize the fluctuations of steam consumption due to batch processing.

An important advantage of computer control is the possibility to modify the temperature profile during the heating and cooking phases, i.e. the termination time of the cooking phase, the cooking temperature and the slope of the heating curve (i.e. the steam-consumption profile). These parameters are closely related to production rate, pulp quality and steam consumption, respectively. Computer control algorithms detailed in the next chapter are able to determine the most appropriate temperature profile for each cook individually, in order to achieve a specified control target. That means that the control system has to shorten the cooking time by cooking at a higher temperature if higher production rate is desired, or to cook at a lower temperature and more time if better pulp quality is the desired target. As it is shown in the example, the individual targets are tightly interconnected and in most cases also contradictory. So the problems arise when different targets are combined together in order to find an optimal solution. The developed control algorithms are based on a priority system: the target with a currently lower priority is being improved as much as possible in relation to higher priority targets and in relation to constraints.

The problem of pulp quality variation can be further improved by the use of a mathematical model of pulp cooking. The aim of this model is to predict the most appropriate cooking time for each batch according to the cooking conditions and the desired final pulp quality. The methods for identification and parameter estimation of the model must be very effective in order to perform successfully the modification of cooking temperature profiles. The work in the area of model reference control and parameter estimation of time-varying model has been published in [3,4] and is still in progress.

4. CONTROL ALGORITHMS

The basic control strategies which are used to coordinate different control targets were designed and tested by simulation, modelling and control techniques. First, simulation was used to examine the nature of the problem and then to design modular control algorithms which were also tested by simulation under different operating conditions. Relations between the discused control strategies and the pulp digester house are shown schematically in Fig. 1.

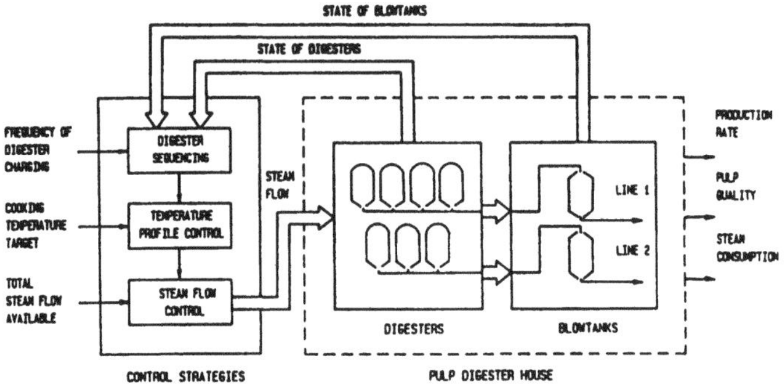

Fig.1.Control scheme of batch pulp digester house.

The algorithm for digester sequencing determines the termination time of cooking for each single digester, so that the desired production rate is being satisfied. Wasted time when waiting with the empty digester or idling chips charging subsystem, is reduced to the minimum. On the other hand, the same algorithm must take care of coordination between the batches on all digesters which are using the same resources, i.e. blowtanks (for digesters on the same line) and the chips charging subsystem (for all digesters in the digester house concerned). This is done by simulating the operation of all digesters for a specified time in advance with the most probable sequence of batches. During such a simulation run, all operational constraints are checked for possible violation.

Having a total time available for heating and cooking, the most appropriate temperature profile is chosen so that the desired pulp quality is obtained at the end of the cook. For control purposes the quality of pulp is predicted by means of already mentioned mathematical model of pulp cooking. It is possible to choose between different temperature profiles to achieve the same quality. Usually, the alternatives are either to choose a higher cooking temperature and slow heating, or a lower cooking temperature and fast heating. The choice depends on the cooking tempera-ture target and the total steam flow available. In situations when more steam is available, it is better to cook at a lower temperature because this gives better quality (the pulp filament is more white and less brittle).

Finally, the algorithm for steam flow control examines and limits, if necessary, the steam flow to selected digesters being at the beginning of the heating phase, in order not to exceed the available total steam flow.

5. RESULTS

From different tests of control strategies obtained by simulating the process and the control algorithms we have chosen to present an example of steam consumption. The time-plot of total steam consumption, the steam-flow and the temperature curves of individual digesters are shown in Fig.2. It can be seen that the steam flow and temperature curves of individual digesters (Fig.2b and 2c) are being adjusted during the heating phase in order to achieve the total steam consumption as smooth as possible (Fig.2a). Fig.2c also reveals how the cooking temperatures are adjusted to improve pulp quality. For this case, the maximum temperature was set to 162°C, whereby the control algorithm has found that this temperature can be lowered and different from batch to batch.

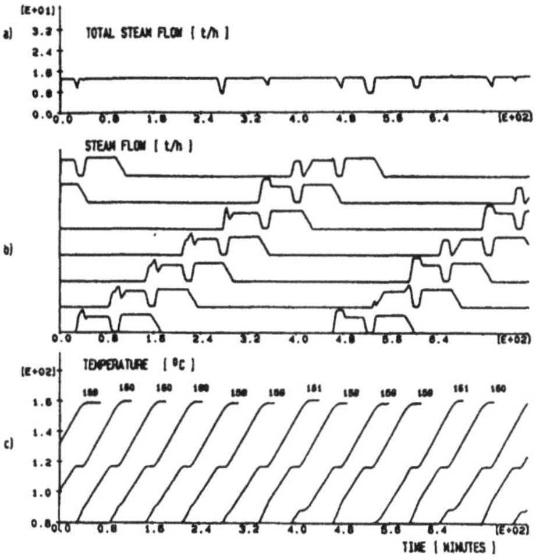

Fig.2. Simulation results:
a) total steam flow,
b) individual steam flows,
c) heating temperature profiles.

6. CONCLUSION

Generally speaking, greater production, better quality and smooth steam consumption are being expected by computer automation of the pulp cooking process. The designed algorithms for digester sequencing, temperature profile control and steam flow control are used to determine the profiles of temperature curves for the heating and cooking phases, in order to achieve specific control targets and to satisfy the limiting operating conditions. Although the simulation tests confirm that the computer control system gives much better results than classical control, some problems are still open. Currently, interactions between control targets and the problem of optimal control are under investigation.

7. LITERATURE

[1] Dumont, G. A.: Application of advanced control methods in the pulp and paper industry-a survey. Automatica 22 (1986) 2, 143-153.

[2] Golemanov, L. A. and K. Leiviska: Methods and algorithms for production and energy management. 6th IFAC PRP Conf.(1986) Akron, Ohio/USA, 184-191.

[3] Hvala, N., J. Černetič, S. Strmčnik and A. Može: Mathematical model of magnefite pulp cooking. Workshop on process automation (1986), Ljubljana, 63-77.

[4] Juričić, Đ.: Parameter estimation of a nonlinear time-varying model of the pulp cooking process. Workshop on process automation (1987), Darmstadt (to be published).

[5] Perron, M. and A. Ramaz: A survey of control strategies in chemical pulp plants. Automatica 13 (1977) 4, 383-388.

MODELLING AND SIMULATION OF A DRIVE MOTOR SYSTEM

Irena Jaworska*, Zbigniew Szulc*, Grzegorz Kryger**

SUMMARY

The paper deals with modelling and simulation of a laboratory motor stand. The motor of 1.5 KW, two regulators PI and the observer of the angular speed are taken into consideration in the research trial. The computer simulation makes a preliminary investigation to utilize the speed observer in a high power industrial drive system.

1. INTRODUCTORY REMARKS

The thyrystor drive systems are widely utilized in industry and municipal facilities due to the high efficiency and easy control. The authors are interested in the motor system driving water pumps in a pipeline [1],[2]. The drive system consists of the asynchronous motor of the power more than 1 MW, transformer, direct current circuit with inverter and rectifier. The idea scheme is shown at Fig.1. There are well known the difficulties occuring in the high power systems effected with the unreliability of the electromechanical elements used to measure directly the angular speed in the feedback loops. The electromechanical elements introduce noises, failures, and need a special structure and instalation working in a direct touch with an environment. To avoid the difficulties the speed observer has been proposed [3]. To make the practical experiment under industrial conditions (water pipeline) the preliminary investigations have been carried on. The trial has been prepared in the laboratory stand equipped with 1.5 kW power motor.

The aim of the paper is :

-to formulate the mathematical description of the subsystems

-to make the computer simulation of the system

-to analize the computer results taking into considerations the behaviour of the speed observer.

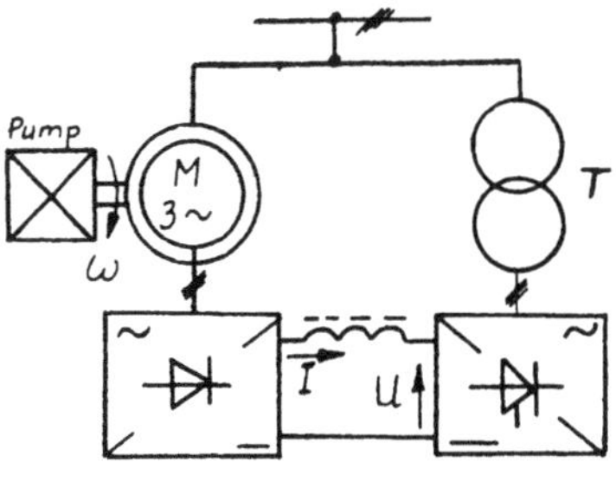

Fig.1 The idea scheme of the asynchronous motor system.

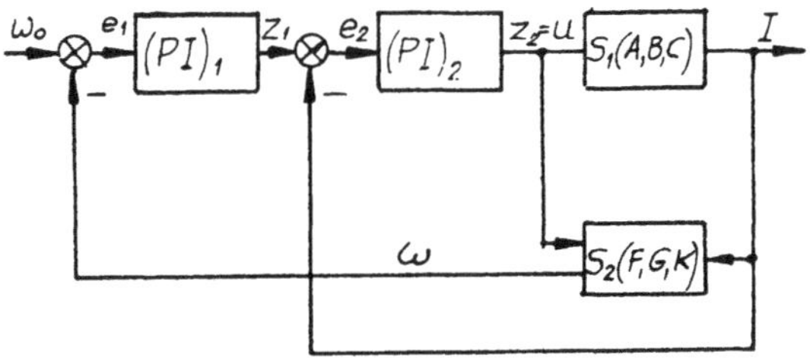

Fig.2 The block scheme of the constant torque cascade.

* Warsaw University of Technology, Koszykowa 75, 00-62 Warsaw, Poland

** Chemical Factory, "Wloclawek", Torunska 222, 87-810 Wloclawek, Poland

2. MATHEMATICAL MODELS OF THE SUBSYSTEMS

The paper deals with the asynchronous motor system with the observer of the angular speed. The block scheme of the system considered is shown at Fig.2. There is so called constant torque cascade with two controllers working according to the PI control algorithms in the feedback loops. The mathematical models are formulated for the following subsystems: motor system, observer, controllers.

2.1 **Motor system** of 1.5 KW : $S_1(A, B, C)$

The model accepted has the form [3]

$$\dot{x}(t) = A\, x(t) + B\, u(t) + v(t) \qquad (1)$$
$$y(t) = C\, x(t)$$

where

$x(t) \in R^2$ state vector (ω, I), $y(t) \in R^2$ output vector, $v(t) \in R^2$ considered as a known disturbance vector - nonlinear function of the angular speed, $u(t) \in R^1$ input (voltage), and

$$A = \begin{bmatrix} 0 & 11.69 \\ -7.36 & -11.55 \end{bmatrix} \qquad v(t) = \begin{bmatrix} 24.36 + 0.01\,\omega^2(t) \\ 1157.7 \end{bmatrix}$$

$$B = \begin{bmatrix} 0 & 14.1 \end{bmatrix}$$

The model is considered in the vicinity of the operating point

$$x_0 = [\,117.2\,,\ 14.5\,]^T \qquad u_0 = 15.1$$

The measurement accessibility is accepted only for the motor current. It means : $C = [\,0\,,\ 1\,]$.

2.2 **The observer of the angular speed :** $S_2(F, G, K)$

The observer of the state vector of the model (1) has the form [4]

$$\dot{\hat{x}}(t) = F\,\hat{x}(t) + G\,u(t) + v(t) + K\,y(t) \qquad (2)$$

$\hat{x}(t) \in R^2$ state vector of the observer

If the following conditions are complied with

$$G = B \qquad F = A - K\,C \qquad (3)$$

the error of the estimation is described by the differential equation :

$$\dot{e}(t) = \dot{x}(t) - \dot{\hat{x}}(t) = F\,[\,x(t) - \hat{x}(t)\,] = F\,e(t) \qquad (4)$$

The eigenvalues of the matrix F may be found by a suitable choice of the matrix K. For the system considered the matrix K is assumed to make slight less the eigenvalues of the observer than the eigenvalues of the system model. Accepting matrix K as follows: $K = [\,0.157 \quad 0.2\,]^T$ the eigenvalues of the matrix F are equal to $\lambda_{1,2} = -5.87 \pm j7.1$

2.3 Controllers of the motor system : $(PI)_i$, i=1,2

At Fig.2 the asynchronous motor system is displayedconsisting of two **feedback** loops with controllers working according to the proportional-integrating algorithms (PI) with saturations. The speed and current controllers have the algorithms

$$z_i(t) = z_i(t-1) + k_i[\ 1 + (\ T_p / T)_i\ ]\ e_i(t) - k_i e_i(t-1) \qquad (5)$$

where

z_i - output signal, e_i - input signal, k_i, T_i - parameters of regulators, $(T_p)_i$ - sampling time, i = 1,2.

The simulation results are presented for the following parameters :

$k_1 = 0.3$, $k_2 = 1$, $T_1 = 0.1$, $T_2 = 0.06$, $T_{p1} = T_{p2} = 5\ E{-2}$

3. COMPUTER SIMULATION AND NUMERICAL RESULTS

The digital simulation has been arranged in the Pascal and curried out at the IBM PC/Xt. The numerical investigations have been led under various system conditions. The system conditions are gathered in Table 1 and

TABLE 1

No	distance* d	initial cond. of observer(2) $\hat{\omega}(0)$	$\hat{i}(0)$	initial. cond. of system(1) $\omega(0)$	$i(0)$
1	0.1	60	10	90	14.5
2	1	60	10	90	14.5
3	5	60	10	90	14.5
4	0.1	0	0	90	14.5

* distance between eigenvalues of the observer (2) and model (1).

numbered from 1 to 4. At Fig.3 and Fig.4 there are displayed the errors of the current and speed estimations for four experiments respectively.

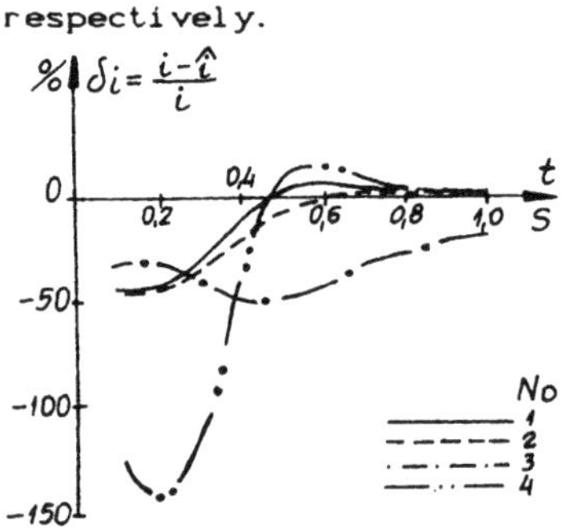

Fig.3 The errors of the current estimation.

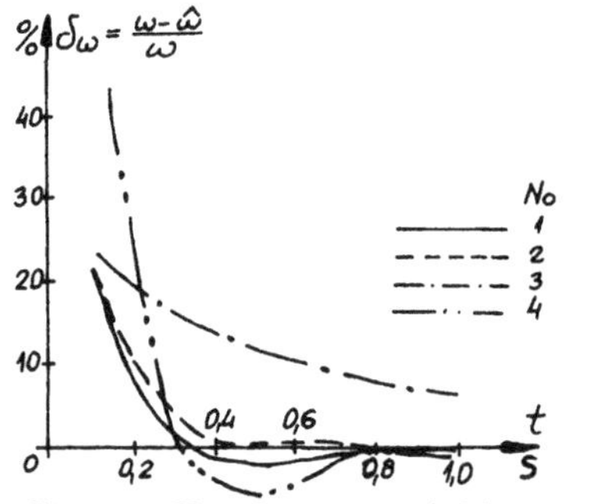

Fig.4 The errors of the angular speed estimation.

The output signals (ω, I) for the experiments No1 and No4 are drawn in Fig.5 and Fig.6. The experiments No1 and No4 show the affect of the initial conditions of the observer the output signals when the distance d between eigenvalues of observer (2) and the system model (1) is the same.

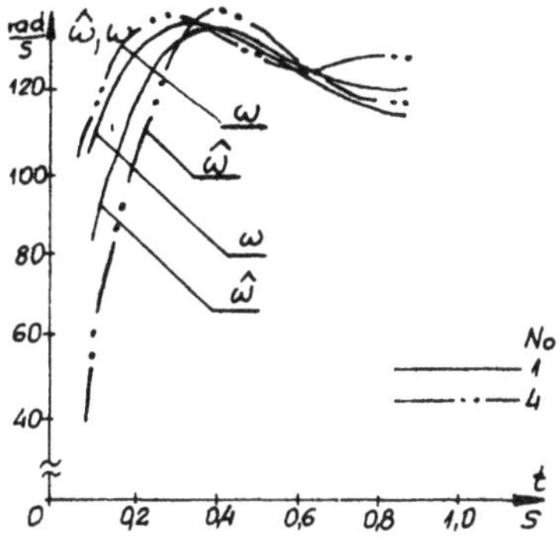

Fig.5 The estimated speeds for the experiments No1 and No4.

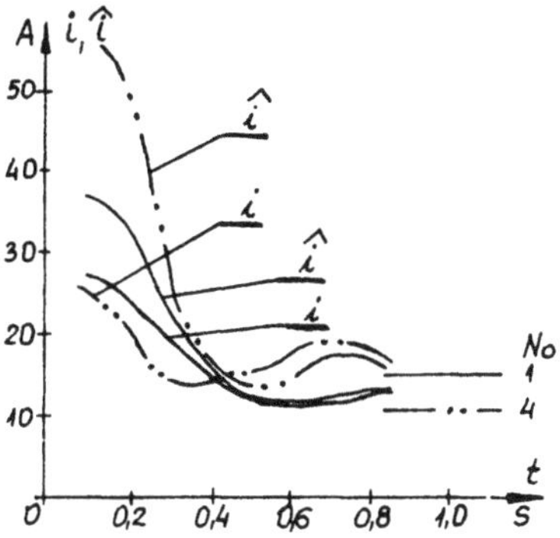

Fig.6 The motor currents for the experiments No1 and No4

4. CONCLUSIONS

In the paper some results of the computer simulations for the laboratory motor stand have been presented. The models of motor system (1) and speed observer (2) have consisted of the nonlinear parts (the square functions of the angular speed). The computer investigations will be carried on by the authors respecting the parametrical changes in the drive system simulated, respecting the other algorithms of the observer and controllers.

To assess the maximum error of the speed estimation the authors' endeavours have been done [3]. The results will be verified with the practical experiment which is prepared at the laboratory motor stand.

REFERENCES

[1] Jaworska I.,Szulc Z. ,The influence of electrical network disturbance on output power of drive motor system,IMEKO TC4Symposium, Como, Italy 1986

[2] Jaworska I.,Szulc Z.,Current measurements in a drive motor system, IMEKO-TC4 Symposium, Warsaw, Poland, 1987

[3] Jaworska I.,Szulc Z.,The measurement technique of the output power in a drive motor system. Report not published. 1987,Warsaw,Poland

[4] O'Reilly , Observers for linear systems ,Academic Press, London,1983

A SIMULATION MODEL FOR ALUMINIUM ELECTROLYTIC CELLS CONCERNING ON PROCESS CONTROL

Laszlo TIKASZ, Miklos ZAYMUS, Andras CSEH *

ABSTRACT

This paper describes the steps and the results of developing work in the field of mathematical modelling and dynamical simulation regarding aluminium electrolytic cells at Hungarian Aluminium Corporation (HUNGALU). Based on the main physical-electrochemical equations, a complex model has been developed. For the examined period of time, the alumina balance formation and the cell voltage alterations are the fundamental processes. The model parameters has been tunned for a real cell. In accordance with the plant situation,the regular operational events as well as the personnel's interactions can determine the simulated process. The computer-based simulator can be applied for test and educational purposes. The research worker or a trained potroom assistant can watch and modify the forming process. It is an effective method to check the results of possible work routines. With the aid of the simulator, the planned control algorithms can be evaluated. Outlines of theoretical background, with examples of practical applications, are given in the paper.

INTRODUCTION

The aluminium electrolytic process is executed by Al_2O_3 electrolysis in molten cryolite at a temperature of 980 °C between carbonaceous electrodes. The process itself is carried out in electrolytic cells. The average cell voltage is about 4.1 - 4.3 [V] and the amperage is in the range of 60 - 80 [kA] in Hungary. The electrolytic cells are serially connected, called potline. The line voltage is in the range of 800 - 1200 [V]. The electrolysis is a continuous process with several scheduled interactions /alumina feeding, metal tapping, interpolar distance adjusting, anode paste feeding, adjusting current carriing elements in the anode, adjusting the chemical properties of the electrolyte, etc./. These interactions are mainly done by manually but there are special machines to make the work easier. An important task is to help the potroom personnel in their work, to generate a reliable, roboust process control

* HUNGALU Engineering and Development Centre,
H-1389 Budapest, P.O.B. 128. Hungary

system. Some years ago, a research project started at HUNGALU. On the basis of specially developed measuring devices, long-term, real-plant data acquisition and processing was done. Although, the analysis of this data gave fine partial results /developing process control algorithms, determination technological states of cells/, a strong demand arised for a more complex understanding of the process. The mathematical modelling of electrolytic cells was begun and this led upto the process simulation.

MODELLING

Preparation of the Model

The fundamental electrolytic process was studied carefully to determine the essential interactions and dynamical properties of the main parameters. The process itself was approximated by submodels calculating the next amounts :

- -cell voltage components,
- -alumina balance,
- -thermal balance,
- -current efficiency.

The basic equations for the cell voltage and alumina balance calculations are given below for demonstration.

The simplified equation for the cell voltage calculation :

$$U = U_o + U_1 + I_s * (R_o + R_1 + R_2) \qquad (1)$$

where

U_0 = sum of equilibrium potential, reaction overvoltage and cathodic overvoltage

U_1 = anodic overvoltage of concentration origin

I_s = line current

R_0 = resistance of mechanical parts

R_1 = resistance of electrolyte between poles

R_2 = resistance of gas layer between poles

The simplified calculation of alumina balance :

$$\frac{d x_1}{d t} + \frac{d x_2}{d t} + \frac{d x_3}{d t} = \Delta m_t - \Delta m_3 \qquad (2)$$

$$\frac{d x_1}{d t} = \Delta m_t - S_1 - S_3 \qquad (3)$$

$$\frac{d x_2}{d t} = S_3 - S_2 \qquad (4)$$

$$\frac{d\,x_3}{d\,t} = S_1 + S_2 - \Delta m_3 \qquad (5)$$

where

x_1 = amount of disperged alumina in the electrolyte

x_2 = amount of settled alumina in the electrolyte

x_3 = amount of dissolved alumina in the electrolyte

m_t = alumina feeded to the cell

m_3 = alumina equivalent of metal production

S_1 = soluted alumina from disperged part

S_2 = soluted alumina from settled part

S_3 = settled alumina from disperged part

The details of electrochemical equations are beyond the scope of this paper.

Structure of the Model

The present model approximates the main parameters of a real cell by interconnected submodels. As a description method, state variables have been choosen for differential equations. State variables have physical meanings to make easier the evaluation of non-measurable parameters.

The state variables are the followings :

-amount of disperged alumina in the electrolyte
-amount of alumina in the settlings
-amount of dissolved alumina in the electrolyte
-temperature of electrolyte
-mass of electrolyte
-temperature of freeze
-mass of freeze
-temperature of molten metal
-mass of molten metal

Input variables are :

-anode adjusting
-alumina feeding
-metal tapping
-line current modification

In actual plant practice, the line current is usually kept at a constant level, rather than used as a real input variable. In domestic plants, however, line current fluctuations are significant. Thus it is inavoidable to include line current in the list of variables. Under normal circumstances, only the cell voltage and the line current can continuously be measured. For test purposes, the next parameters were measured:

-cell voltage,
-alumina concentration,
-temperature of electrolyte,
-temperature of cathode side-wall
-level of electrolyte in the cell
-level of molten metal
-anode-cathode distance

The differential equations are solved by fourth-order Runge-Kutta methods. The calculation is fast enough to work in a user-friendly dialog form.

SIMULATION

Basic Applications

The dynamic model is the basis to attemp to simulate the behaviour of electrolytic cells. The simulation can be applied for evaluating different alumina feeding methods, and for testing control strategies represented by different algorithms. To partially demonstrate the possibilities of the model, the calculated components of alumina balance and the alteration of the cell resistance are shown for wheel breaker operation.

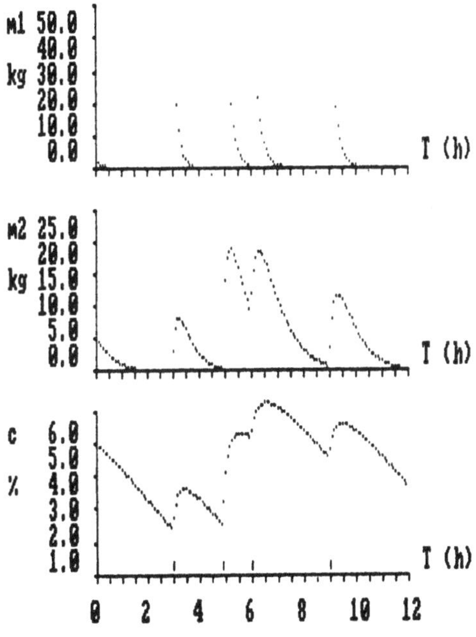

Figure 1. Simulation of alumina balance
/ m_1 , m_2 : amount of disperged and settled alumina,
c :alumina concentration in the electrolyte/

Figure 1. simulates the alumina balance parameters. It is presumed that the feeding process takes place at every 180 minutes accuratelly, but the amount of feeded alumina changes according to the normal operational alterations. Vertical lines sign the feeding actions where about 80 kg of alumina was added.

Figure 2. simulates the cell voltage parameters. For practical purposes, the cell resistance is calculated and displayed. Line current data are generated with statistical characteristics accordingly the real line current by a special submodel. During the simulation, the interpolar distance was altered manually.

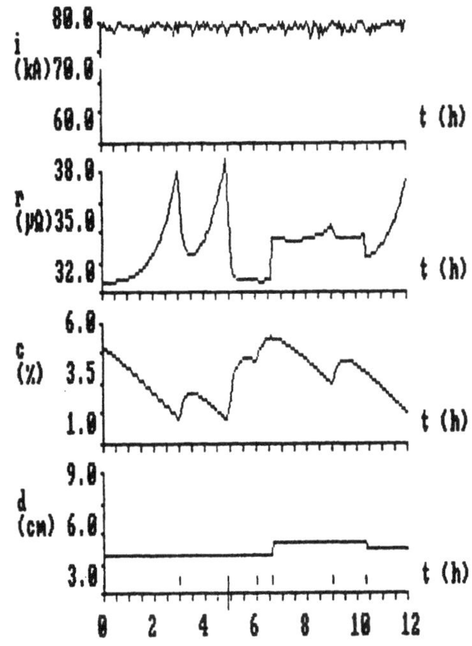

Figure 2. Simulation of cell resistance
/ i : line current, r : resistance of the cell,
d : interpolar distance/

Advanced Applications

A really usefull application of this simulation tool was the developing and tunning procedure of point feeders' algorithms. In the real-plant situation, a two-level computerised control system is in operation. The lower-level computer is connected to the electrolytic cell. It measures the electrical parameters of the cell, it fulfills resistence control by adjusting interpolar distance. This computer contains the different routines for alumina feeding procedures, but it is'nt able to choose the proper routines adequat for the necessary alumina content. The higher-level computer collects data from the lower-level one, analyses them and gives orders initialising the alumina feeding routines.

In our simulation, two computers were interconnected. The simulation model planted into the first computer acted as a cell regarding the main parameters and the responses. The second was nominated to be the higher-level supervisor. The real data communication subroutines were in operation on both computers. During the simulation, the first computer generated the cell's parameters and fulfilled the basic control functions and the data transfer. The second computer accepted the simulated plant data, processed them, sent orders to the first computer and also fulfilled the data transfer.

This arrangement was realistic. The well-tunned cell simulator produced cell parameters of different technological situations. With the aid of this, adjusting of alumina feeding algorithms was done in the laboratory significantly decreasing the time-consuming, expensive work at plants. After it, the feeding routines were put in real-plant operation. The results were excellent.

These and similar simulations can conveniently be done by a personal computer. During the simulation process, the research worker or the trained potroom assistant can watch and modify the forming process. Depending his demand, he can see either the normal, "visible" cell parameters or the "hidden" ones. It is an effective method to check different imaginations about the process.

RESULTS

The close co-operation between experts of different subjects has proved to be fruitful. Improved results have been obtained connecting real-plant data processing and simulation methods. Based on the achieved results, the electrolysis process can be led on a more accurate way.

ACKNOWLEDGEMENT

The authors are thankful to their colleagues for their co-operation. Special thanks are due to Mr. Jozsef Pocze and to Mr. Antal Klemm for their precious advices during the laboratory and real-plant work.

REFERENCES

1. L. Tikasz :
"Modelling and Control of Alumina Reduction Cells"
Dr Techn. Thesis, Technical University of Budapest, 1985.

2. M. Vajta et al. :
"Anode Effect Prediction of Aluminium Reduction Cells"
Report, Dept. of Automation, Technical University of Budapest, 1984.

3. A. Ek, G.E. Fladmark :
"Simulation of Thermal, Electric and Chemical Behaviour of an Aluminium Reduction Cell on a Digital Computer"
Paper presented at the AIME Light Metals, 1973., pp. 85-104.

4. O.A. Asbjornsen, J.A. Andersen, A. Larsen :
"On the Use of Mathematical Models for the Phisical Interpretation of Current and Voltage Fluctuations in an Alumina Reduction Cell"
Paper presented at the AIME Light Metals, 1979., pp. 517-539.

5. P. Entner et al. :
"Investigation of the Dynamic Behaviour of Aluminium Pots"
Paper presented at the AIME Light Metals, 1984., pp. 701-710.

6. M. Vajta, L. Tikasz :
"Adaptive Prediction of Anode Effects in Aluminium Reduction Cells"
Paper presented at the IFAC/IMACS International Symposium on Simulation of Control Systems Vienna, Austria, 1986., pp. 117-121.

7. L. Tikasz et al. :
"Electrical Measurements, Process Control and Electromagnetic Interaction in Aluminium Electrolysis"
UNIDO Course Material, 1984.

8. H. A. Oye ed. :
"The 2nd International Course on Process Metallurgy of Aluminium"
Course material, Norwegian Institute of Technology, 1983.

9. J. Horvath et al.:
"Physical Chemistry and Practice of Aluminium Electrolysis"
UNIDO Course Material, 1984.

10. L. Tikasz, J. Horvath, I. Horvath, M. Vajta :
"Computer Control System for Aluminium Electrolytic Cells Model Studies and Plant Operation"
Paper presented at the AIME Light Metals, 1988., pp. 583-588.

A COMPUTER SYSTEM FOR DECISION SUPPORT IN THE SUGAR CANE AND ITS DERIVATIVES INDUSTRY

Raúl Sabadí Díaz[1], Eugenio de la Vega, Diego Fernandez

Introduction

In the last years, sugar prices has been decreasing in the world market and it seems this situation will not change. Under this condition, the sugar cane derivatives industry plays an important role in the national economy of producer countries. With aims of analysis of integrated productions of sugar and cane derivatives, the computer package SANPAD is designed; which, based on the Revised Simplex algorithm of linear programming, allows the technical and economical optimization of production strategies. In this paper, the version 1.1 of this package is presented.

The production scheme

The production scheme incorporated into the system includes plants which produce animal feeding and/or raw materials for this production, selected because of their importance to our country[3,4]. The plants included are those for production of torula yeast, alcohol, proteic molasses, pamusal[2], hydrolised bagasse, hydrolised green residuals, mixed cattle feeding, a formulation of molasses and urea and the sugar factory with its cleaning center. There are defined all the relationships between the sugar mill and the other factories taking into account the flows from each one which are raw material for the others.
In the sugar mill, a three massecuites processing is considerated. The production of alcohol, yeast and proteic molasses can be also analysed in the no harvest season.
The possibility of using several combinations of raw materials in each factory gives enough flexibility to the system and the user can stablish different technologies.

The mathematical model

In the mathematical model, calculations associated to the material, steam

---

1) Computer Systems Dpt. / Engineering Division / ICIDCA / P.O.Box 4026 / Havana / Cuba
2) Bovine cattle feeding from harvest green residuals. Formulation developed in ICIDCA.

and electricity balances for the integrated scheme are included. The first version of this package only included material balances and had another limitations[5]. There are also included other technological restrictions such as crystals content, fermentation yield, protein balances, steam generation parameters, plant capacities, steam and electricity compsumption rates, etc. All this information is enclosed in a set of linear equations, which can be used as a linear programming model for optimization of alternatives.
The model for the scheme which includes all the factories is made up of 213 equations with 137 variables.

The computer package SANPAD

The computer package SANPAD (ver. 1.1) is implemented for IBM compatible microcomputers. It is written in Turbo Basic language, from Borland International, and runs under MS DOS 3.2 operating system with ramdrive feature installed. It needs also color graphics adapter and a graphic 80 columns printer for optional output. It can run using the math co-processor 8087 improving its speed significantly.
There are four programs in the system:

- Program SAN1: which carries on the system initialization. It works as a combination of graphical menu and data editor which allows the modification of information associated to the scheme.
- Program SAN2: for variable bounds and objective function definition. It's in charge of the mathematical model generation.
- Program SAN3: performs the revised simplex algorithm of linear programming.
- Program SAN4: for results output on display and printer.

The user accesses the system through the command SANPAD11 and, after initialization and a brief description of the objectives of the package, is given a graphical menu, which is a representation of the production scheme. In this menu, the user has some choices: to move over the plants with the up and down arrows keys, to select a plant with the <ENTER> key and to ask about information with the <F2> key. To select a plant means that the user includes it in the scheme to be analysed.
If the <F2> key is pressed, the user can get information about the plant where the cursor is on. This information may be the diagram of the plant (<F4>) or about the variables (<F5>), i.e., the physical characteristics of the materials (<F6>) and technological relations in the plant (<F7>). The information about these variables is displayed in user-oriented window data editors (WDE) and the user can modify these values in order to adjust the scheme to a particular case. These WDE operate associated to each plant and, because there is common information between them, the system updates the modified parameters for the whole scheme. Pressing the

<ESC> key the user can go back to the main menu and go on along the program with the <F3> key, being possible to print all this information. The next step is to define the upper and lower bounds of the variables, i.e., of the flows which are raw materials in each plant. It is done through another easy-to-use data editor. This option allows the final identification of the scheme with the case being analysed. After this definition, it's possible to select the objective function from a menu or set up a new one using edition facilities of the program. If the choice was one of the built-in economical objective functions, the user can look at the basic economic information (costs, prices) and modify it.
In this moment the particular case is already defined and the system generates the appropriate mathematical model on disk. The optimization is carried on using the Revised Simplex algorithm of linear programming and the results are saved on disk also. If the calculation fails, i.e., if the solution is infeasible or no bounded, a message is displayed and the user can redefine the problem or end the job. The user can see the results of the problem on screen using a menu which permits to look at each plant. It is possible also to print all the information.

Conclusions

This package has been used for the analysis of a sugar mill and cattle feeding complex[1] and in the analysis of integrated production of sugar and alcohol[2]. The results has demonstrated that SANPAD is an efficient tool for decision support in the sugar cane and its derivatives industry. This PC based software was developed for being easily handled by the user, without needing a computer specialist and to avoid the formulation of complex models for strategy analysis.
Future developments of this system will include other sugar cane derivatives industries such as pulp and paper, particleboards and furfural plants.

Literature

[1] De la Vega, Eugenio; Sabadí Díaz, Raúl. "Engineering study in a sugar cane mill and cattle feeding complex". 9th International Congress CHISA'87. Prague, Sept. 1987.

[2] De la Vega, Eugenio; Sabadí Díaz, Raúl. "Modelación matemática del proceso de producción azúcar-alcohol". Revista ICIDCA, vol.XXII, no.1, 1988.

[3] "El desarrollo de la alimentación animal a partir de los subproductos y productos de la industria azucarera". ICIDCA. Informe interno. Febrero 1987.

[4] "La industria de los derivados de la caña de azúcar". ICIDCA. Editorial Científico Técnica. La Habana, 1986.

[5] Sabadí Díaz, Raúl; de la Vega, Eugenio; Fernandez, Diego. "Sistema de analisis de la producción de azúcar y derivados. SANPAD (ver.1.0)". INFOAZUCAR'87. C.Habana, 1987.

SIMULATION MODEL OF HIGH PRESSURE GAS PIPELINE NETWORKS[1)]

Ryszard Klempous, Jan Nikodem, Jedrzej Ulasiewicz[2)]

1. Introduction.

High pressure simulation of gas pipeline networks is one of the factors which enables to aid the system dispatcher decisions. Because of its complexity, this problem can be solved effectively only by using a computer techniques. As a basic fact one should recognize the dynamic phenomena appearing in the network, accumulating of the gas and a great number of admissible decisions. Moreover in order to elaborate proper control decisions it is necessary to perform a large number of simulations. Each of this simulation is connected with the necessity of the iterative solutions of nonlinear differential equation systems describing the dynamic phenomena appearing in the network. In the papers [4], [5] the solution methods of the above mentioned problem in the case of quasi static flows are described. This algorithms ensure a satisfactory concordance between the results obtained from simulation with the measuring data carried out on a real network especially in case of low pressure gas networks. Only in such situation it is possible to neglect in the model of network the phenomena of gas accumulation.

In the paper a model which takes into considerations the above mentioned problem adopted to the requirements resulting from its aim to aid the dispatcher decisions will be described.

2. Quasi static model description.

The system considered in this paper is the nonlinear gas distribution network with the complex loop structure. This network structure is described by the graph:

$$G = \langle W, L, f \rangle \tag{1}$$

where:

W, L are sets of nodes & arcs and $f: W \times W \longrightarrow L$.

The static gas flow in the network are expressed based on the following principles [1]:

[1)] This research was supported by the Scientific Program RP.I.02 " Theory of Control and Optimization of Continuous Systems and Discreet Processes "

[2)] Institute of Engineering Cybernetics, The Wroclaw Technical University, Wroclaw, POLAND.

a). <u>equations of the continuity of the gas flow:</u>

$$\sum_{j\in S(t)} Q_j = q_t \quad ; \; t=1,2,\ldots,w \tag{2}$$

where: Q_j - the quantity of the gas flow through the j-th arc,

q_t - the gas consumption in t-th node,

$\sum_{j\in S(t)}$ - an algebraic sum for all the arcs connected with j-th node,

b). <u>path equations:</u>

$$\sum_{j\in S(k)} \Delta p_j = 0 \quad ; \; k=1,2,\ldots,l-w+1 \tag{3}$$

where: Δp_j - the pressures difference between two ends of the j-th arc,

$\sum_{j\in S(k)}$ - algebraic sum for all the arcs belonged to the k-th loop.

c). <u>Herming's formula</u> [2]

$$(p_{2j})^2 - (p_{1j})^2 = k_j Z_j Q_j^2 \; (p, T) \quad ; \; j=1,2,\ldots,l \tag{4}$$

where:

p_{2j}, p_{1j} - pressures in the initial and final nodes of the j-th arc, respectively,

k_j - resistance coefficient of the j-th arc.

The relations (2), (3) are equivalent to 1-st and 2-nd Kirchhoff's law for nonlinear networks, while the formula (4) defines the relations between the quantity of the gas flowing through the pipeline. There are also taken into consideration the varying pipeline resistance dependent on gas compressibility. This coefficient Z(p,T) appearing in the formula (4) has been proposed by Redlich & Kwong and the method of its evaluating has been described in the paper [4].

Let us consider the pipeline of l-th lengths and k-th resistance coefficient. Let us denote as follows: p^* - the pressure at input node, Q - the flow in the considered pipeline, x - the distance from the pipeline beginning. Now we can define, according to Herming's formula (4) the pressure value in the pipeline at point x :

$$p^2(x) = p_*^2 - x \frac{kZ}{l} Q^2 = (A - Bx) \tag{5}$$

Based on the Boyle-Marriotte equation as well as the formula (5) the quantity of the gas accumulated within the pipeline is given as follows:

$$N = \frac{d}{4RT} \int_0^l p(x)\, dx = \frac{2C}{3B} (Bx - A)^{-3/2} \Bigg|_0^l \tag{6}$$

The algebraic transformation of the formula (6) allows us to obtain the following expression:

the following expression:

$$N = \frac{2V}{3RTkZ}[p_* + (p^*)^2/(p^* + p_*)] = \frac{V}{RTkZ} \cdot p_r \qquad (7)$$

where: $p^* = p(1)$,
V - the capacity of the pipeline,
p_r - mean working pressure in the pipeline.

The quantity (7) of the gas accumulated in the pipeline of a V volume with the mean working gas pressure p , enables us to determine the total amount of the gas accumulated in the network:

$$N_C = \sum_{t=1}^{w} p_t \cdot V_t \qquad (8)$$

The time dependent changes of the quantity (8) should be taken into account in the balance equations (2). This one results from changes of pressures in network pipelines. The solution method of the above problem will be described and the influence of some simplifications on the accuracy of the obtained solution will be discussed below.

3. Description of the method.

Now we describe the proposed high pressure gas network simulation method and finally present the manner on which the equations (7)-(8) have been included into the balance equations (2).

Assuming that the approximated mean working pressure p_R is expressed by the formula:

$$p_R = 0.5 (p_* + p^*) \qquad (9)$$

and basing on the formula (7) we obtain:

$$N = \frac{1}{RTkZ} (p_* V/2 + p^* V/2) \qquad (10)$$

In the last formula the parameter V is a steady one hence the quantity of gas accumulated in the pipeline depends on the pressures on its nodes. Therefore a total quantity of gas accumulated in the all pipelines of the network is related to the pressures in their nodes according to the formula:

$$N_C = \sum_{t=1}^{w} p_t V_t/2 = \sum_{t=1}^{w} p_t \cdot \sum_{j \in S(t)} V_j /2 \qquad (11)$$

In general case the node pressure vector $p=[p_1,p_2,\dots,p_w]$ is time varied. Therefore the system of equations (2)-(4) must be solved many times.

Let us assume that the whole period of time $[0,T]$ can be divided into intervals $[0,T_1],[T_1,T_2],\dots[T_{v-1},T_v=T]$ in such a manner that the values of the pressure are constant in each of this subperiods. Then the simulation problem of high pressure gas pipeline network in the $[0,T]$ period of time can be formulated as follows:

a) for the first interval $[0,T_1]$ we should solve the system of equations (2)-(4) and determine the vector of node pressures p,

b) for each of intervals $[T_1,T_2],\dots,[T_{v-1},T_v]$ we should solve the system of equations (2)-(4) assuming that in the formula (2):

$$q_t' = q_t + D(p_t^{m-1} - p_t^m)\, V_t/2 \quad ; \quad t=1,2,\dots,w \quad m=1,2,\dots,v \ ; \ D\in R \tag{12}$$

and we should determine the vector p .

4. The final remarks.

It is necessary to justify two following aspects strictly connected with proposed above method:

- in which conditions the mean working pressure (8) can be approximated using the formula (9),
- in which manner it is possible to determine the value q_t of the flow (12) because of the fact that its value should be obtain concurrently with computing the elements of vector p.

From the carried out numerical experiments results that if the differences between pressures p_r, p_R does not exceed 30% of p_* value, then the accuracy of the mean working pressure approximation (determined by the formula (7)) using equation (9) is better than 0.9% of the p_* value.

The solution algorithm of equation system (2)-(4), which takes into account the formula (12) is presented in our paper [3].

5. References.

[1] Cross H.,Analysis of flow in networks of conduits or conductors, Bulletin University of Illinois Engng. Exp. Stat., No. 286, 1936.
[2] Herming F.;Grundlagen und Praxis der Durchflussmessung, VDI- Verlag, Duesseldorf 1967.
[3] Klempous R., Nikodem J., Ulasiewicz J., Simulation of the flow in the gas network with accumulation, Proceedings of the II Conference on theory of control and optimization of continuous systems and discreet processes, Kazimierz Dolny, 1988.
[4] Nikodem J., Ulasiewicz J., Statical flow optimization in nonlinear gas network based on modified Cross method, Journal of Gas, Water & Environment Engineering, No. 8-9, Warsaw 1987.
[5] Osiadacz A., Gas network simulation methods, Proceedings of the II Conference on Industrial application of computers, Szczecin 1981.

CHOSEN PROBLEMS OF AIR TRAFFIC CONTROL SYSTEMS MODELLING

Andrzej Urbanik Ph.D [1)]

Air traffic control systems /ATC systems/ belong to the class of complex systems with hierarchical structure of control. ATC system consists of two main parts : Air Traffic block /AT block/ representing air traffic carried in the air space and Air Traffic Control block /ATC block/ which includes everything which concerns air traffic control. Air traffic is represented by a set of aircrafts moving in limited air space. Each air-craft is characterized by its specific technical characteristic. The structure of the limited air space is specified by adequate international law regulatins. Air traffic constitutes controlled plant. The control in ATC systems depends on controlling of the current traffic situation and making up appropriate decisions aiming at safety of all air flights carried on in the space being under control. Process of the air traffic control consists of the following partial processes :

1. calculation of optimal flight trajectory, which is possible using aircraft technical characteristics and taking into account cost of fuel used by airplane during its flight in specific conditions;
2. detection of conflict situation in the air space, such conflict situations can be determined as situations in which any reaction of air traffic controller is necessary to avoid dangerous situation i.e. breaking of separations between different airplanes;
3. analysis and solving of conflict situations i.e. issuing of control decisions in order to solve existing conflicts;
4. elaboration of control decisions and transmitting of these decision to the aircraft captain.

Problems mentioned in points 2 and 3 have fundamental significance for the researched problem. That is why special, original method of detection, analysis and solving of the conflict situations has been elaborated. Proposed method consists in that safety management of air traffic is equivalent to finding of admissible allocation of flight levels and aircraft velocities. Notion of allocation used in the theory of graphs and networks has been applied. Such allocation problem has been stated. It has become evident that application of any previously published algorithm is in practice impossible, so new original algorithm, partly heuristic, has been worked out. Defining of the quality index was one of the most complicated tasks in the presented problem. Proposed algorithm was implemented and tested. Completely positive results were obtained so that algorithm has been used in the simulation model of ATC system.

When modelling ATC system firstly one has to identify modelled air traffic control system. Precise identification of ATC system has been performed. Simulation process of air traffic can be understood as servicing of aircraft flow according to the appropriate international regulations. Appearance generates specific control decision elaborated by ATC system. These decisions are dependent on the current situation in the controlled air space.

Simulation model of air traffic in the controlled air traffic area /FIR/ has been prepared. The simulation model has been implemented in Proffesional Fortran on personal microcomputer compatible with IBM AT.

Bibliography

[1] Gładyś S., Malarski M., Urbanik A. : Criteria for Evaluation of Quality of Air Traffic. in The Proceedings of the Fourth Scientific Conference "Science and Practice in Transport", Warsaw 1985.

[2] Korzan B. : Elements of the Theory of Graphs and Networks. Methods and Applications. WNT, Warsaw 1978, in Polish.

[3] Skorupski J., Urbanik A. : Method of Flight Levels Allocation Aiming at the Solution of the Conflict Situations in the Limited Air Space. to be published in the Research Works of the Institute of Transport, Warsaw University of Technology. Warsaw 1988.

---

[1] Institute of Transport
Warsaw University of Technology
Koszykowa 75, 00 662 Warsaw
Poland

DESIGN OF MULTIVARIABLE CONTROL LOOPS FOR DISTILLATION COLUMNS USING DIGITAL SIMULATON

M.Wagner,Dr.2) ; Ch.Hille,Dr.1)

Abstract:
Building a dynamic nonlinear model of a distillation column using sub-models for the equipment of chemical technological processes. The sub-models are envolved in a library of a digital simulation system for continuous processes. Simulation experiments with the nonlinear dynamic model for to find out the most suitable structure of the multivariable control loops. Derivation of a linearised dynamic model for the use of a program for design linearised multivariable control loops. Realization of the control loops with the TDC 2000 control system.

1. Description of the problems

The distillation column separates a binary mixture of solvents by demand of a high purity (0.9995) for bottom and overhead products. The column is a component of a polyacrylnitril - fibre plant.
During the startup of the plant there were some difficulties with the stabilization of the temperature and the concentration profiles along the column using the designed usual control loops. One reason of these problems is the heat feedback beetween bottom product flow rate and feed flow rate.
In the time of startup the column it was required to design multivariable control loops in a short time to stabilize the column process. The design took place by using a program of digital simulation continuous processes. The technological flowsheet of the column process and the structur of the calculated multivariable control loops as results from the use of digital simulation are represented in figure 1.

2. Model building

The simulation system for continuous processes DIGSIM /1/ envolves a library of nonlinear dynamic submodels for chemical engineering equipment and automatic control devices. These submodells consist of ordinary differential equations and algebraic equations.
The nonlinear dynamic model of the column with 32 practical plates was built using the following types of dynamic submodels:
. plate for binary mixtures with enthalpy balances and with or without side streams as material or enthalpy streams
liquid level changes in the bottom
reboiler with control valve
types of controllers and transfer functions

1) VEB Chemiefaserwerk "Friedrich Engels" 1832 Premnitz
2) VEB Chemieanlagenbaukombinat Leipzig - Grimma
Betrieb Anlagenbau, 7010 Leipzig

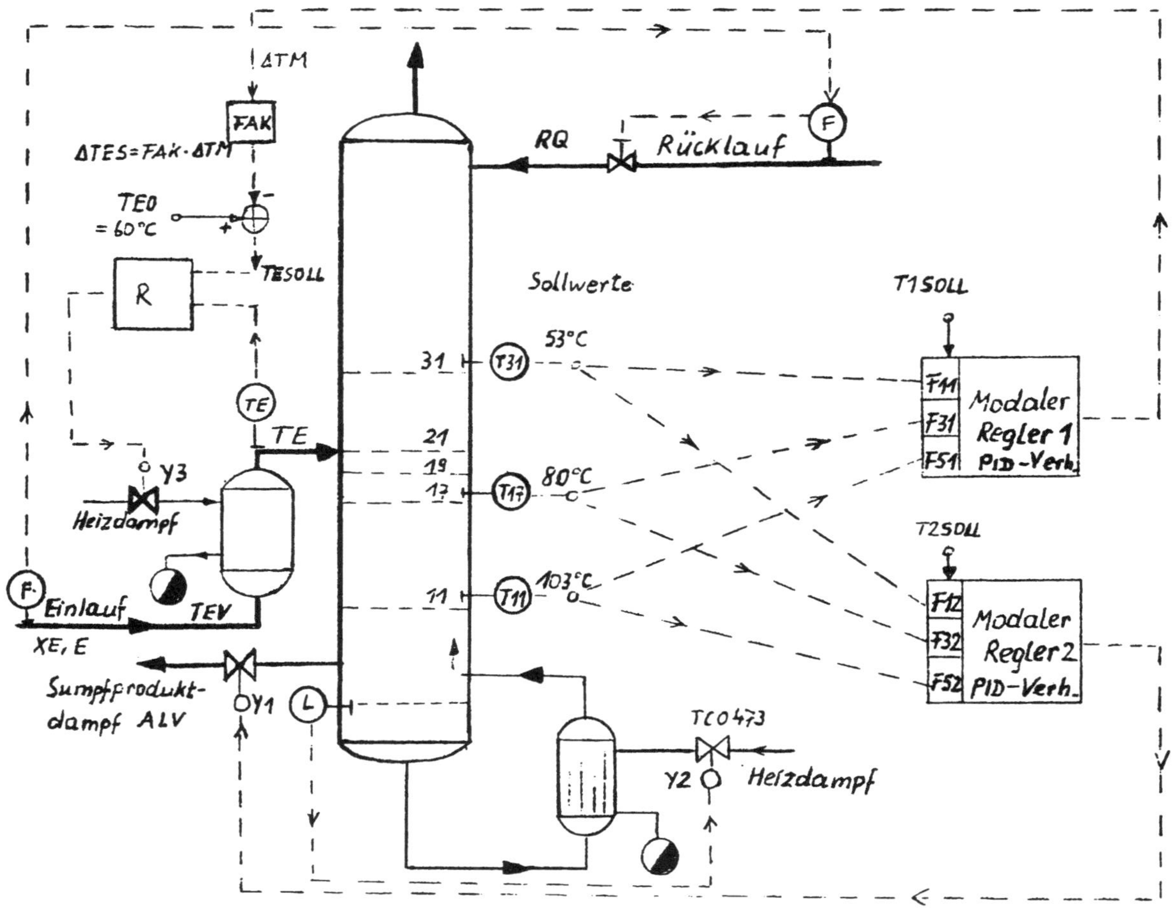

Figure 1
Technological flowsheet and multivariable Control loops

The plate submodels involve ordinary differential equations for material balances and algebraic equations for enthalpy balances and for the equilibrium conditions beetween liquid and vapor phases /2/.

3. Simulation experiments

A great number of simulation runs was necessary for the lot of possibilities of control structurs to calculate the most suitable structures of the multivariable control loops.
These structures are characterised by the position of the temperature measurement points, the weighting factors for summation, and the controller parameters.

The number of variants can be reduced by study the model step responses with reference to changes in feed flow rate, heating steam flow rate and others.
In this way it was possible to find out the most suitable control variables and positions of temperature measurement points to create multivariable control loops.
A linearised dynamic model of the column follows from the model step responses at the working point. This linearised model in the form of state variable equations was applied as input data for a program to design linearised multivariable control loops /3/. This is a second way to calculate the most suitable multivariable control loops.

4. Startup of the multivariable control loops

The control algorithm were realized with the TDC 2000 control system using the involved special controller algorithm and the real time programming language SOPL for to write down the connections beetween the temperature measurement points and the controller inputs.
There are some problems during realization too; e.g. the real transfer factors of the controlling valves had to be determind experimentally at the plant, for to correct the amplification factors of the controllers. These factors were calculated with digital simulation at assumption of a linear behaviour of the controlling valves.
The control system could be realized in a short time of some weeks using digital simulation only, which guarantees a stable run of the column process.

5. References

/1/ Anwenderinformation DIGSIM
VEB CLG, Betrieb Anlagenbau Leipzig , 1979

/2/ M.F.Doherty, J.D.Perkins:
On the Dynamics of Distillation Processes
Chem. Eng.Science 37 (1982), S.381-392

/3/ Dokumentation zum Programmsystem IMGRE
(Entwurf und Simulation von linearen Mehrgroessenregelungen)
Zentralinstitut fuer Kybernetik und Informationsprozesse
(Institutsteil Dresden) 1984

AN INTEGRAL LOOK AT AUTOMATION FUNCTIONS IN MODERN PRODUCTION SYSTEMS

H. Ehrlich[1)]

1. Introduction

Today's production systems are continually being upgraded as regards productivity, added-value content and efficiency, with the accent increasingly on cutting costs, saving materials, energy and human effort, and using ecologically sound methods with a minimum of waste. The extraordinary capital outlays involved require that plants be used to maximum capacity, while greater demands are being made on mobility and flexibility. There is not only a chance but also an urgent need to solve these problems by utilizing microelectronics and informatics to the largest possible extent and optimizing automation strategies.

2. Automation of continuous processes

New qualitative and quantitative requirements must be met by production control systems, one of these being an integrated approach to partial functions /1/. Integrated automation systems are fairly large and must therefore be viewed in their entirety. If they are to provide flexible production control, they must themselves be flexible, which generally makes them time-variable or non-stationary systems.

The following is a closer look at some aspects of automating production processes /2/, /3/ which involve the conversion of materials and energy and yield a product on a continuous basis. This is in contrast to discontinuous processes which only yield a product at the end of a cycle.

Conventional automation of continuous production emphasizes working point stabilization and the suppression of disturbance variables by way of closed-loop control, while the overriding principle in discontinuous production is logic (run-off) control.

The specific differences between these two approaches tend to become less conspicuous with growing levels of integration in the automation of

- start-up, shut-down and reversing operations and safety functions in continuous production, and of
- closed-loop control and arithmetic processing functions in discontinuous production.

1) Prof. Herbert Ehrlich, Technische Hochschule Leipzig, Sektion Automatisierungsanlagen

This trend is also reflected in the available range of automation equipment, where process controllers and memory-programmable control systems are increasingly designed for universal application. This versatility is further enhanced by bus link-ups using standard interfaces.

There is a growing demand for flexibility also in continuous processes and their control systems to enable the following operations:

- reversals made necessary by different load stages, inputs and initial conditions;
- start-up and shut-down control for changeover and adjustment to new production conditions, and for maintenance, repairs and breakdowns; and
- control action to deal with various faults (that can not be corrected) so that normal operation is restored (process reliability).

Diagnostic aids for the process and plant condition, and for determining trends and the flexible control steps required, are important instruments to ensure reliable operation /2/.

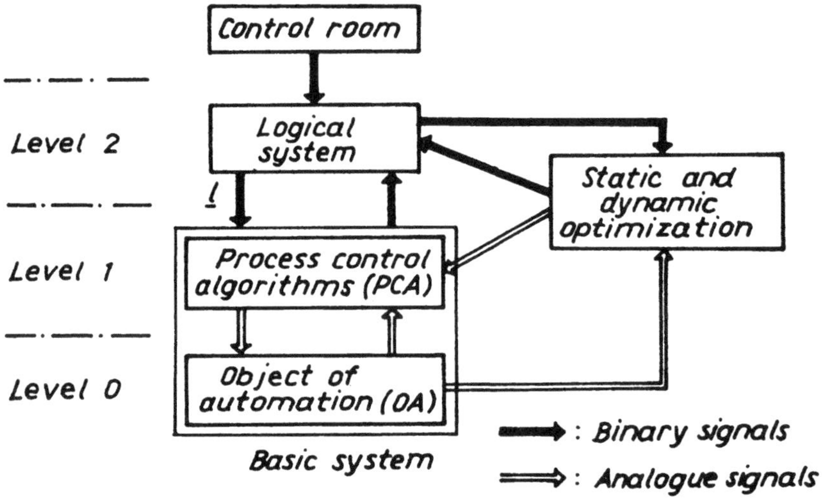

Fig. 1: Integrated automation system

Flexible control and automation not only permits but at the same time requires the integration of all important modes and their interaction over a certain time span (cycle, time between two standstills, life cycle) /3/. This is the second dimension in considering integrated automation systems.

When it is assumed that both the object of control and its algorithms are modular, then flexibility is mostly structural in nature, and the master logic must provide structural control in response to particular needs and situations. Fig. 1 is the aggregate version of a crude schema showing an integrated automation system.

3. Models of dynamic systems having controllable structures

Systems consisting of two interacting partial systems, one dynamic and the other logical, are classified as logical/dynamic systems /4/.

The following discussion is limited to the dynamic (basic) system and the structural relations resulting from the outputs of the logical system (vector $\underline{l}$, see Fig. 1). It is assumed that,

1. the elements are linear and not time-variable; and
2. the behavior of the elements is described in a continuous state space.

Fig. 2 shows the elements and basic structural relations.

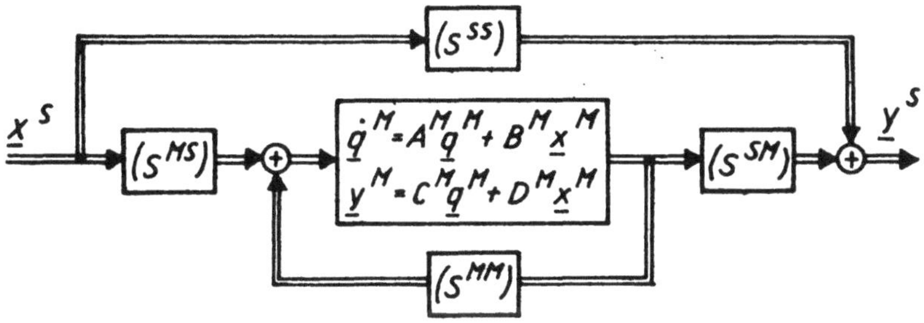

Fig. 2: Signal flow diagram of a structure-variable system

Based on the state space representation of the elements

$$\dot{\underline{q}}_i = A_i \, \underline{q}_i + B_i \, \underline{x}_i; \quad (1) \qquad \underline{y}_i = C_i \, \underline{q}_i + D_i \, \underline{x}_i; \quad (2)$$

which, in compact form, is

$$\dot{\underline{q}}^M = A^M \, \underline{q}^M + B^M \, \underline{x}^M; \quad (3) \qquad \underline{y}^M = C^M \, \underline{q}^M + D^M \, \underline{x}^M; \quad (4)$$

where

$$A^M = \text{diag } A_i \; ; \qquad \underline{q}^{M'} = (\underline{q}_1', \underline{q}_2', \ldots, \underline{q}_m');$$

$$B^M = \text{diag } B_i \; ; \qquad \underline{x}^{M'} = (\underline{x}_1', \underline{x}_2', \ldots, \underline{x}_m');$$

$$C^M = \text{diag } C_i \; ; \qquad \underline{y}^{M'} = (\underline{y}_1', \underline{y}_2', \ldots, \underline{y}_m');$$

$$D^M = \text{diag } D_i \; ;$$

the model of the complete system results acc. to Fig. 2 as

$$\begin{aligned} \dot{\underline{q}}^S &= \left\{A^M + B^M(\underline{I} - S^{MM}D^M)^{-1} S^{MM} C^M\right\} \underline{q}^S + \left\{B^M(\underline{I} - S^{MM}D^M)^{-1} S^{MS}\right\} \underline{x}^S \\ &= \qquad A^S \qquad \underline{q}^S + \qquad B^S \qquad \underline{x}^S \end{aligned} \tag{5}$$

$$\begin{aligned} y^S &= \left\{S^{SM}(\underline{I} - D^M S^{MM})^{-1}\right\} \underline{q}^S + \left\{S^{SM}(\underline{I} - D^M S^{MM})^{-1} D^M S^{MS} + S^{SS}\right\} \underline{x}^S \\ &= \qquad C^S \qquad \underline{q}^S + \qquad D^S \qquad \underline{x}^S \end{aligned} \tag{6}$$

where $\underline{q}^S = \underline{q}^M$.

The structural matrices for the signal flow diagram of a structure-variable system or the related signal flow graph can be easily found with the aid of a graphic scheme /6/. The elements of the matrices A^S, B^S, C^S and D^S change their numerical values as a function of $\underline{1}$ and are structure-variable.

4. References

/1/ Williams, Th. J.: The integrated approach to Industrial Control. 10th World Congress of the IFAC, Munich, July 1987

/2/ Ehrlich, H.; Ostapenko, Yu.: Erweiterte Steuerungskonzeptionen für kontinuierliche technologische Prozesse Chemische Technik ... (1988) manuscript submitted for publication)

/3/ Ehrlich, H.: Steuerungskonzepte zur Automatisierung von An- und Abfahrvorgängen sowie von Umsteuerungsprozessen 16th Autumn Course on Technical Cybernetics, Leipzig 1985 in: Wiss. Berichte der Technische Hochschule Leipzig, no. 15, 1985

/4/ Zhuk, K.D.; Timchenko, A.A.; Dolyenko, T.I.: Issledovaniye struktur i modelirovaniye logiko-dinamicheskich sistem Naukova dumka publishers, Kiev 1975

/5/ Ehrlich, H.: Beschreibung und Analyse von dynamischen Systemen mit steuerbarer Struktur (in Russian) 7th Internat. Conference on Control Systems and Computer Science, Bucharest, May 1987 (Preprints)

/6/ Ehrlich, H.; Fiedler, K.: Logisch-dynamische Systeme zur Struktursteuerung großer Antomatisierungssysteme msr 31 (1988) (manuscript submitted for publication)

DYNAMIC SIMULATION OF CORN GROWTH

Z. Dunkel[1] and M. Hunkár[2]

There are several aspirations to model the soil-plant-atmosphere system. They are different efforts according to the order of magnitude of resolution of physical and physiological processes and requirements of computer technique. Our main purpose was to build up a corn model using a personal computer. Because of the importance of maize /Zea mays L./ in Hungary the model was fitted and tested by maize data /c.v. SZTC-255/ derived from the Agrometeorological Research Station at Keszthely /46°44'N; 17°15'E/. The model is based on the Sirotenko model [1], but it is a simpler one. The structure of the model can be seen in Fig. 1. The time

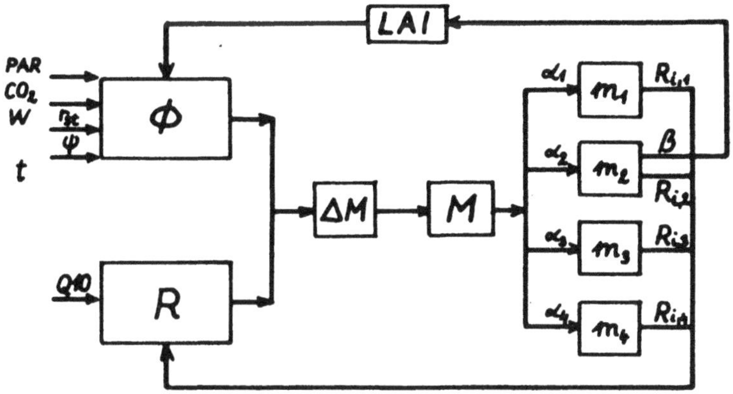

Fig. 1. The structure of the dynamic model

unit of our model is one day. The photosynthesis and the respiration were calculated every day and the plant growth was considered as sum of these two terms. The rate of phosynthesis was described by means of the Chartier formula

$$\Phi^j = \frac{\tau^j \psi_T^j L^j}{2(r^j - r_c)} \left[C + \alpha I^j r^j - \sqrt{(\alpha I^j r^j - C)^2 + 4 r_c I^j C \alpha} \right] \quad (1)$$

1) Central Meteorological Institute, Budapest P.O.Box 38
H-1525 Hungary

2) Agrometeorological Research Station, Keszthely P.O.Box 80
H-8361 Hungary

where

τ^j is the length of the photo-period, [s]

ψ_T^j is the photosynthesis-temperature efficiency function, the curve of this function is showed in Fig. 2.

L^j is the surface of photosynthesizing organs, in our model it's only the leaf area index /LAI/, $[m^2\ m^{-2}]$.

r^j is the leaf resistance, it is taken as sum of three terms, namely the stomatal, the mesophyll and the so called carboxylation resistance. The mesophyll and the carboxylation resistance are supposed constant during the vegetation period, $[s\ m^{-1}]$.

r_c is the carboxylation resistance, $[s\ m^{-1}]$.

C is the carbon dioxide concentration, $[kg\ m^{-3}]$.

α is the initial light use efficiency for fixing CO_2 in leaves, in C_3 plants $1.1\cdot10^{-8}$, in C_4 plants $1.4\cdot10^{-8}$, $[kg\ J^{-1} = kg\ W^{-1}s^{-1}]$.

I^j is the photosynthetically active radiation, PAR, $[W\ m^{-2}]$.

The respiration is considered the sum of two terms, namely the maintenance and the growth respiration. Both terms contain the respiration rate of every organ. The growth respiration depends on the growth rate. The total respiration can be written in this form

$$R = \sum_{i=1}^{4} R_{m,i}\, m_i + \sum_{i=1}^{4} R_{g,i}\, \Delta m_i \qquad (2)$$

where the 1th term is the maintenance respiration. The factors are independent of time, they are supposed constant. The values are given in [2].

The environmental conditions like photosynthetically active radiation and carbon dioxide concentration - are expressed directly in the photosynthesis formula. The influence of the air temperature is considered using efficiency functions, ψ_T^j for the phosythesis is shown in Fig. 2. and the well-known Q10 function for the respiration. The only independent variable is expressed indirectly is the water supply. It is taken into account in the water supply. It is taken into account in the stomatal resistance,

$$r_{st} = r_{min}\, exp(a_1 \psi_s^j) \qquad (3)$$

where ψ_s^j is the water potential, a_1 empirical constant. We haven't been able to measure the water potential directly, so we had to calculate as it was given by Sirotenko [3],

$$\psi_s^j = b_1\, exp\left(b_2 \frac{W_i^j - W_i^{WP}}{W_i^{FC} - W_i^{WP}}\right) \qquad (4)$$

where

W_i^j is the actual water content of the soil layer was examined, [mm]

W_i^{fc} is the field water capacity of this layer, [mm]

W_i^{wp} is the wilting point value in water [mm],

b_1, b_2 are constant, depending on the type of soil.

The change of soil water content can be described using a submodel which calculates the evapotranspiration of plant canopy or we can give its values directly by time series.

In our model realization not only the soil water content was taken from experimental data series but the carbon dioxide concentration and the photosynthetically active radiation [4] values on the maximum photossynthesis rate level too [5].

The model simulates the time course of the dry matter production of maize canopy from the emergence till maturity in dependence of environmental factors as it was shown in Eq. (1), (2) and (3). The total dry weight comes from the difference of phosythesis and respiration. The increase in the total dry weight, in kgm^{-2} day^{-1}, of the crop is partitioned over the plant organs: leaves, stems, roots and storage organ. The ratio of the weight of individual organs were determined experimentally [6]. The partitioning factors are introduced as a function of degree-day. In our model the biological minimum temperature is $10^{o}C$. The curve of the partitioning factors - growing functions - is shown in Fig. 3. The increase in photosynthesizing surface, i.e. green area of the leaves, the stem and finally the storage organ, follows directly from the growth rate of the organs. According to our experimental measurements the photosynthesizing area - dry mass of the organs is not constant. The Fig. 4. shows the curve of this function.

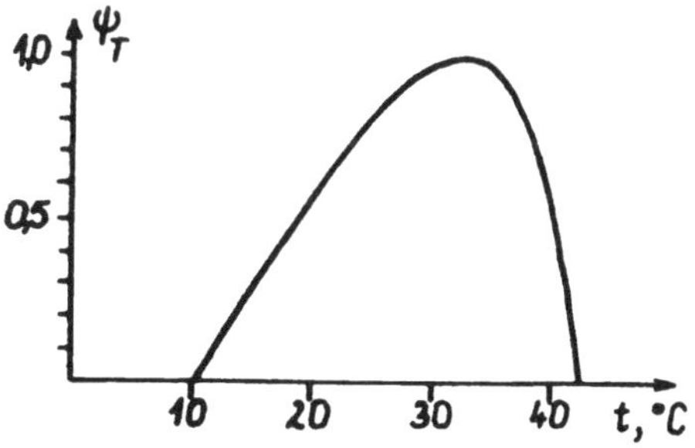

Fig. 2. Photosynthesis-temperature function

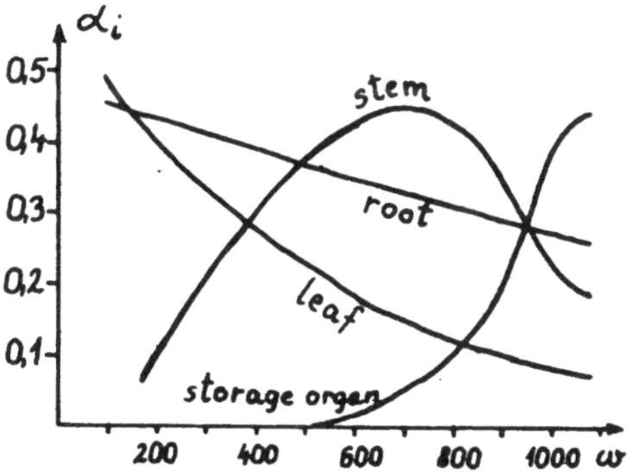

Fig. 3. The growing functions

As an illustration of the computed results of our model, the cumulative crop production and the calculated leaf area index are shown in Fig. 5. Few experiments were made changing the environmental factors in the model.

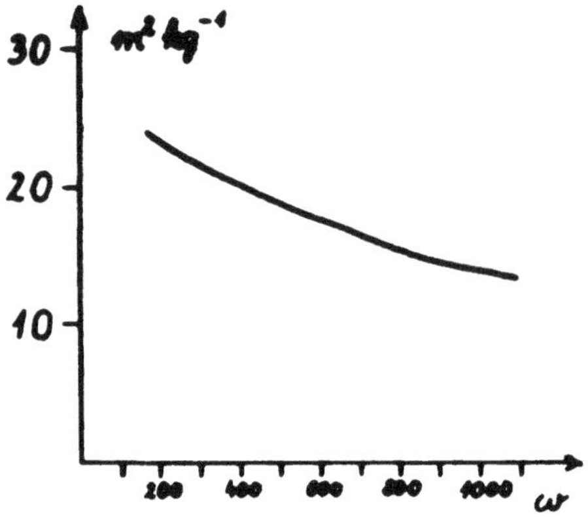

Fig. 4. Leaf area-leaf mass function

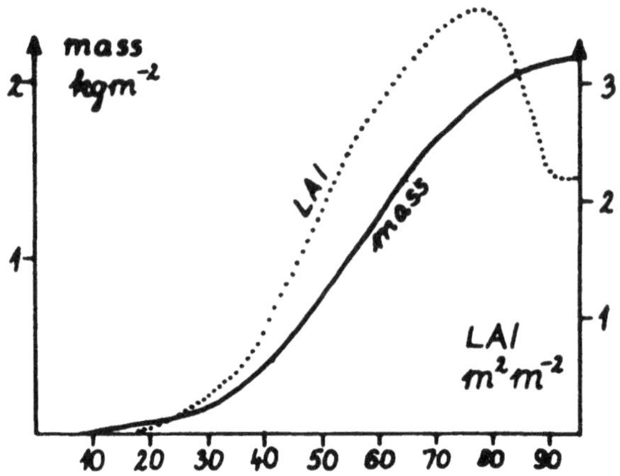

Fig. 5. The time course of the calculated total dry mass and leaf area index

For example substituting the carbon dioxide concentration for a constant value we have got a smaller gross production when this average was the climatic average of measured data. Further research on the applied variables and parameters is needed. Our model have to be further tested and if it is necessary adjusted. Knowledge of the parameters and the interaction of parameters have great importance for planning investigations and experimental projects.

LITERATURE

[1] Sirotenko, O.D., Boyko, A.P., 1975: On construction of a closed system of equations describing energy and mass exchange aimed at calculating the biomass of crops. /in Russian/ Meteorologija i Gidrologija, 78-87.

[2] Penning de Vries, F. W., van Laar, H.H., 1982: Simulation of plant growth and crop production. 308 p. Wageningen.

[3] Sirotenko, O.D., 1981: Mathematical Modelling of Moisture and Heat Regime and Productivity of Agricultural Ecosystem. 176 p. Leningrad.

[4] Hunkár, M., 1984: Simple calculation of the vertical profile of average PAR flux density within a maize stand. Időjárás 88, 139-144.

[5] Dunkel, Z., 1987: Measurements and estimates of CO_2 and PAR flux in and above maize canopy. Vortragstagung Agr. Met. Halle/Saale/ 1987/17 /S60/

[6] Hunkár, M., 1986: Growing functions used in the dynamical model of plant development. /in Hungarian/. Rep. on Scient. Res. in Hungary. 111-120.

STUDIES ON THE ADAPTATION OF A WINTER WHEAT MODEL

Hunkár, M.[1/] and Zemankovics, F.[2/]

Abstract

The TRITSIM model was tested under Hungarian climatic condition. Winter wheat above ground biomass was simulated using measured and evaluated environmental variables. The results were compared to field experiments.
Keywords: winter wheat, simulation, environmental variables.

Introduction

The SONCHES simulation system and the models ofwinter wheat agroecosystem (AGROSIM-W) were developed by the Central Institute of Cybernetics and Information Processes of Academy of Sciences of the GDR and several other institutes of the GDR. The system and models were performed at the 2nd International Symposium on Systems Analysis and Simulation in 1985 [3], [4], [6]. These software tools were installed at the University of Agricultural Sciences of Keszthely in Hungary during 1986.
The TRITSIM model which is a part of the AGROSIM-W winter wheat agroecosystem model describes the growth and yield formation of winter wheat. The present study investigated the possibility of the adaptation of the TRITSIM model to the Hungarian conditions.

Simulation experiments

The adaptability of TRISTSIM was investigated from two different aspects. At first creations of meteorological input tables were examined since the model requires special meteorological input variables. On the other hand after finding the suitable environmental inputs the model outputs were compared to field experiments. The field experiments were carried out by the Institute for Plant Breeding of University of Agricultural Sciences in winter wheat cv. MV-4 in 1985 and 1986.

The original model structure of TRITSIM has not been changed. Initial values and environmental variables were chosen according to local condition. Input tables of meteorological elements such as mean temperature of photoperiod (TP), mean temperature of dark period (TN), daily amount of photosynthetically active radiation (QP) and the quotient of the

---

1/ Agrometeorological Research Station, H-8361 Keszthely, P.O.B. 80. Hungary
2/ University of Agricultural Sciences Keszthely, H-8361 Keszthely, P.O.B. 71. Hungary

actual to potential evapotranspiration (EQ) were derived from special agrometeorological measurements carried out at the Agrometeorological Research Station. Temperature was recorded continously by thermograph in a Stevenson screen, photosynthetically active radiation was measured by LI-190SB quantum sensor, and EQ was determined by an empirical formula of Antal [1] in which available water content of the soil is the measured characteristic:

$$EQ = w \frac{w + b}{1 + b} \tag{1}$$

were w - the relative water content of the soil (it is measured every ten days), b - a plant factor, expresses the effect of transpiration on evapotranspiration ($0 \leq b \leq 1$) [5].

The daily values of EQ come from interpolation.

These meteorological variables are not available at any place but there is a common way in meteorology to derive required element from other meteorological elements which are available more frequently in time and space. Therefore TP and TN can be calculated from daily maximum, T_{max} and daily minimum, T_{min} temperatures. Star (*) means evaluated value.

$$TP^* = T_{max} - 1/4\ (T_{max} - T_{min}) \tag{2}$$

$$TN^* = T_{min} + 1/4\ (T_{max} - T_{min}) \tag{3}$$

$$QP^* = 0.4\ G \tag{4}$$

where G is the total incoming shortwave radiation (global radiation).

$$EQ^* = E_a^*/E_p^* \tag{5}$$

where E_a^* is the actual and E_p^* is the potential evapotranspiration [2].

$$E_a^* = k \frac{w_o + P}{\sqrt{1 + \frac{w_o + P}{E_p^*}}} \tag{6}$$

$$E_p^* = \frac{1 - f}{2 - f}\ T \tag{7}$$

where w_o is an initial value of the available water content of the upper 20 cm layer of the soil determined at the beginning of the given period (in the end of February in Hungary w_o usually is equal to the field capacity).

P - is precipitation, f - is the daily mean of air humidity, T is the daily mean temperature, k - is an empirical factor depending on soil type.

E_a^* and E_p^* are issued for ten days periods by Agrometeorological Information Service for the standard meteorological stations in Hungary. Evaluated EQ^* came from these data.

Discrepancies of the measured and evaluated values in the case of TP and TN show normal distributions, $\overline{TP}^*$ is a little bit higher than $\overline{TP}$, $\overline{TN}^*$ is a little bit lower than $\overline{TN}$ ($\overline{TP} - \overline{TP}^* = -0.72$; $\overline{TN} - \overline{TN}^* = +0.62$).
QP^* is mostly higher than QP but the discrepancies shape an $\cup$ distribution. The average of the evaluated EQ^* for the simulation period in 1986 is almost the same as the measured one ($\overline{EQ}^* = \overline{EQ}$). In 1985 the evaluated EQ was significantly higher than the measured one ($\overline{EQ}^* > \overline{EQ}$). The discrepancies in the daily values show uniform distribution.

Simulation experiments were carried out with the measured and evaluated meteorological inputs as well. The effect of replacing of individual meteorological elements by evaluated values was also examined in numerous simulation scenarios.

The weather in the growing seasons of 1985 and 1985 was quite different, especially in May. In 1985 there was cool and rather wet, but in 1986 there was a serious drought. Monthly precipitation and mean temperatures are shown in table 1.

Table 1.

The monthly mean temperature (t_k) and amount of precipitation (P) at Keszthely, Hungary

		April	May	June	July
1985	t_k °C	10.5	16.6	16.8	20.9
	P mm	54.9	79.7	76.2	81.0
1986	t_k °C	11.6	18.3	18.4	19.7
	P mm	54.1	32.6	79.4	42.2

From model outputs the total above ground biomass was taken into account because these data were available also from field experiments. Sequentially harvested and measured dry matter can be seen in figure 1. Simulated above ground biomass using measured meteorogogical inputs are shown in figure 2. The great difference between the two years is reflected in the curves. The effect of input evaluation can be seen in figure 3 a/ and b/ for 1986 and 1985 respectively. In 1985 evaluation of EQ resulted more biomass, but evaluation of TP and TN resulted less biomass than the simulated biomass with measured inputs. In 1986 all of evaluated inputs resulted less biomass than the simulated one with measured variables.The final differencies can be significant: ± 25 %. It means that the model is highly sensitive to the meteorological input variables. A little change (in average no more than 10 %) in inputs may cause great differences in the final production.

The crucial meteorologicalinput seems to be the EQ. In 1985 when EQ^* was higher than EQ in the majority of the days, the final production became the highest among the simulation experiments, and came near to field experiment. The applied formulas determining EQ seem to be too rigorous especially in the point of plant water stress. This refers to the fact that all simulated biomass (where EQ was calculated by Antal formula) were significantly less than the measured dry matter in both years. The Antal formula results water stress when relative water content of the soil falls bellow 0.86. It is well known that the optimum relative water content (w) is about 0.7, therefore water stress may occur when $w < 0.7$. A new formula had to be constructed assuming that in a bare soil the evaporation is potential only when it is saturated with water but in a developed plant stand the evapotranspiration remain potential until the available water content exceeds 0.8 of field capacity. This effect is expressed by a plant factor. The new formula based on the assumptions mentioned above is as follows:

$$EQ = (0.25b + 1)\,w \tag{8}$$

If EQ determined by (8) were more than 1.0; EQ = 1.0
Simulation experiments with the new EQ values gave much better results comparing to the field experiments. See figure 4.

Conclusions

The TRITSIM model seems to be adaptable to Hungarian conditions. Nevertheless it has to be assesed that the meteorological inputs have great importance. Evaluation of that variables may cause 20-30 per cent change in biomass production. Especially the role of EQ is expressed. Examinations pointed out that researches ought to be conducted to develop a more sophisticated method to get it.

References

(1) Antal, E.: A new method for computing evapotranspiration (in Hungarian). 1967, OMSZ. Hiv. Kiadv. XXXIV, Budapest, 414-423.

(2) Dunay, S.; Pósza, I.; Varga-Haszonits, Z.: A simple method for determination of actual evapotranspiration and the moisture content of soil. Part I. 1968,Meteorology and evaporation (in Hungarian). Öntözéses Gazdálkodás, VI. 2, 39-48.

(3) Ebert, W.: Modelling and simulation of agroecosystems - the winterwheat agroecosystem "AGROSIM-W" (1985). Systems Analysis and Simulation, Akademie-Verlag Berlin, 64-77.

(4) Matthäus, E.; Flechsig, M.; Bellmann, K. and Wenzel, V.: Use of task facilities of the simulator "SONCHES" for agroecosystem-model investigations and control (1985). Systems Analysis and Simulation 1985. II.,Akademie-Verlag Berlin, 133-140.

(5) Pózsa, I. and Stollár, A.: Plant constant evaluated by several years measurements for the calculation of effective evaporation (in Hungarian) 1983. Időjárás (87.) No. 3., 170-177.

(6) Wenzel, V.; Bellmann, K.; Matthäus, E. and Flechsig, M.: SONCHES - an interactive simulation language for design, validation and usage of ecological models (1985). Systems Analysis and Simulation 1985 II., Akademie-Verlag Berlin, 129-132.

Figure 1.

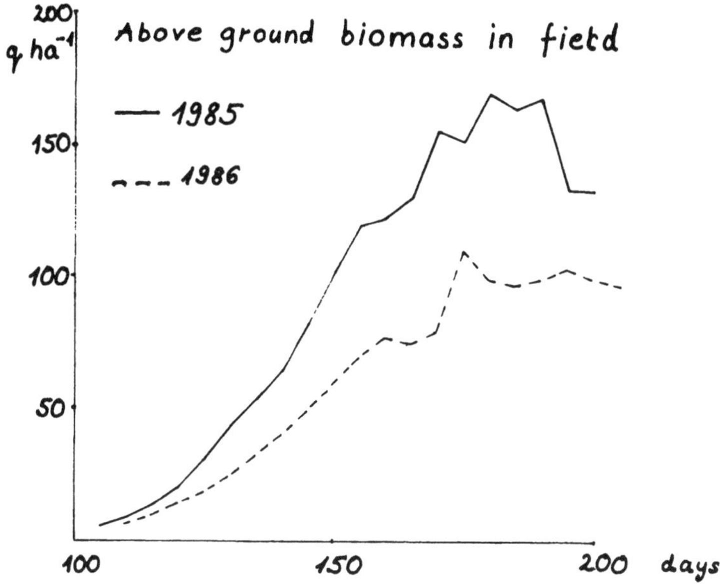

Figure 2.

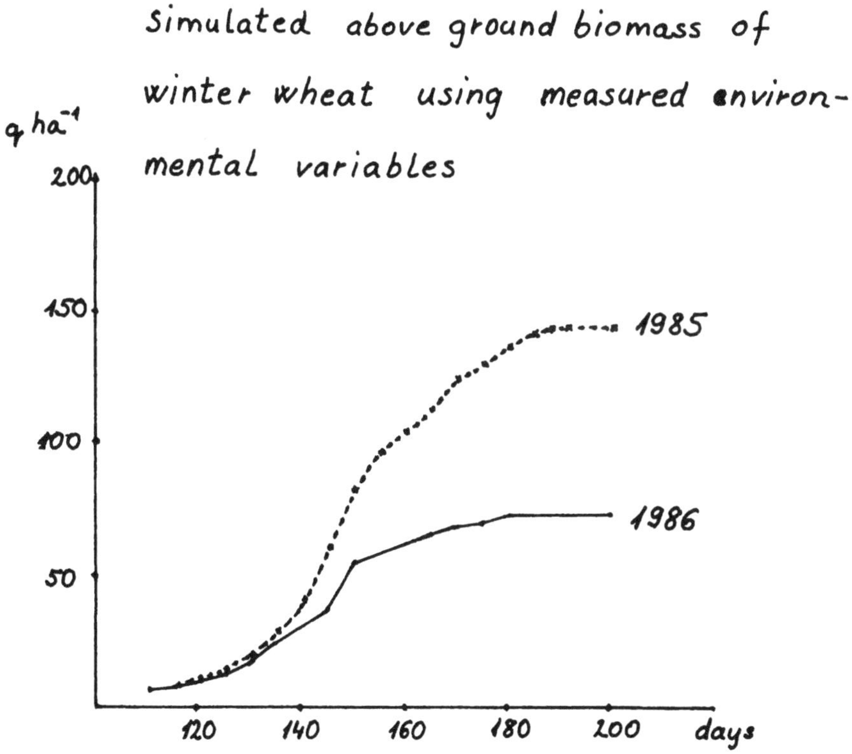

Figure 3.

Above ground biomass of winter wheat

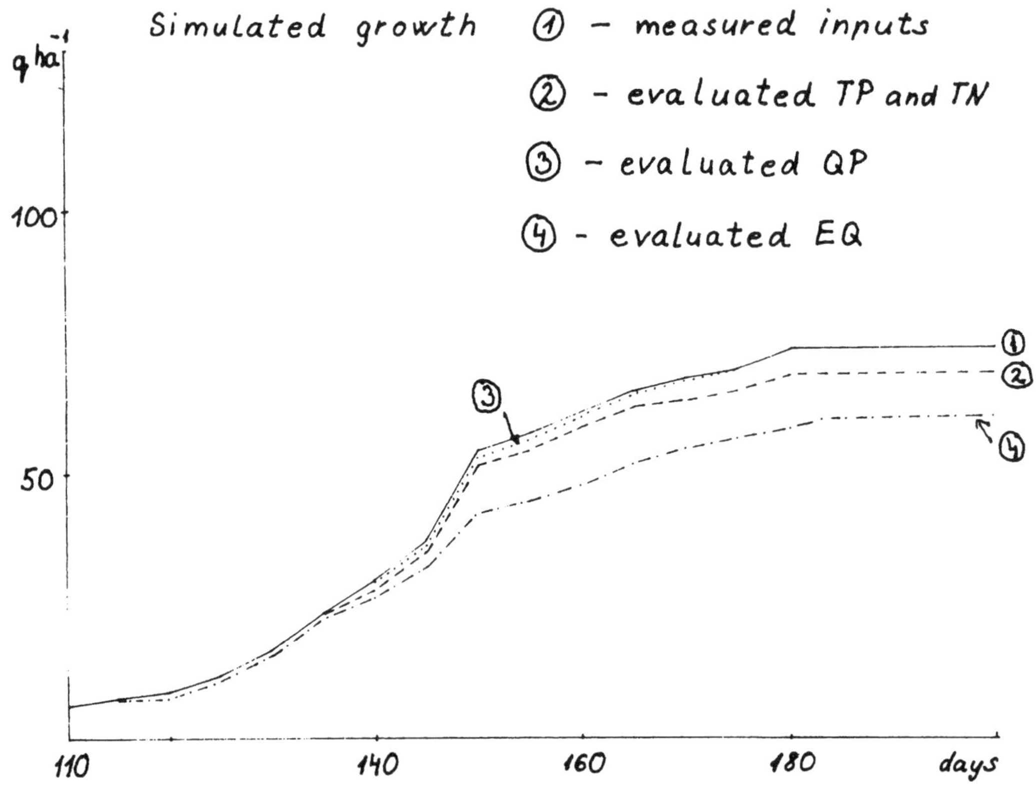

a) - 1986

Figure 3.

b) - 1985

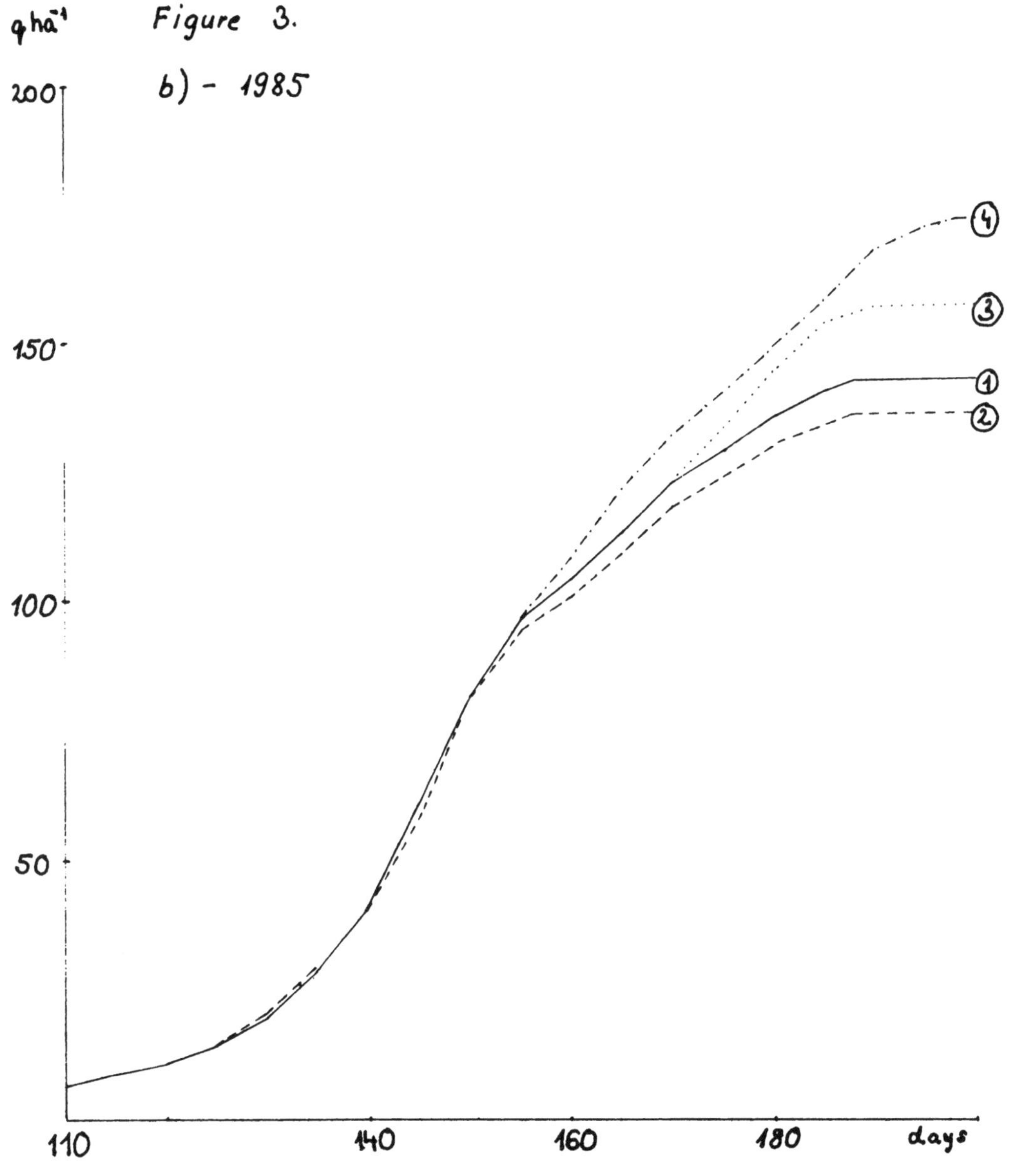

Figure 4. Above ground biomass of winter wheat

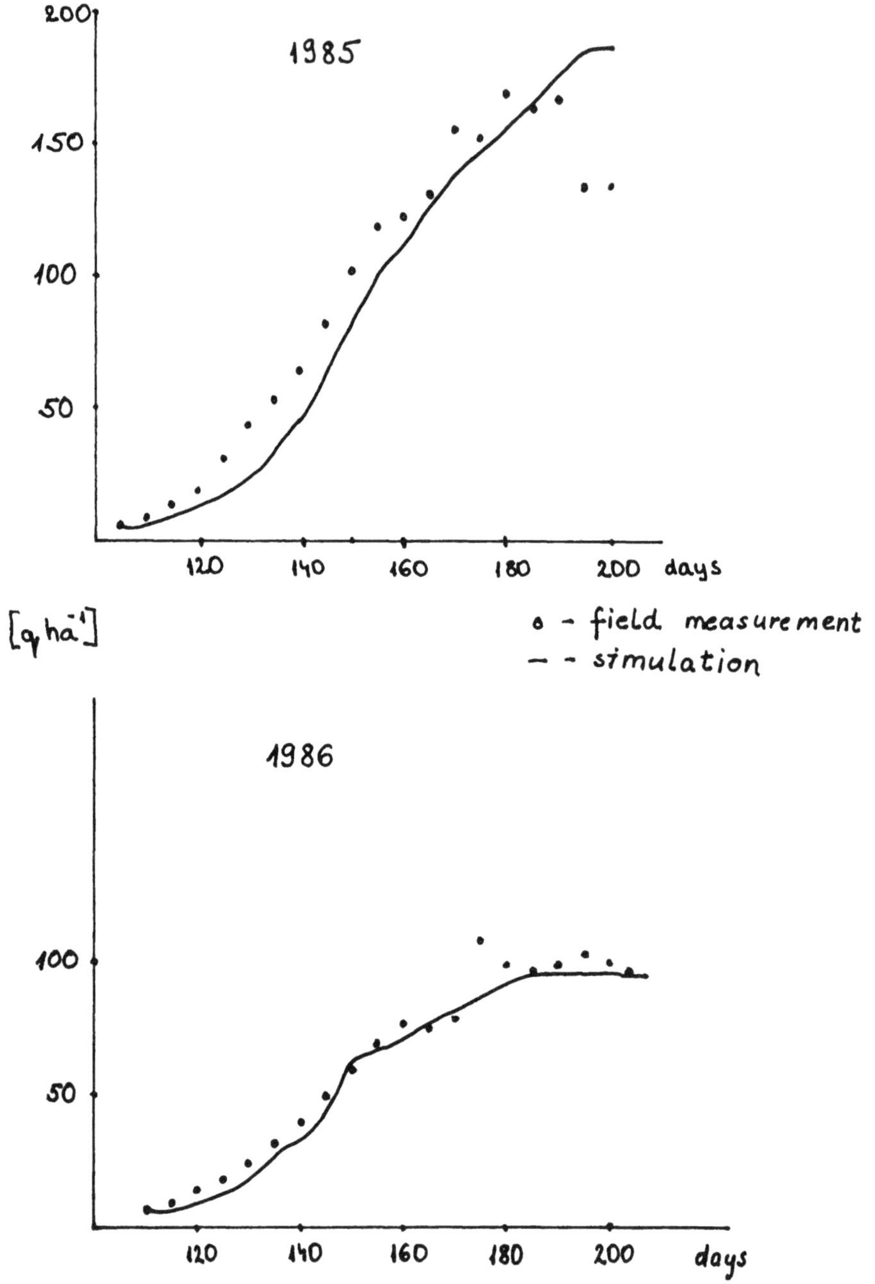

Modelling of Growth Processes in Broiler Production Using the Simulation System SONCHES

K. Grosskopf[1], *E. Matthäus*[2], *M. Flechsig*[2]

Abstract: The basic structure of a mathematical growth model for computer aided controlling of the broiler production process is described. The core of this model is represented by the mathematical description of the growth of the single animal under optimal environment. The whole model contains additionally the influence of the animal stock and climatic control. Model building and validation is carried out by means of SONCHES.

Keywords: Broiler production modelling and simulation, single animal model, stock model, climat model, use of simulation software

The development of a mathematical growth model is aimed at a software solution for computer aided control of the broiler production process. Such a model has to be based on the causal factor-product-relationship and interrelationships forming the yield and expenditure which are to be modelled in such a way that it will be possible to reflect (to simulate) the behaviour of the real system (production process) by means of a computer (yield programming). Figure 1 shows:

- the aims of growth modelling in broiler production
- the most essential factors influencing these aimes
- the interrelations of these factors

The material result of broiler production is usually measured by the live weight which is produced in a certain time interval T^* on a certain production area (Fig.5). It is determined by

- the live weight per chicken at slaughtering which depends on the daily live weight gain (<W[3]>) and on the length of the production period
- the number of chickens for slaughtering per unit of production area per T^* determined by the initial number of chickens per unit of production area, the number of production periods within T^* and by the losses of animals during the production period.

In order to reduce the complex real system to the essential relationships the following factors were selected from the enormous number of potential factors of influence:

(1) length of production period and/or ontogenetic age of the chicken
(2) sex of the chicken
(3) energy and protein content of a food ration fed ad libitum
(4) actual live weight per unit of production area as an expression of the stocking density
(5) temperature inside the poultry house
(6) relative air humidity inside the poultry house.

---

[1]Karl-Marx-University Leipzig, Section of Animal Production and Veterinary Medicine, Institute of Socialist Farm Management Schillerstrasse 6, Leipzig, DDR-7010

[2]Academy of Sciences of the GDR, Central Institute of Cybernetics and Information Processes, Kurstr.33, Berlin, DDR - 1080

[3]for SONCHES syntax see contribution of WENZEL in this proceedings

The analysis of the causal relationships between influencing factors among each other and between them and the aims has shown that the model subject can be devided into three subsystems which are interdependent and for which special criteria subordered to the aim of the whole model can be defined. These are:

- growth of a *single animal* expressed by <W
- development of the *stock* ignoring the influence of climatic conditions inside the poultry house with the losses of animal
- ensurance of suitable *climatic conditions* inside the poultry house corresponding to temperature and air humidity.

The mathematical description of the growth of the *single animal* dependent on the above mentioned influencing factors (1,2,3), i.e. under an optimal environment, represents the core of the simulation model of the broiler production process. Therefore a mathematical description of the genetically determined live weight gain (DW), protein retention ability (DP) and lipid retention characteristic (DL) in dependence on the ontogenetic age of the animal is required. It will be used for the determination of a daily food energy (DE) and food protein requirement (FPR):

$$DE = CEP_{norm} * DW * DPDW + DL * CEL + CEM * W^{0.75}$$

$$FPR = (DP + CPM * W^{0.667})/PCR$$

with DPDW - genetically determined protein retention per unit of <W

CEP_{norm} - energy requirement per unit of protein retention under the condition of a genetically determined growth

W - actual live weight per chicken

CEL - energy requirement per unit of lipid retention

CEM - energy requirement for maintenance per unit of metabolic W

CPM - protein requirement for maintenance per unit of metabolic W

PCR - protein conversion rate

These requirements are met by the daily food intake. Its prediction under the condition of ad libitum feeding is a central problem of growth modelling. Different control functions (Fig.2) can be used alternatively. An upper bound for daily food intake (MFI) caused by a limited intake capacity of the digestive tract and by its passage ability is involved:

XFOOD = DE/FEC - food intake, controlled only by energy requirement

or

XFOOD = Max(DE/FEC,FP/FPC) - food intake, controlled by energy as well as by protein requirement according to the limiting one

IF (XFOOD.GT.MFI) XFOOD = MFI

with FEC - energy content of the ration

FPC - crude protein content of the ration

Based on such a daily food intake the real daily intake of metabolizable energy and protein will be obtained and the chicken performs protein and lipid retention (Fig.3) and hence live weight gain. These values internally enlarge the respective values of the previous time-step and a next simulation step can start:

- $AVALE=XFOOD*FEC-CEM*W^{0.75}-EDL$ (available food energy for growth)

  $AVALP=XFOOD*FPC-CPM/PCR*W^{2/3}$ (food crude protein for protein retention)

- IF ($AVALP*PCR*CEP_{act}$.LT.AVALE) THEN

  P = Min (AVALP*PCR,DP) (actual protein retention under the condition of protein limited protein synthesis

ELSE

$P = \mathrm{Min}\,(\mathrm{AVALE}/\mathrm{CEP}_{act}, \mathrm{DP})$ (actual protein retention under the condition of energy limited protein synthesis)

ENDIF

$- L = (\mathrm{AVALE} - \underbrace{(P_t - P_{t-1})}_{\hat{=}\mathrm{PG}} * \mathrm{CEP}_{act} + \underbrace{0.1 * (<W * \mathrm{CEL})}_{\hat{=}\mathrm{EDL}}) / \mathrm{CEL}$ (actual lipid retention)

$\hat{=}$PG - daily protein retention

$\hat{=}$EDL - energy for minimum lipid retention

$- <W = (\mathrm{WCP} * \mathrm{PG} + \mathrm{WCL} * (L_t - L_{t-1})) / (1 - \mathrm{WCA} * 0.045)$

with WCP,WCL,WCA - specific contribution of the different retention components to <W

A *single animal* growth model can't be used for practical production control unless it is extended to the level of *stock* and takes into consideration the influence of *climatic conditions*. These facts mean to describe the development of a spacially limited stock in numbers and weight in dependence on the stocking density, climatic conditions and on factors involved in the *single animal* growth model. The description of animal losses per time-step is the prerequisite for mapping the time behaviour of number of animals per stock. During the starting period (the first 10 days) losses of animals are a function of time and after that they are related to the level of exceeding a critical live weight per unit of area which itself is a function of W and of number of animals per unit of area (Fig. 4). Modelling of influence of the thermal environment on yield and expenditure of a stock is related to the relationships between temperature and air humidity on the one hand and food intake, <W and requirements for ventilation and heating on the other hand. All these processes have to be controlled to guarantee a permanent coincidence between heat production and heat losses of a stock, because of homeothermy. Heat production is inseperably connected with processes of metabolism. Under the condition of a thermoneutral environment the energy requirement of the chicken for maintaining its body temperature is completely met by the basic metabolism and the thermogenius effect of the nutrients in the maintenance metabolism. If the ambient temperature is below thermoneutrality the chicken can maintain its body temperature only by producing more heat. Increasing ambient temperature decreases the energy requirement for thermoregulation. Hence, the balance between heat production and heat losses of the chicken and the influence of food intake on this balance is the main point of mapping thermal environmental factors in broiler production. This is reflected in the model by self-regulating the energy maintenance requirement and consequently the food intake in dependence on the degree of violating the thermoneutrality and by computation of the amount and the kind of heat losses of the chicken in dependence on temperature, W and <W. The heat output of the stock is simultaneously an input to the submodel *climat inside the poultry house*. Its aim consist in determining such measures which guarantee the maintaining of a physiologically and economically founded range of inside temperature considering the defined interrelationships between temperature and relative air humidity under different outside conditions. These measures can be

first of all ventilation but also heating and if necessary cooling of the ventilated air. The adequate measure is elected and its required amount is calculated.

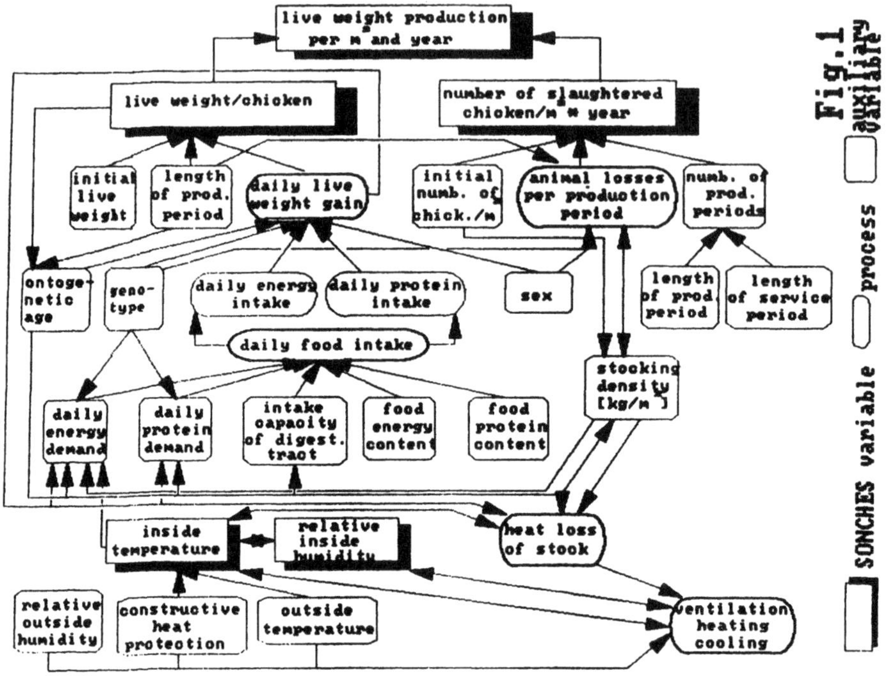

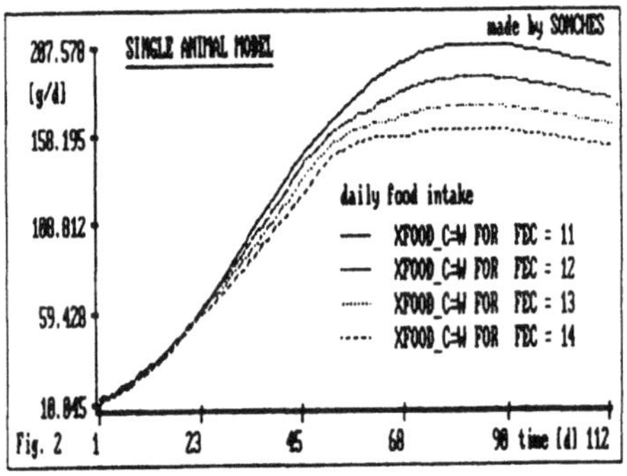

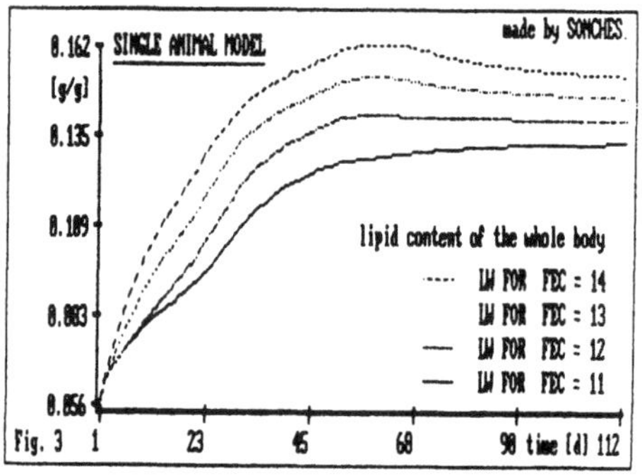

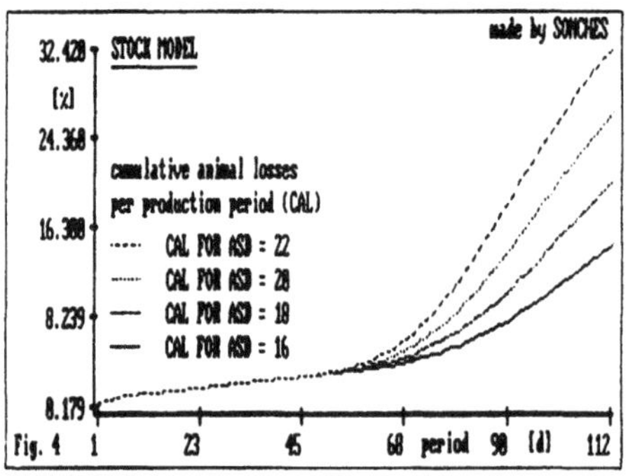

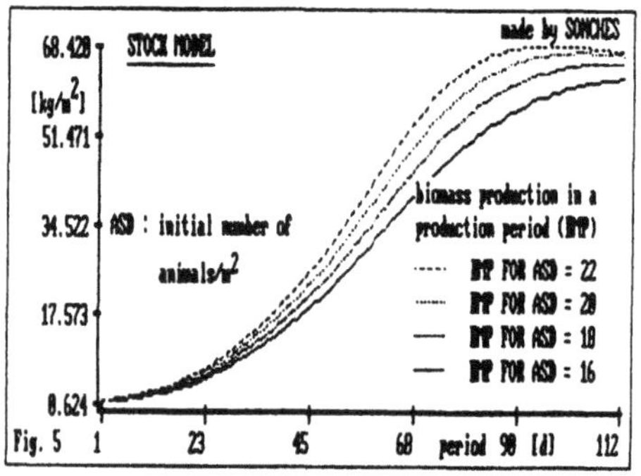

COMPUTER AIDED MODELLING AND ITS APPLICATION IN ECONOMY AND ECOLOGY

J.A.Müller[1]

Abstract

Two possible trends in computer aided modelling are considered:
1. Use of the dialogue to consider the a priori information available on the side of user
2. Automatic model selection in the way
   a. of generation of model variants with growing complexity and selection of models of optimal complexity and
   b. of selection of the best model as a special case.

Keywords: model building; models of optimal complexity; dialog systems; computer aided modelling; economic models; ecological models

1. Introduction

In modelling the following problems have to be solved:
- inadequate a priori information
- methodological problems in connection with modelling
- problems of epistemology.

There exists at present a lot of several mathematical, cybernatical, statistical and other methods for modelling. Experience of the application of this tools, methods and techniques have only a few specialists. In the practice of economic and ecological decision making they play a small role, the necessary knowledge for a sucessful application does not exist by the specialist of these sciences.
In order to solve these problems and to deal with ill defined systems in particular the necessity arises to seek for ways of how it is possible with the help of the developing information engineering to reduce the great share needed by model formation for the solution of task (according to Shannon the share is estimated to be 75 - 90 %) by means of a computer aided design of mathematical models. There are two possible trends that become obvious:
a. Use of the dialogue to consider the a priori information available on the side of user
b. Automatic model selection.

1) Hochschule für Ökonomie "Bruno Leuschner", Sektion Wirtschaftsinformatik, 1157 Berlin, Hermann-Duncker-Str. 8, GDR

2. Use of dialogue systems

The aim is to create highly qualitative programming systems with a maximum degree of automation of the phase of construction, solution and analysis of models, which can either qualify as econometric models or also algorithmic models. It is typical of both trends that due to interactive model formation (dialog system) the indeterminacy of model formation available on the side of user shall be reduced, i.e. that user shall be enabled to make a correct and clear-cut choice of a suitable, or even best model in the indicated sense. During interactive operation user can at the same time step-by-step formulate his own problem and can react quickly to intermediate results and to peculiarities from his immediate knowledge of problems. As a result, the expert information available, the knowledge of facts and experience existing, which are difficult to formalize, or are not at all formalizable, flow into model formation by subjective evaluation a.o.

The own experience by the development of a

- dialog system for computer aided construction of system-dynamic models as a tool for simulation of complex economic and ecological systems
- dialog system for construction of ARIMA-models, of trend, regression and other functions as tools for analysis and prediction of economic an ecological processes

and several results by application of these tools for modelling of national economy, analysis and prediction of economic and ecological processes (for instance: monthly electric energy consumption, national economy, Cl-discharge of seeping water and drainage outflow) has shown: the dialog-assisted model selection serves users who seek insight in the process of model selection and possess the mathematical, statistical, cybernetical knowledge required and have time for dialogue and, apart from possibly good results on selected positions, want to have information on lawful regularities in the investigated process.

3. Automatic model selection

Automatic model selection serves those users who take only interest in the model results (for instance analysis, forecasts) proper and have hardly any knowledge on mathematical, cybernetical and statistical techniques, nor any time for the computer dialogue, yet have to process problems of a large scope of time series or systems.

This mode of approach is based on experimental systems analysis. As was proved by Kalman this task has no clear-cut solution. Since

on the basis of theoretical systems analysis this indeterminacy can only be insufficiently reduced due to the lack of a priori information, a clear-cut solution is achieved by means of automatic model generation and adequate evaluation of model variants. Below deliberations are devoted to this approach.
There are two possible trends that become obvious

a. Generation of model variants with growing complexity and selection of models of optimal complexity
b. selection of the best model as a special cases.

In both directions it is necessary proceeding from the a priori information available as a result of the theoretical systems analysis especially on the system and on the aims of model formation to select a model class containing all possible special cases. Obviously this model class becomes the more comprehensive, the smaller is the a priori information.

3.1 Selection of models of optimal complexity

The task of experimental systems analysis is substituted for the following tasks.

Given is the following:

1. There is a clear reproduction of the factors of influence $\underline{x}$ in the scalar output variable y: $y = f(\underline{x},\underline{a})$.
2. Given is a quantity F of functions $f_r(\underline{x},\underline{a}^r)$, with $\underline{a}^r$ being the adequate model parameters.
3. Given is the following structure on a quantity F of functions, which have been selected for the description of the reproduction of the input variables in the output variables: proceeding from a partial quantity F_1 of elementary functions there are partial quantities existing F_i $(i=2,3,\ldots,q)$, which contain the preceding partial quantity F_{i-1}. Thus the following is valid $F_1 \subset F_2 \subset \ldots \subset F_q$. The partial F_q coincides with the quantity F of functions.
4. The structure shall be given in a way that the complexity k_i of the functions of the partial quantity F_i is smaller than the complexity of functions contained in the partial quantity F_{i+1}, i.e. $k_1 < k_2 < \ldots < k_q$.
5. Given is a set of observations $X = \{x_1,\ldots,x_T\}$.
6. Given is a T-dimensional vector $\underline{y}$ of observations of the initial variable in the points of quantity X.

Sought

is the model of optimal complexity, i.e. such an element of the partial quantity $F_r, r = 1,2,...,q$, with the complexity k_r, which minimizes a given selectiv criterion Q.

In such a way autoregressive models, polynoms, several dynamic models (equation of Kolmogorov-Gabor, systems of interdependent equations) can be produced. In all these cases a system of elementary models is necessary, which can be designated as basis functions. For example the user can formulate an orthogonal function system or polynoms or Volterra-Lottka-equations.

By means of several program packages, which realize these algorithms, systems of interdependent equations for description of national economic processes, of meteorological processes and of ecological processes, were produced. Also a sucessful forecasting of electric energy consumption of fish distribution centres (autoregressive models) was achieved.
Results achieved in such a manner allow the conclusion that these algorithms are appropriate for the synthesis of models (systems of interdependent equations for description macro-economic, meteorological and ecological processes) in principle.

3.2 Selection of the best models as a special case

In special cases it is possible to select the best model on the base of the model class, containing all possible special cases. Evidently, most various principles of the selection of the best model are thinkable, for instance by the system-dynamic modelling.
It could be shown, that a class of growth processes can be described by Volterra equations. They include the most types of trend functions used in economy. If it is possible to identify unknown parameters of the Volterra system on the base of observations x(t) (t=1,2,...,N) the solution of the Volterra system gives the best model for the given observations.

In order to estimate the unknown parameters of the equation it seems reasonable to combine
- the numerical solution $x^M(t)$ of the Volterra system for given parameters by means of the 4.order Runge-Kutta techniques
- the search algorithm of Powell to minimize the square (or other error function)

$$Q = \sum_{t=1}^{N} g_t \, (x(t) - x^M(t))^2.$$

g_t denotes weights, which can be estimated on the base of a priori information.
Several districtions and heuristical extimation of initial values were necessary for a two step search procedure.

POLICY SIMULATIONS AND OPTIMAL CONTROL WITH A SMALL ECONOMETRIC MODEL FOR AUSTRIA

Reinhard Neck[1)]

1. Introduction

In order to assess the effects of government stabilization policies upon an economy, it is necessary to have some quantitative estimates of the relations between the main macroeconomic aggregates for the country under consideration. Usually this information is provided by an econometric model estimated on empirical data. Moreover, if the question of policy design is addressed, it is desirable to have a standard of comparison for the evaluation of different policies. This can be obtained from simulation and optimization experiments performed with the econometric model; for the latter purpose, an objective function has to be formulated. Policy-makers or their advisers can then be informed about the consequences of particular policy measures and can even be guided such as to choose those measures in an optimal way. The present paper intends to provide a first basis for such informations concerning the Austrian economy by estimating a small econometric model for this country and reporting about the results of some policy simulations and an optimal control experiment obtained with this model.

2. The Econometric Model

For the present purpose, a small linear demand-side macroeconometric model in the IS-LM framework with exogenous prices has been built. The model consists of six endogenous variables: C (real private consumption), I (real fixed investment), M (real imports of goods and services), R (real rate of interest, constructed as the yield on newly issued bonds minus the GDP-deflator inflation rate), Y (real gross domestic product at market prices), and V (real total final demand, the sum of Y and M); three control (instrument) variables, assumed to be at the disposal of the policy-makers: G (real public consumption), T (real net tax revenues of the public sector, constructed as GDP minus disposable income) and M1 (real money supply M1, deflated by the GDP price index); and one non-controlled exogenous variable: A (real autonomous expenditures, consisting of exports of goods and services and inventory charges including errors and omissions according to the national income accounts). Using annual data for Austria from 1955 to 1986, several specifications of single equations have been estimated and subjected to various diagnostic checks; tentative simulations have been performed with some alternative versions of the entire model. The version presented here was chosen mainly for its comparatively satisfactory tracking capability with respect to the developments of the endogenous variables.

The final model consists of a consumption function, an investment function, an import function, a money demand function, and two identities; it was estimated by three stage least squares. It contains the following equations:

$$C_t = \underset{(0.15314)}{0.40141}\, C_{t-1} + \underset{(0.13521)}{0.45379}\, (Y_t - T_t) + \underset{(0.09813)}{0.41783}\, M1_t - \underset{(5.03937)}{9.50237} \qquad (1)$$

$R^2_c = 0.99814$, SE = 4.78091, MAPE = 0.96392, DW = 1.96277, $\rho = 0.01411$

$$IR_t = \underset{(0.01513)}{0.95861}\, IR_{t-1} + \underset{(0.04525)}{0.12707}\, (V_t - V_{t-1}) - \underset{(0.54799)}{1.40099}\, R_t + \underset{(3.30484)}{11.17847} \qquad (2)$$

$R^2_c = 0.98991$, SE = 5.30393, MAPE = 2.69528, DW = 1.60608, $\rho = 0.17697$

$$M_t = \underset{(0.09595)}{0.66962}\, M_{t-1} + \underset{(0.05059)}{0.18647}\, Y_t - \underset{(14.1238)}{43.12065} \qquad (3)$$

$R^2_c = 0.99188$, SE = 9.74238, MAPE = 4.68438, DW = 1.73024, $\rho = 0.10896$

$$M1_t = \underset{(0.00682)}{0.13192}\, Y_t - \underset{(0.89466)}{2.20979}\, R_t + \underset{(4.98696)}{35.68886} \qquad (4)$$

$R^2_c = 0.91623$, SE = 7.97347, MAPE = 5.60772, DW = 0.64701, $\rho = 0.6978$

---

1) Institut für Volkswirtschaftstheorie und -politik, Wirtschaftsuniversität Wien, Augasse 2-6, A-1090 Vienna, Austria. Financial support by the "Fonds zur Förderung der wissenschaftlichen Forschung" (research project no. P 5846) is gratefully acknowledged.

$$Y_t = C_t + I_t + G_t + A_t - M_t \tag{5}$$

$$V_t = Y_t + M_t \tag{6}$$

Numbers in parentheses below coefficients are their estimated standard deviations; R^2_c denotes the coefficient of determination, corrected for the degrees of freedom, SE the estimated standard error of the equation, MAPE the mean absolute percentage error, DW the Durbin-Watson statistic for serial correlation, and ρ the estimated first-order autocorrelation coefficient of the residuals. Similar coefficients result from ordinary least squares and two stage least squares estimations.

The consumption function (1) is modelled in accordance with the permanent income or the habit persistence hypothesis and includes a real balance effect (Pigou effect). The rate of interest does not become significant when added to the other variables. The investment function (2) follows the flexible accelerator hypothesis. In spite of its low significance in an OLS regression, the interest rate is retained here to model a Keynes effect of monetary policy. Using Y instead of V or levels instead of first differences for these variables results in equations with lower fit. For the import function, relative import prices have been tried as additional regressors, but get the theoretically "wrong" sign. Unplausible coefficients arise from using V instead of Y or first differences of these variables in (3). The demand-for-money function (4) uses the real instead of the nominal interest rate to keep the model linear. Exchange rates as additional regressors remain insignificant, and Y is to be preferred over V as scale variable for statistical reasons. Although (4) exhibits significant serial correlation, including the lagged M1 variable does not provide a remedy as this results in unplausible and partly insignificant coefficients. Equation (4) assumes equilibrium in the money market; it is solved for the rate of interest in simulating the system. Estimating a reduced-form equation for R results in similar coefficients as those derived from (4).

The ex-post predictive performance of the entire model can be assessed from the results of error analyses obtained with dynamic simulations of the model. Both the results for the estimation period (1955-1986) and for the period of interest for the policy experiments (1971-1986) are given in table 1, which shows the means, standard deviations, root mean square errors, lower and upper bounds of the simulation errors for the endogenous variables, both in absolute and (with the exception of the interest rate) in percentage terms.

Table 1: Results of Error Analyses

	1955 - 1986					1971 - 1986				
	mean	std.dev.	RMSE	low.bd.	upp.bd.	mean	std.dev.	RMSE	low.bd.	upp.bd.
C	3.714	19.714	20.061	-26.549	54.321	6.668	26.599	27.422	-29.137	53.333
	0.5%	5.0%	5.1%	-7.0%	12.7%	1.2%	6.3%	6.5%	-7.7%	12.5%
I	4.682	16.976	17.610	-23.658	46.334	11.615	19.235	22.470	-24.109	44.797
	0.4%	11.3%	11.3%	-20.0%	24.3%	6.0%	10.0%	11.6%	-11.7%	23.5%
M	3.640	13.811	14.283	-16.190	35.190	8.201	15.531	17.563	-16.578	34.485
	-1.8%	10.4%	10.5%	-24.6%	12.5%	3.0%	5.3%	6.1%	-4.4%	12.0%
R	0.286	3.221	3.233	-5.661	6.151	0.421	3.921	3.944	-5.734	6.124
Y	4.755	25.766	26.201	-25.331	69.670	10.080	34.197	35.652	-31.108	68.461
	0.4%	3.7%	3.8%	-5.8%	8.9%	1.1%	4.6%	4.7%	-4.9%	8.7%
V	8.393	34.058	35.076	-32.179	100.648	18.279	42.684	46.434	-35.431	98.123
	0.2%	3.9%	3.9%	-7.0%	9.8%	1.6%	4.2%	4.5%	-4.3%	9,6%

3. Policy Simulations and a Control Experiment

With the above model, several policy simulations can be performed to determine the policy multipliers of the instrument variables and to get informations about the effects of alternative hypothetical stabilization policies during the past. The time paths of the endogenous variables obtained by simulations with historical values of instrument and exogenous variables are used as control solution, which is "shocked" in various ways. The time period considered is always 1971 to 1986, which were the years of the one-party government of the Socialist Party (SPÖ) until 1983 and of the "small coalition" of the SPÖ with the Liberal Party (FPÖ) from 1983 to 1986.

First we calculate the effects of sustained shocks of the instrument variables over the simulation period for each instrument separately and compare the resulting paths of the endogenous variables to those of the control solution to determine the policy multipliers of the model. The results are as follows: An increase of real public consumption by 1 billion AS as compared to the control solution results in an increase of real GDP, which has its maximum value of 1.502 billion AS in 1971; then the multiplier decreases steadily to 0.416 in 1986. This is due to crowding-out effects induced by a rising rate of interest and hence declining real fixed investment from 1972 onward. The negative multiplier on investment rises to -0.652 in 1986. Multipliers on real private consumption and on real imports are positive, with maximum values of 0.961 in 1973 and 0.567 in 1974, respectively. Opposite effects obtain from an increase in real net tax revenues of 1 billion AS, but they are weaker, with a maximum GDP multiplier of -0.961 in 1973. Here the strongest effects are exerted on private consumption, where the multiplier has its maximum of -1.397 in 1974. Imports are reduced and investment is raised in this simulation. On the other hand, monetary policy is much more effective than fiscal policy in our model: A sustained increase of M1 by 1 billion AS over the period of simulation raises investment (via the Keynes effect of a lower interest rate) and consumption (via the Pigou effect), with multipliers increasing steadily to 3.599 and 4.762 in 1986, respectively. Although some of these additional expenditures are absorbed by rising imports, the multiplier on real GDP rises to a maximum of 5.395 in 1986. Thus we can conclude that real money supply has vigorous cumulative effects on the real variables of our model.

A similar impression can be obtained from simulations of one-time shocks of policy variables. Specifically, we consider raising public consumption and lowering net tax revenues by 5% from their historical levels on the one hand, and raising money supply by 5% on the other, both in the year 1975 where the first oil-price shock led to a decline of real GDP in Austria. The combined effect of the two fiscal policy measures increases real GDP by 18.115 billion AS in 1975; however, already in 1977 the effect on GDP becomes negative and remains so until 1986, with a maximum negative effect of -2.769 in 1979. Similar developments are exhibited by private consumption, fixed investment, and imports. Thus a one-time fiscal policy expansion has only transitory effects on the economy. On the other hand, the increase in M1 leads to positive effects on GDP over the entire period of simulation; the maximum effect is 10.211 billion AS in 1975, but even in 1986 the effect is still 0.536 billion AS. Similar effects obtain for private consumption, fixed investment, and imports. Combining the fiscal and monetary policy shocks in 1975 leads to dampened oscillations in additional GDP. Thus one-time fiscal policy measures have strong immedate effects in our model, which evaporate, however, soon and are even reversed, whereas monetary policy effects last for a much longer period. A combination of fiscal and monetary expansion may be appropriate to deal with a transitory exogenous decline of real GDP in our model.

Although policy simulations provide much insight into the dynamic properties of the model and its reaction upon changes in instrument variables, for the purpose of using the model to derive policy recommendations or evaluations of the performance of past stabilization policies additional information is required. One possibility consists in formulating an explicit objective function and performing optimization experiments in order to determine how "optimal" stabilization policies would have looked like under the particular objective function if the model were exactly true. The following optimization experiment has been performed with our model, using the optimal control algorithm of Chow ([1]): Historical values for all variables were assumed to be given for 1970, and the optimization period was 1971 to 1986. Desired ("ideal") values of endogenous and control variables were specified as smooth paths, starting from their historical values in 1970 and growing with the constant rates of 3.5% (the actual average growth rate for real GDP was about 3% during the period under consideration), except for the real interest rate whose desired value was kept constant at 2%. This specification serves to express the objective of achieving a balanced growth of the real aggregates of the model. A quadratic objective function was formulated to penalize deviations of the objective variables from their "ideal" values. The weighting matrix of the objective function was specified as a diagonal matrix, with the following weights attached to deviations of each variable:

variable	C	I	M	R	Y	V	G	T	M1
weight	5	5	5	2.5	10	0	5	5	1

The results of this optimization experiment can be summarized as follows: The optimal path of real public consumption exhibits a move discretionary policy behavior than suggested by its "ideal" and by its historical values; in most years (exception: 1975) it remains below the historical path. Optimal real net tax revenues follow the "ideal" path rather closely; the historical 1975 drop in this variable is not reflected in the optimal path. Optimal monetary policy, on the other hand, is much more active: In 1975, M1 is above its "ideal" value; otherwise it remains considerably below them (particularly 1979 - 1980 and 1984 - 1985). Compared with the historical path, optimal M1 is much more volatile, too, remaining below historical values until 1980 and above them thereafter. This combination of a rather active monetary policy with mildly discretionary fiscal policy results in an optimal path for real GDP, the main objective variable, which is much more close to the "ideal" path than that of the control solution. Optimal GDP is below the control solution path until 1981 and above it since 1982. Although during the eighties there is much stronger growth in optimal GDP than in the control solution, it still remains short of the "ideal" 3.5%, which must be considered as too optimistic for those years. Similar developments occur to private consumption, fixed investment, and imports, whose optimal values are in general below those of the control solution during the seventies and above them during the eighties. Thus monetary (and to some extent fiscal) policies in this model are effective in smoothing the time paths of the main aggregate variables and generating higher growth than occured actually in the stagnation period of the eighties. An interesting side-effect of this is the fact that optimal imports and hence total final demand are considerably above their "ideal" values. The interest rate, which is regarded as an intermediate target in our objective function, exhibits an optimal path which is slightly less volatile than that of the control solution. The main effects of monetary policy on real GDP work through the real balance effect and not through the Keynes effect.

4. Concluding Remarks

We have shown that in the present model fiscal and especially monetary policies have effects on the real economic variables considered, particularly on real GDP, and that business cycles can be smoothed out to a considerable extent. Monetary policy is much more effective than fiscal policy due to the real balance effect, which is a striking contrast to an otherwise very similar model considered in [2]. It would be premature to derive from these simulations policy conclusions to be used directly for actual recommendations to policy-makers. One of the reasons for this is the neglect of stochastic elements in the simulations and the optimization experiment. Moreover, further experiments with more sophisticated models are required, which have to introduce, inter alia, flexible prices and the supply side of the economy to give a more realistic picture of Austrian economic developments. This will be the subject of further research in the next future.

References

[1] Chow, G. C.: Econometric Analysis by Control Methods. John Wiley & Sons, New York 1981.

[2] Neck, R., and U. Posch: On the Sensitivity of Optimal Macroeconomic Policies to Econometric Models: An Empirical Analysis for Austria. In: T. Basar and L. F. Pau (eds.), Dynamic Modelling and Control of National Economies 1983, Pergamon Press, Oxford 1984, 91-98.

TOTAL POLLUTION EFFECT AND TOTAL ENERGY COST PER OUTPUT OF DIFFERENT PRODUCTS FOR POLISH INDUSTRIAL SYSTEM

Henryk W.Balandynowicz[1]

1. Introduction

For many years a broad use has been made of the indices of total energy requirements in the whole large production system corresponding to unit output of particular goods (Boustead I., Hancock G.F.,1979). These indices satisfy well the requirements of system analysis. Because of their efficiency, an attempt was made to solve a similar problem of the determination of the indices of total pollution generation requirements emerging as a result of complex multi-stage (multi-object) production processes.

2. Method

The method which was elaborated for the needs of the research uses some elements of the theory of the graph of flows in a network (Bibrowski Z.,1981). The main features of this method consist in representing the processes of the production and consumption of products in the production system considered in the form of a graph. This graph is characterized by the fact that it is directed, labelled and weighted. Each apex of it corresponds to the process of production of one object, while its label expresses the amount of yearly production. Each arc of the graph represents the flow of a given product to the production of another object or outside the system considered. The direction of the arc indicates the product consuming, while its weight shows the flowing annual amount, used up, e.g. in the production indicated. The magnitude of flows in the graph is expressed in physical units. Such a graph can be described in mathematical form with the following equation:

$$(I - A)\ X = C \qquad (1)$$

where:

I - unit matrix,

A - matrix of the relative technical coefficients (in natural units),

X - vector of the sough indices of total pollution generation requirements,

C - vector of known indices of direct pollution generation requirements.

For this model of the system it is possible to determine the indices of total requirements of a broad class of factors (including e.g.total employment requirements

[1] Institute of Fundamental Technological Research of the Polish Academy of Sciences, 00-049 Warsaw, 11/21 Swietokrzyska St., POLAND

total water rquirements, total emission of SO_2 requirements):

$$X = (I - A)^{-1} C \tag{2}$$

if only the respective indices of direct consumption or generation per unit production (in natural unit) of total output, are known in the form of the vector C.

3. Calculation scheme

Calculations of indices of total pollution generation requirements were carried out for the Polish production system consisting, to a much simplified extent, in the production of 36 industrial goods using many technologies (Balandynowicz H.W., et all.,1987). In all, 86 producto-technologies were considered for the state and conditions of this system in 1985. The study considered:

1) consumption of energy,

2) emission of SO_2, NO_x (recalculated to N_2O_5), CO, C_xH_y, volatile dust and metallurgical dust,

3) generation of liquifed waste containing organic and nonorganic substances reducing the biochemical oxygen demand (so-called BOD), suspended pollutants,

4) generation of solid waste: ashes and power-plant slags, rock waste from mining.

4. Model

To carry out calculations, a computer model, called CUMUL, was elaborated. Fig.1 shows its software structure. It consists of:

1) master program CUMUL which contains orders for controlling procedures.

2) procedure READ-G implementing the names of identifiers: of branch, section and product.

3) procedure READ-D serving to read and sort out numerical data. Within this procedure, the matrix A is generated.

4) procedure ORDER ordering the matrix A generated in the preceding matrix.

5) procedure SIMQ-3 serving to invert rare matrices.

Further procedures of the CUMUL model are added depending on what kind of pollutants is considered. These procedures are two-fold:

6) procedure BALANCE-N including two groups of partial balances.

7) procedure PRINTOUT-N generating printouts of two tables.

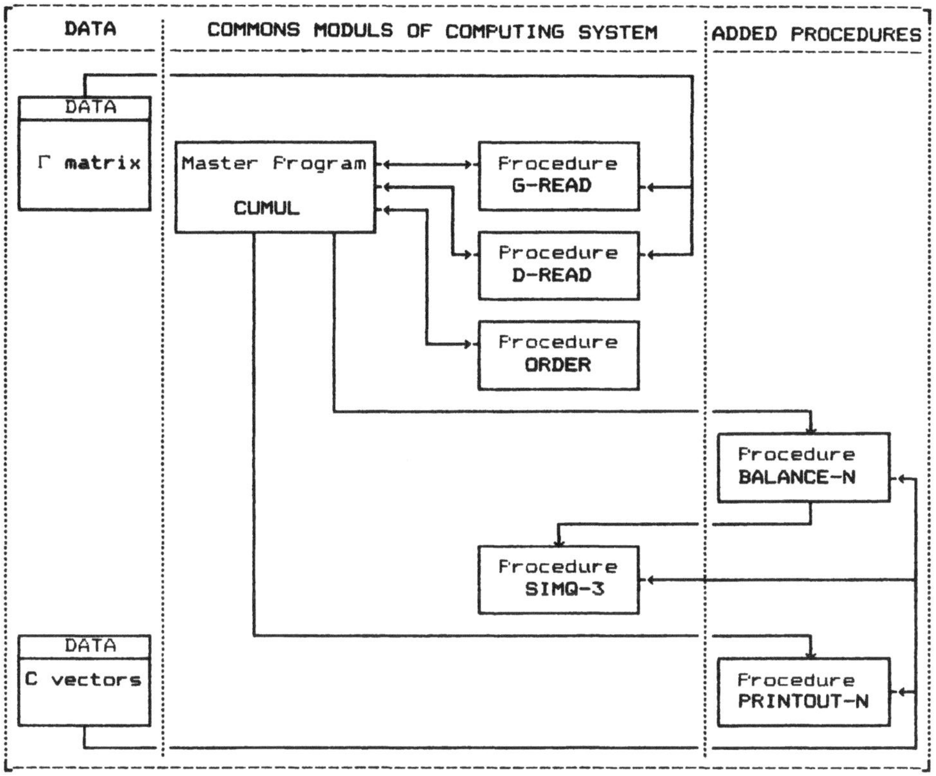

Fig.1 The modular structure of computing system CUMUL

5. Results

The calculation results are given in tables 1 and 2. It is evident, e.g., that although the making of some products is not related directly to the generation of some pollutants, this can, indirectly (by using other goods), cause generation. This phenomenon is well illustraded by results on the generation of mining waste and power-plant slags. Despite the fact that the direct generation of this waste involves small amounts of goods, all of them are characterized by the values of the indices of total pollution generation requirements of these pollutants.

The values of total pollution generation requirements of mining waste also indicate the coal-absorptive structure of the Polish production system. In practice, the production of each of the considered goods in Poland causes an increase in waste on dump heaps near mines.

References

[1] Balandynowicz H.W.,et all: Total Energy and Pollution Generation Requirement of Industrial Products for Poland. Department of Energy Problems of IFTR, Warsaw 1987
[2] Bibrowski Z.: Studies on the Indices of Total Energy Requirements. The Method of Signal Flow Graph. Archiwum Energetyki,4,1981
[3] Boustead I., Hancock G.F.: Handbook of Industrial Energy Analysis. Ellis Horwood, New York-Chichester-Brisbane-Toronto, 1979

Table 1. Indices of direct and total pollution generation (energy) requirements of industrial goods in Poland

No.	Industrial goods	Units	Energy GJ/unit		SO_2 kg/unit		NO_x kg/unit		CO kg/unit		C_xH_y kg/unit		Volatile dust kg/unit	
			D	T	D	T	D	T	D	T	D	T	D	T
1.	Steam coal	tce	0.5	1.5	0.0	1.4	0.5	0.7	0.1	0.2	0.03	0.06	0.0	0.5
2.	Coking coal	tce	0.4	1.1	0.0	0.9	0.0	0.2	0.2	0.3	0.01	0.02	0.0	0.3
3.	Lignite	tce	0.5	2.3	0.0	2.4	0.0	0.3	0.0	0.4	0.00	0.03	0.0	0.7
4.	Coke	tce	35.1	37.6	1.3	4.2	0.5	1.2	1.2	1.9	2.79	3.16	0.0	6.2
5.	Crude tar	t	32.4	34.6	1.2	3.9	0.5	1.1	1.1	1.8	2.58	2.92	0.0	5.7
6.	Crude benzol	t	72.5	77.5	2.7	8.7	1.1	2.4	2.5	3.9	5.77	6.53	0.0	12.8
7.	Coke-oven gas	tce	35.1	37.6	1.3	4.2	0.5	1.2	1.2	1.9	2.79	3.16	0.0	6.2
8.	Natural gas[1]	tce	0.2	0.4	0.0	0.3	0.0	0.1	0.0	0.0	0.00	0.00	0.0	0.0
9.	Crude oil - domestic	tce	1.2	1.9	0.0	0.9	0.0	0.2	0.0	0.1	0.00	0.05	0.0	0.2
10.	Crude oil - import[1]	tce	0.1	0.2	0.0	0.2	0.0	0.1	0.0	0.0	0.00	0.00	0.0	0.1
11.	Petroleum products	tce	33.1	36.2	0.9	5.5	0.1	0.6	0.9	1.8	1.13	1.37	0.0	0.3
12.	Electricity[2]	MWh	10.8	12.3	12.2	16.1	1.5	2.1	0.1	0.4	0.00	0.03	2.1	3.5
13.	Electricity[3]	MWh	11.9	15.7	13.5	19.1	1.5	2.1	0.1	0.6	0.00	0.04	3.1	4.2
14.	Electricity[4]	MWh	10.6	12.8	7.9	9.9	1.4	1.8	0.1	0.5	0.34	0.69	1.4	1.7
15.	Heat	GJ	1.3	1.5	0.9	1.2	0.2	0.2	0.3	0.4	0.11	0.15	0.5	0.5
16.	Iron ore - import[5]	t	0.4	0.8	0.0	0.6	0.0	0.1	0.0	0.0	0.00	0.01	0.0	0.1
17.	Agglomerate	t	2.6	4.3	1.5	2.9	0.3	0.6	4.9	5.3	0.73	1.02	0.0	0.8
18.	Outside scrap	t	0.0	0.1	0.0	0.1	0.0	0.1	0.0	0.0	0.00	0.00	0.0	0.0
19.	Pig iron	t	15.9	28.8	0.1	8.9	0.7	3.1	120.0	130.0	0.02	4.44	0.0	7.7
20.	Steel	t	1.8	24.7	1.2	9.6	1.3	3.9	1.5	100.5	0.59	3.96	0.0	6.2
21.	Hot rolled products	t	6.5	40.2	0.3	15.6	0.1	5.8	0.6	125.4	0.00	5.68	0.0	10.1
22.	Steel bars&sections	t	5.6	38.9	0.0	14.2	0.0	5.4	0.0	127.2	0.00	5.44	0.0	9.1
23.	Seamed tubes	t	3.9	55.2	0.0	21.3	0.0	7.7	0.0	154.4	0.00	7.60	0.0	14.2
24.	Cold rolled products	t	1.4	51.2	0.0	23.4	0.0	7.7	0.0	143.3	0.00	6.63	0.0	13.2
25.	Cu ore concentrate	t	6.3	13.1	6.9	26.7	1.4	5.1	0.0	1.5	0.08	0.19	0.0	4.9
26.	Cu-Pb ore concentrate	t	13.9	45.8	6.9	54.6	0.1	8.2	3.3	8.4	0.00	0.35	0.0	11.4
27.	Zn-Pb ore concentrate	t	6.3	13.1	4.8	14.4	0.1	1.6	3.3	8.6	0.00	0.38	0.0	2.3
28.	Commercial lead	t	28.6	81.3	9.1	74.9	0.0	10.5	7.7	27.8	1.01	2.03	0.0	18.7
29.	Electrolytic Cu	t	35.3	78.8	62.4	126.0	1.6	12.3	756.0	771.3	9.08	11.50	0.0	16.7
30.	Electrolytic Zn	t	45.4	119.0	14.5	107.0	0.1	15.5	12.2	37.1	0.02	1.33	0.0	27.8
31.	Sulphur	t	7.5	9.6	0.0	6.4	0.0	0.4	0.0	0.3	0.00	0.03	0.0	2.8
32.	Carbon disulphide	t	9.5	18.1	0.6	3.8	1.1	2.6	0.0	0.2	0.00	0.03	0.0	0.7
33.	Sodium chloride	t	0.1	0.5	0.0	0.4	0.0	0.1	0.1	0.4	0.00	0.02	0.0	0.4
34.	Ammonium sulphate	t	32.3	35.1	0.1	3.2	0.1	1.1	0.1	0.6	0.26	0.61	0.0	5.9
35.	Sulphuric acid	t	4.1	7.2	12.9	17.4	11.3	11.8	0.0	1.2	0.00	0.04	0.0	1.7
36.	Cement	t	5.9	7.8	9.1	11.2	1.3	1.7	16.1	16.2	0.00	0.02	0.03	0.7

Table 2. Indices of direct and total pollution generation requirements of industrial goods in Poland

No.	Industrial goods	Units	Metallurgical dust kg/unit		LIQUIFED WASTES: BOD 10^{-3} kg/unit		LIQUIFED WASTES: suspended 10^{-3} kg/unit		SOLID WASTES: ashes&slags kg/unit		SOLID WASTES: rock waste kg/unit	
			D	T	D	T	D	T	D	T	D	T
1.	Steam coal	tce	0.00	0.01	16.4	17.4	28.0	34.8	0.0	23.8	476.8	490.7
2.	Coking coal	tce	0.00	0.01	22.8	23.5	37.5	43.1	0.0	13.3	710.0	719.0
3.	Lignite	tce	0.00	0.03	10.7	12.2	291.5	313.9	0.0	39.3	0.0	3.5
4.	Coke	tce	0.00	0.02	15.5	47.1	2.6	60.3	0.0	29.2	0.0	878.6
5.	Crude tar	t	0.00	0.02	14.3	43.4	2.4	55.5	0.0	26.9	0.0	809.7
6.	Crude benzol	t	0.00	0.04	32.0	97.1	5.4	124.3	0.0	60.2	0.0	1812.
7.	Coke-oven gas	tce	0.00	0.02	15.5	47.1	2.6	60.3	0.0	29.2	0.0	878.6
8.	Natural gas[1]	tce	0.00	0.00	0.0	0.2	0.0	1.7	0.0	4.1	0.0	1.3
9.	Crude oil - domestic	tce	0.00	0.00	0.0	0.6	0.0	10.0	0.0	10.8	0.0	3.5
10.	Crude oil - import[1]	tce	0.00	0.00	0.0	0.1	0.0	1.2	0.0	3.0	0.0	1.1
11.	Petroleum products	tce	0.00	0.06	3.1	4.8	128.4	158.3	0.0	9.9	0.0	7.1
12.	Electricity[2]	MWh	0.00	0.03	1.1	8.0	4.8	19.9	165.0	190.3	0.0	193.0
13.	Electricity[3]	MWh	0.00	0.03	2.8	9.8	5.5	105.3	210.0	310.0	0.0	3.7
14.	Electricity[4]	MWh	0.00	0.00	2.8	8.8	5.0	18.2	165.0	184.0	0.0	64.8
15.	Heat	GJ	0.00	0.00	0.8	1.7	1.5	3.6	20.0	24.0	0.0	25.0
16.	Iron ore - import[5]	t	0.00	0.00	0.0	0.4	0.0	4.9	0.0	8.5	0.0	2.7
17.	Agglomerate	t	1.50	1.78	0.1	4.5	0.2	11.9	0.0	18.3	0.0	75.1
18.	Outside scrap	t	0.00	0.00	0.0	0.1	0.0	0.9	0.0	2.1	0.0	0.7
19.	Pig iron	t	0.20	2.45	0.2	37.7	0.3	66.4	0.0	89.7	0.0	661.6
20.	Steel	t	1.80	6.43	0.4	29.9	1.4	61.2	0.0	93.1	0.0	511.6
21.	Hot rolled products	t	0.00	7.98	0.2	46.0	0.4	100.6	0.0	167.8	0.0	783.8
22.	Steel bars&sections	t	0.00	8.17	0.0	44.3	0.0	93.6	0.0	146.5	0.0	760.1
23.	Seamed tubes	t	0.00	9.75	0.0	62.9	0.0	139.4	0.0	240.4	0.0	1068.
24.	Cold rolled products	t	0.00	9.10	0.0	55.9	0.0	147.5	0.0	283.1	0.0	929.4
25.	Cu ore concentrate	t	0.23	0.34	320.0	332.0	2890.	3012.	0.0	316.0	66.6	183.5
26.	Cu-Pb ore concentrate	t	0.79	1.17	80.3	111.1	2500.	2791.	0.0	813.3	4.0	310.9
27.	Zn-Pb ore concentrate	t	0.79	1.23	80.3	90.0	2500.	2552.	0.0	188.7	4.0	166.0
28.	Commercial lead	t	0.09	2.36	14.9	185.8	37.2	3870.	0.0	1128.	0.0	747.5
29.	Electrolytic Cu	t	7.93	8.52	0.4	513.6	1.1	4336.	0.0	774.9	0.0	1061.
30.	Electrolytic Zn	t	0.15	2.75	23.7	220.8	59.1	4062.	0.0	1638.	0.0	1051.
31.	Sulphur	t	0.00	0.02	2176.	2258.	641.8	683.7	0.0	95.6	0.0	101.9
32.	Carbon disulphide	t	0.00	0.01	0.0	15.6	0.0	393.6	0.0	43.2	0.0	183.3
33.	Sodium chloride	t	0.00	0.02	11.5	11.9	20.3	23.0	0.0	6.4	474.6	478.9
34.	Ammonium sulphate	t	0.00	0.01	1.4	24.3	0.2	44.9	0.0	36.5	0.0	581.5
35.	Sulphuric acid	t	0.00	0.01	0.0	514.8	5.1	154.2	0.0	62.9	0.0	56.9
36.	Cement	t	0.00	0.01	7.4	12.1	25.6	45.7	0.0	34.2	0.0	94.9

* NOTE ON THE TABLES:
D-direct
T-total
[1] pipe-line only,
[2] power stations-hard coal,
[3] power stations-brown coal,
[4] industrial power&heat generating plant,
[5] railway transport only

A MULTISECTORAL ECONOMETRIC INPUT-OUTPUT-MODEL OF THE G.D.R. FOR PREDICTION AND DYNAMIC SIMULATION OF THE NATIONAL INCOME

Hans-Peter Höschel [1]

For analysis, prediction and planning of the national income of the GDR a complex mathematical model of the national economy has been developed in the division of national-economic accounts and analysis of the Central Statistical Office (CSO). By use of time series of the national income statistics the productive part of the economic reproduction cycle has been represented with its complicated subsystems, interactions and feed-backs. A simplified representation shows the following figure:

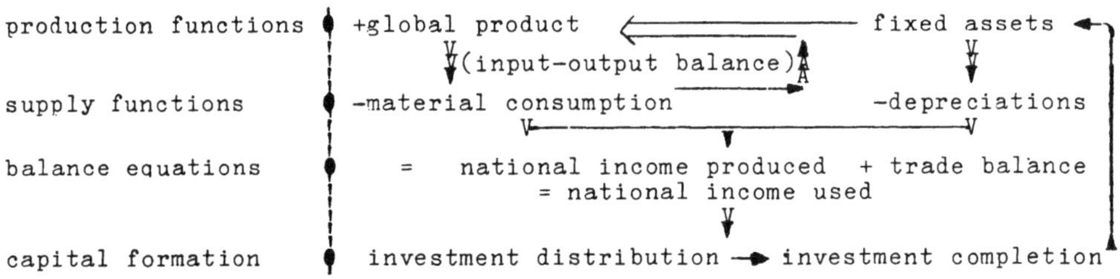

Compared with this simplified picture the true model comprehends of course all relations in much more detail. It is a multisectoral linear simultaneous equations model including several difference equations and a complete quasi-dynamic input-output model for the 15 producing industrial and other sectors of the material sphere. Altogether it consists of about 250 equations with 350 variables. The cycle of completion of investments from gross fixed capital formation to new fixed assets put into use has been investigated especially carefully by mathematical statistical methods. Exogenous factors as the conditions of vegetation and lignite-mining have been included. There are only few basic control variables delivered by the state planning commission :

- the gross investments in the material sphere (gross fixed capital formation) and their planned share among the industrial and the other producing sectors
- the gross investments in the non-material sphere and the private and social consumption
- the exports and imports with their components capital goods, consumer goods and materials .

The main results by the model are of course the forecasted values of the produced and used national income. However, another model variable has turned out as important for the evaluation of the results obtained. Namely, on the one hand, there is the predetermined final use on the basis of the abovementioned planned control variables. On the other hand, one has the final product which is the national income plus depreciations. The difference between the predicted final product and the planned final use is the final-use-balance, respectively in the positive case called final-use-reserve. If this final-use-balance is negative, then the

1) Staatl. Zentralverwaltung f. Statistik, H.-Beimler-Str.70, Berlin, 1026

tentatively planned final use can not be covered by the growth of the national income. To ensure in this case a balanced development of the national economy, there are essentially two possibilities. At first one could reduce the tentatively planned final use. If in this case it is not possible to reduce especially the consumption then a reduction of the productive investments would be nessecary. By a negative feed-back this would lead to a diminution of the economic growth in later periods. However, keeping the predetermined investment growth would reduce the share of the final consumption and of the external trade. On the other hand, one could also raise credits to purchase new equipment. Then, however, in later periods one has to supply additional exports for discharging the external trade obligations.

With respect to the construction of the model all relevant experiences show that for macroeconomic models one should use only such data and timeseries which have been centrally collected and maintained and which are accounted for by the CSO. The model conception had to ensure that with the available data a sufficiently detailed representation of national economic processes will be possible. On the other hand one cannot include too small economic units. The model should consist of relatively independent submodels. From all macro-economic models known to the author the most suitable concept had been proposed in [1]. The set of 15 industrial and other producing sectors reflects important structural processes sufficiently detailed and stable. In an extension of the approach in [1] and the former version of our model in [2] now a complete quasi-dynamic input-output balance has been included.

1. SUBMODEL : PRODUCTION

The applied production functions include as production factors material consumption and fixed assets. With this tool the essential processes of improving the efficiency of material consumption by new fixed assets can be represented. On the contrary, the "classical" production functions include fixed assets and employees - "capital" and "labour". However this yields no suitable model for socialist economies. Since there a direct substitution of labour by capital is impossible for social reasons. The substitution of workers is a complicated process. They cannot simply be replaced by equally productive equipment and then be "fired". Thus these form of production functions is too simple a model for this purpose and the data-fitting will be poor. This is a result of a computational analysis for the sectors of the material sphere of the GDR (cf. [2] for references and a detailed description of the whole model).

However, if one uses, in a similar production function, material instead of labour, the results are much better, because here a substitution by the use of efficient equipment is not restricted for social reasons. Even more, the saving of material has strongly been stimulated by the newly introduced state measures adopted with a view to further improving management, planning and economic accounting, especially with the broad use of the net product as scoring index. Since the model framed had to be linear for several reasons also the production functions included had to be so as well. The linear difference equations used were of the following form:

DGP = a * DMC + b * f(FA) + CORR(year)

where Dx := x - lag(x), GP - gross output, MC - material consumtion, FA - fixed assets and CORR is a time-dependent correction term. CORR consists of 0-1-dummy variables for year-intervals. The parameters a, b,.. of the regression function have been determined by a weighted-least-squares estimation. For the function f one of the following variants may occur

FA, L1FA :=lag1(FA), L2FA :=lag2(FA), D12_FA :=lag1(FA) -lag2(FA), D02_FA

The optimal variant is determined by an extensive mathematical model choice together with the best form of CORR. The appearence of lag-terms means that new fixed assets are generally not working with full capacity in the first year. The conditions for the classical production functions are not required for the present ones, such as the intersection of the origin or the monotony. As a production function in a wide sense one can regard any function of certain production factors. This is sensible with respect to the aim of modelling the behaviour of the system over a restricted area sufficiently exactly. Thus such regression functions may be interpreted as approximations of the unknown "true" production function over the range observed.

2. SUBMODEL : MATERIAL CONSUMPTION

The aim of this submodel is to reflect the interdependences of the sectors from mutual supplies. The main idea for this model was to include all dependences in a complete quasi-dynamic input-output model

MCyy = a*GP1 + b*GP2 + ... + c*GP19 + d*IMPMAT + CORRyy(year)

where MCyy denotes the material consumption and GPxx the gross output of the units 1,..6,10,..,19 (two-digit classification number of the sectors) and IMPMAT denotes the imported materials. The parameters a, b,.. are the output coefficients as shares of the total product of the suppliers for a fixed year. Their sum for each xx is one. They are computed as output coefficients from basic input-output tables, but they do not coincide with the usually applied input-coefficients. The optimal form of CORR is determined by automatic model choice. Since CORR is time-dependent we obtain a quasi-dynamic input-output model.

3. SUBMODEL: DEPRECIATIONS

In this submodel the depreciations had been represented by estimated average depreciation rates for the fixed assets with some time-dependent correction term of the form abs(year - 1975).

4. SUBMODEL : NATIONAL INCOME BALANCES

This submodel connects the national economic main indicators. The formulae include the main widely known balances of the system of national accounts within the frame of the dynamic system model. To emphasize the use of the MPS system we apply the denotation global product for the national gross output. Finally all main balances of the system of national accounts (MPS system, cf. [3]) are included which connect the produced and the used national income.

6. SUBMODEL : CAPITAL FORMATION

By this submodel the cycle of reproduction of the national income included in the model will close. Here the process of investments is represented where fixed assets are put into use and become completed and finally productive fixed assets. In comparison to other macro-economic models (cf.e.g. [1]) it was tried to capture this process more thoroughly in the present model For the different economic sectors have been detected significantly different characteristic time-lags in the completion of investments by statistical model choice. It was one of the aims for the model construction to get in by that the quantitative consequences of alternative investment strategies. The basic form for representing the investment completion as the absolute increase of the fixes assets is

$$DFA \longleftarrow f(\, L0IG,\, L1IG, \ldots,\, L5IG) + \text{completion shocks},$$

As alternative functions f of the investments (gross) in the model choice we used all moving-average combinations from 0 up to 5 years. Therefore one can interpret the corresponding regression coefficient as an average completion rate. In addition to this model of an average completion, in some cases it turned out that an additional shock variable would considerably improve the data fitting. The mathematical representation of these put-into-use waves may be recognized from the following typical cases (The terms in brackets are logical variables with value 1 for "true" and 0 else. The symbol 'mod' denotes the remainder function.):

$u52 = \text{mod}(year - 1 - \underline{2}, \underline{5})/(\underline{5} - 1), v52 = (u52 = 1) - (u52 < 1)/4, w52 = (u52 = 1).$

Changed production conditions of the future may be dynamically simulated by suitably chosen exogenous controlvariables. Therefore realistic forecasts are possible up to 20 years in advance. On the basis of different variants of the basic economic control variables one obtains several medium- and long-term simulative predictions. For simulation there is some specified year where the forecast is to be started. Until this year static simulation is done, that is actual endogenous values are found and fed into the lagged endogenous terms. Then dynamic simulation begins with predicted values being fed into lagged endogenous terms.

[1] Bilow, W.; Grahl, B.; Walter, D. (1981): Das ökonometr. Modell DÖM-2 - ein Modell f. d. Analyse und Prognose der Volkswirtschaft der DDR. In:'Planungs- und Prognosemodelle'.Akademieverlag Berlin. p.56-63.

[2] Höschel, H.-P. (1986): Ein ökonometrisches Modell für Analysen und Vorausberechnungen des Nationaleinkommens der DDR. Österreichische Zeitschrift für Statistik und Informatik 16, Nr.4, 212-239.

[3] UNO (1977): Comparisons of the system of national accounts and the system of balances of the national economy. Part one: Conceptual relationships. U N Dept. of Economic and Social Affairs, Statistical Office, Studies in Methods, Series F, No.20. Sales No. E.77.XVII.6.

STATISTICAL METHODS FOR SUPPORTING THE SELECTION OF STRATEGIES IN AUTOMATIZATION BY REGARD OF SOCIO-ECONOMIC FACTORS

Sigrid Zeipert [1)]

In introducing new technologies, not only technological, economical and organizational aspects should be in the focus of our considerations, but also the position of people within the labour process, a position which is effected by and effecting new technologies.
In selecting, implementing and valuaring strategies of automatization, the consideration of labour- and socio-economical factors is everytime a valuation of human activity, mode of action, conditions, abilities, skills and of other factors. In the valuation and quantification of such factors,in connection with technical and economic aspects of automatization e.g. the following issues should be investigated:

- Which influence have solutions of automatization and on an increasing level also solutions of flexible automatization, towards socio-economic factors?
- How do socio-economic conditions effect in implementing of solutions of flexible automatization?
- Which kind and how dense is the interrelation between technical, economic and social factors?

Herewith, we have to consider, that often just labour- and socio-economic aspects are complicated if we want to take them into account for quantifications.
Mathematical statistics, esp. the theory of correlation, gives a serie of methods, supporting the catching of interrelations between factors, which we cannot quantify or can do it only worse.
In the process of using the theory of correlation in the field of social sciences, the assumption of a normal distribution of figures in many cases is not senseful. Rather it becomes obvionsly, that for a large number of fields the conditions for an implementation of parametic correlation coefficients are not given. It is even true, then, if the economic process to be investigated are normally distributed with a really good approximation. As a criterion proves the circumstance, that often only a relatively small number of character values can be involved for correlation analyses. Moreover, many settings of problems can be

---

1) Doz. Dr. sc. Sigrid Zeipert, Karl-Marx-Universität Leipzig,
Sektion Wirtschaftswissenschaften,
DDR - 7010 Leipzig, Karl-Marx-Platz 9

investigated only with a relatively reduced information basis, because the structure of the data material results in a great proportion from qualitative graduations and valuations, and from the beginning we have to take into account the appearance of predominantly binary features.

Thus, correlation coefficients (CC) free of parameters have a primary importance just for labour- and socio-economic investigations. CC free of parameters can be divided in two groups, i.e.
- CC with a basis in frequency data,
- CC with a basis in rank data.

Those CC free of parameters which are basing in frequency data can be divided in association coefficients and coefficients of contingency.

They can be used, if the phenomena to be investigated cannot be expressed in figures, but can be divided only conceptionally. Such characters are e.g.
- the profession,
- sex,
- the subordination in the management pyramid,
- the experience,
- job satisfaction,
- the qualification,
- the readiness for efforts,
- the work with computer aided remedies.

The term "assoziation" is implemented in particular for the interrelation objectively existing between conceptual valencies of characters, e.g. between job satisfaction (yes/no) and the existing working conditions (good/bad) or the use of information processing techniques (yes/no). Contingency analyses the interrelations between more than two variations of characters, e.g. between job satisfaction (very good/good/bad) and profession respective job (production worker, scientific-technological staff, scientific assistant, ...) or qualification (0, 1, ..., 6).

In such cases, if economic investigations require the finding out of interrelations between two phenomena of which the phenomenon of the one character considered arises in several metrically scaled variations, while these of the ofher character are binary variations, the biserial respective lag correlation can be of use. This concerns e.g. the investigation of the interrelation between the level of automatization (0, 1; ...; 1) and the job satisfaction (yes/no).

Association, contingency and biserial correlation open on the one hand in principle the opportunity of measuring interrelations between characters, but on the other hand, there are a serie of problems in interpre-

ting, e.g. the relatively strong simplification of the variations of characters by fixing them on binary characteristic values.
In analysing correlative interrelations between charakters in their diversity we should take into consideration the rank correlation.
The rank correlation coefficients from S p e a r m a n and K e n - d a l l give the chance of interpreting qualitatively the interrelations between conceptual characters, withont the given precondition of a normal distribution of characters. The rank correlation is above all useful for such analyses, in which quantitative figures on the phenomena to be investigated are indeed impossible, but a rank will be uniquely defined, e.g. for the qualification level (0, ..., 6), the quality of the work (1, ..., 5), management rank (1, ..., 5).
Within the rank correlation the coefficient of concordance is a suitable measure. In difference to the rank correlation coefficient of K e n - d a l l and S p e a r m a n , it measures the interrelation between several rank series and should likewise be used for economic investigations.
The correlation theory allows not only the calculation of coefficients of the mutual interrelation resp. correlation, but permits moreover by use testing methods, to point out whether the interrelation stated is a significant and by provable causes influenced phenomenon or only of accidental character. The correlation analysis, however, does not disclose causes for the interrelations pointed out.
Thus, on the basis of an e.g. high correlation between continuity of production and qualification level of production personal, it should not be drawn the conclusion, that allone an improvement of the qualification structure will improve the continuity of production. It has to be considered, that notwithstanding the mutual (non - causal) interrelation between the figures analyzed, only tendencies are shown, and also other factors have influence.
In selecting and implementing solutions of flexible automatization, the mathematical-statistical instruments mentioned above should be involved in finding out background informations for the decision process.

The paper in continuing the thougts mentioned above shall present a mathematical-statistical and computer-aided method of simple handing, which can take part to the qualification of the decision process in selecting strategies of automatization, and that by its use for interpreting the collective preparation of a decision by inquiry of experts.

MODEL ANALYSIS OF ENERGY-ECONOMY INTERACTIONS

Janusz Cofala, Krzysztof Ostromecki [1]

1. Introduction

The proper social and economic development of both individual countries, as well as of the whole world is impossible without the adequate supply of energy carriers. Energy is the indispensable for all branches of economic activity production factor. The manufacturing of fuel and energy is becoming more and more expensive because of the necessity to reach for more costly primary resources, application of more costly technologies , and also because of the increasing demands in environmental protection. The interactions between the energy system and the economy are becoming ever more complex. Therefore investigation of the development of energy system i.e. the complex of production, transport, distribution and utilization of energy, must be closely bound with investigation of the superior system of the national economy.

This work presents a method of investigating the relationships between the energy sector and the whole economy by use of a set of computerized models designed for application under Polish conditions. The application of these models has been discussed from the point of view of searching for energy system development paths, consistent with macroeconomic development. Although the method is highly formalized, heuristic methods need to be used, too. This is just an outcome of the problem itself, so difficult to be modelled. There are significant problems in adapting adequately the models to reality, caused by many covert mechanisms affecting structure of the investigated system, uncertainty of input data, etc . The models are usually just a simplified image of the reality. However, those simplifications are necessary to make the whole procedure transparent and to receive solution in a short enough period of time. Irrespective of the simplifications, it is purposeful to apply these models reasonably in analyzing the energy-economy interactions. Guidance of experts should be, however, sought in collecting and verifying model input data, selection of data

---

[1] Institute of Fundamental Technological Research of the Polish Academy of Sciences, 00-049 Warsaw, 11/21 Swietokrzyska St. POLAND

sets for analysis within models, and, finally, to assess the results and, eventually, to change the assumptions.

2.Modelling technique

The research into mutual relationship between the energy system and economy is carried out by use of the appropriate set of models called SPSEK. This set consists of four mathematical models : macroeconomic model STRUK, final energy demand model PROSK III and energy supply models DORSEK and SALDO. Interactions between these models and sequence of calculations together with non-formalized (heuristic) operations are illustrated in Fig. 1. The SPSEK set serves for formation and analysis of various development scenarios for the next 20-25 years. General rules of the set functioning are presented below. For more detailed description of separate models one should refer to [1,3].

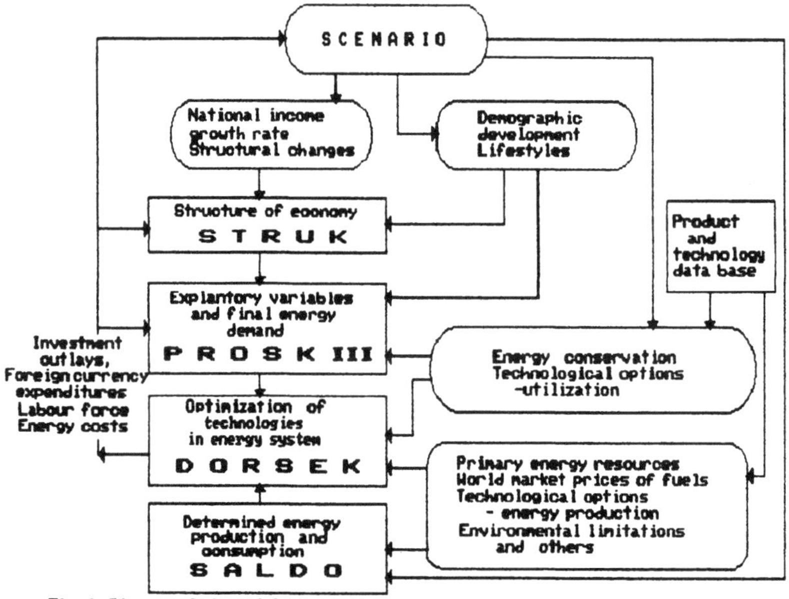

Fig 1. Diagram of the models for medium-term programming of national energy system development S P S E K

The analysis of the energy system development begins with the formulation of the assumptions for the scenario. The term "scenario" stands for a set of mutually interrelated quantities, the values whereof will be decisive in future periods with respect to country social and economic development. The work on formation a scenario begins with the assumptions, i.e. with preliminary fixing of the quantities depicted in it with defining trends and rate of future changes by the scenario-builder. At the beginning those trends are depicted in a verbal form. They are attributed concrete numerical values only during the formation of the scenario. This is performed by use of formalized mathematical models which assure the consistency in calculations.

The first of models entering the structure of the set SPSEK is the STRUK model. The task of STRUK is to generate general macroeconomic proportions for the scenario. STRUK is an aggregated dynamic input-output model. It distinguishes five to six 5 years subperiods. The model specifies 12 sectors, including 3 branches of the energy industry (EI)-coal, non-coal fuels and electricity and 9 non-energy sectors , i.e. metallurgy (MI)-ferrous and non-ferrous, chemical industry (CHI) , building materials (BMI) , equipment goods industries (EGI) , other industrial branches (OIB) , construction (CON) , agriculture (AGR) and other sectors (EOS). The scheme of relationships analyzed in the model is presented in Fig.2.

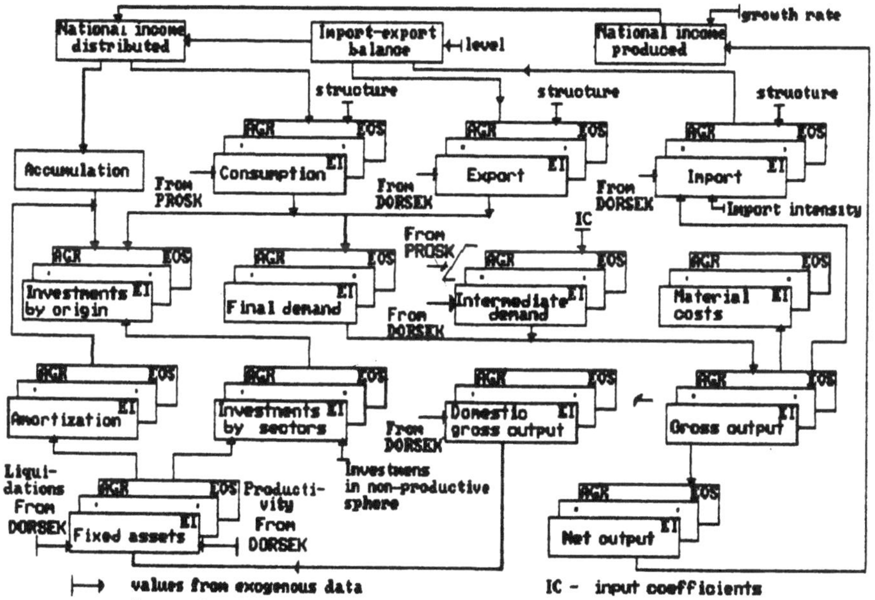

Fig 2. Main relationships among macroeconomic quantities in model S T R U K

STRUK provides a representation of process of formation and expenditure of national income in a form which makes it possible to study the development of the energy system. Modelling is also used for the fixed capital formation, so as bring about the realization of the projected economic development. STRUK operates for the pre-determined growth rate of national income produced and has to take into account several structural limitations describing the permissible shares of sectors in exports, imports, consumption etc. The model generates values of sector output for each 5 year subperiods, maximizing the weighted sum of consumption in the period under study.

The model STRUK is combined with models PROSK III and DORSEK within the system SPSEK. Therefore some input data to the macroeconomic model are endogenous to the whole system SPSEK. It refers first of all to the future trends of sector energy intensities, which are supplied by models PROSK and DORSEK. Also data on energy exports, imports and on productivity changes of fixed assets and immobile property liquidation coefficients for energy industries are derived from SALDO and DORSEK. Other input data for STRUK are the result cooperation between scenario-builders and experts on macroeconomic issues. The experts prepare data using results of analyses and simulations carried out in other research institutes dealing with macroeconomic problems.

Information obtained from model STRUK serves to determine the values of variables which explain the energy consumption. The explanatory variables denote the activity levels of the most energy-consuming economic processes in the production of goods and services. They are measured in natural or monetary units. The natural units are used for measuring output for energy-intensive sectors (for instance steel or cement), or the levels required to meet the energy-intensive needs of the population (space heating, transportation needs etc.). For each variable, an analysis of specific energy consumption is carried out with the help of model PROSK III. It is a simulation model, based on engineering data.

The final energy demand serves as input information for the dynamic linear optimization model DORSEK. DORSEK chooses the most advantageous technological structure in energy production, as well as in exports and imports of fuels. The model also takes into account the vital technological options in the sphere of energy utilization. DORSEK enables to select the technology only for newly built or subjected to up-dating objects for which a decision whether to built has not yet been taken. The balance of energy production and consumption in the already existing objects at the initial horizon, and objects under construction is calculated in the simulation model SALDO. The solution of the model DORSEK makes it possible to provide a fuel-energy balance sheet for the country at the level of primary energy.

Some of the model results are compared with the values assumed at the initialization of the scenario, and with values calculated in other places of the whole procedure through a different model. First of all, comparison is being made of investment expenditures for development of the energy industry received from DORSEK with the expenditures resulting of STRUK. Similarly, comparison is made of fuel export and import values received in these both models. Checked are also the assumed in STRUK tendencies to reduce the energy consumption in individual economic sectors. These tendencies should correspond with conclusions received from energy models. In this way a check can be made to ensure that no glaring inconsistencies exist between the results and the assumptions. If considerable discrepancies do appear, the calculations are repeated, with the necessary corrections in the input data. This assures a compatibility between energy system development path and macroeconomic scenario.

3.Applications

The above described set of models has been used to analyze the expansion of the national energy system up to the year 2005 for several qualitatively differing developmental assumptions. Calculations for a scenario assuming an average growth rate of the economy - similar to that adopted in the 1986-1990 5 year plan have been shown in Table 1. During the initial phase of the scenario writing process a significant increase in non-productive investments was assumed, to satisfy the growing housing demands. The moderate reduction in the energy intensity of the whole economy was also assumed, employing several means of energy conservation. By those assumptions primary energy demand has increased by 16% for the year 1990 and by 52% for the year 2005. To satisfy such a demand an intensive investment program was required resulting in doubling the investments in energy system to compare with previous years. In consequence the relative contribution of investment expenditures in the national income has increased, and only a small increase of consumption was possible. Although the investment expenditures for the energy industry were high, the scenario has shown a rapid decline in hard coal exports, further deepening the deficit in foreign fuel trade.

After discussing the results with the experts it was decided to change some of the input assumptions. First of all, consequences of reducing housing activities were analyzed, what led to reduction in non-productive investments, and it was assumed that more stringent energy conservation methods would be employed. Such changes reduce the national primary energy demand for the year 2005 by some 5% (see table 1) , still permitting to export hard coal. Also current consumption could be increased over the whole period by 4% . The finally accepted plan for 1990 assumes further limitation of energy investments, what is an outcome of an assumption that more pronounced reduction of the energy intensity of the economy would take place by deep restructuring of the economic mechanisms.

Table 1. Characterization of scenarios analyzed 1/

No	Category	Unit of measurment	Year				
			1986	Initial variant		Modified variant	
				1990	2005	1990	2005
1.	Population	10^6 people	37.4	38.6	41.7	38.6	41.7
2.	National income prod.	1985=100	100	119	185	119	186
3.	Per capita consumption	1985=100	100	108	137	112	143
4.	Five-years investment outlays 2/ of which:	1981-85=100	100	133	276	123	260
4.1	non-productive invest.	%	38.0	35.0	35.3	32.7	31.8
4.2	investments in energy industry	%	13.0	25.2	12.1	24.5	12.5
5.	Final energy demand	10^6 toe	124	140	180	134	170
6.	Primary energy demand	10^6 toe	176	204	268	197	255
7.	Energy intensity of national income	1986=100	100	97	82	94	78
8.	Foreign trade balance in energy industry 3/	10^9 US $/yr	-3.3	-4.9	-7.9	-4.3	-7.3

1/ All economic values in constant 1982 prices

2/ Outlays in the five-year period, the last year whereof is specified in the table heading

3/ Foreign currency difference values in exports and imports of fuels in constant prices

4. Conclusions

The up-till-now applications of the SPSEK set of models have confirmed its practical usefulness in analyzing the expansion of fuel and energy complex. SPSEK is not a system giving a "press button" one and only correct answer. It is used rather to back up the analyses of energy sector development including the assessment of the influence of various model parameters on the final result. Cooperation with technical and economic experts enables to analyze a wide range of future changes which - by way of discussing intermediate results and the overall effect - may be modified to create a family of feasible states of the energy system, consistent with the defined directions of social and economic growth.

The research of this type should be carried out in a systematic way, taking into account the changing developmental possibilities, technical progress and changes in the external situation. The results received for scenarios should help to define what needs to be done today to shape the energy future in the desired direction. Model results may be the starting point for discussions and analyses of the energy policy. Such discussions may become more open and concrete. The usefulness of model investigations rests just in creating the foundation for such discussions and not in providing compulsory policy recommendations. The here to now applications of SPSEK were aimed just in that direction and seem to be in line with suggestions by Keepin and Wynne [2] concerning application of model analysis .

5. References

[1] Cofala J.: A Model of the Polish energy system .[in:] Kydes A. , Geraghty D.M: Energy Markets in the Longer-Term: Planning under Uncertainty. Elsvier Science Publishers B.V. Amsterdam 1985.

[2] Keepin B., Wynne B.:Technical Analysis of IIASA Energy Scenarios. Nature Vol. 312, (1984) pp. 691-695.

[3] Ostromecki K., Cofala J.: Macroeconomic Modelling in Relation to Energy System Development [in:] Macromodels`86.Polish Scientific Publ.(PWN) Warsaw. (forthcoming)

COST MINIMIZATION IN A FIRM'S POWER STATION

Günter Fandel and Joachim Reese*

1. Introduction

Energy represents for every industrial site a fundamental process input which can only be substituted with great difficulty and without which industrial manufacture in general is impossible. The requirement is hereby concentrated on the utilization of distinct types of thermal energy for the process steam and heating, as well as for power and light. According to WOLF (1987) the total demand in 1981 for energy for industrial purposes in the Federal Republic of Germany was divided as follows - 75.9% for the manufacturing process, 10.9% for space heating and 13.2% for power and light. Although several energy resources were used equally for the production of heat, power and light were generated almost without exception by means of electricity. Heat as well as power and light, or rather electricity, are all commodities which have only a limited storage capacity. This leads to considerable procurement problems when the energy requirement fluctuates.

In the case of the external supply of all types and amounts of end-use energy the difficulties often lie in peak demands, because they lead to considerable extra cost. Even if the peak values are only sporadic, i.e. they occur for relatively short periods of time, they nevertheless make a substantial contribution to the overall cost in the firm in that they necessitate standby quantities at the supplier which cause in turn an according standby tariff. This is why the main concern must be to subject the period requirements to regular checks with a suitable operational policy in mind. According to this suitable instruments are either temporary delivery cuts or special energy usage programs. In spite of the many proposals in this area (e.g. CONSTANTOPOULOS et al. 1983, FATTI 1983 and WOLF 1987) the results are often usually unsatisfactory because breaks in production occur or unforeseeable demand cannot be calculated with sufficient accuracy.

For this reason it is often to be recommended that the problem basis be broadened to include the self generation of energy. The efficiency of a firm owned power-station attached to a plant and used to supply energy for production in a medium-sized firm in the electrotechnical industry has been tested here with the aid of the classical stage-oriented calculation (see e.g. ALBACH 1962). As the power-station is integrated within the organization of the firm as a cost center, within

*Prof. Dr. G. Fandel and Prof. Dr. J. Reese, Fernuniversität Hagen, Chair of Business Administration, P.O. Box 940, 5800 Hagen, F.R.G.

which the internal energy requirements are to be guaranteed at the most advantageous costs possible, and does not have to produce any market performance of its own, the analysis has accordingly to limit itself to an investigation of the costs incurred. Certainly, the outside supply of electricity will be included in the study, but the possibility of releasing energy into the public network is left out of the considerations. The data used in the study were collected over the period of a calendar year. Because of the extreme seasonal and shiftbased fluctuations in energy demands the data were at first arranged according to suitable periods. Finally, special calculations for each period of the year observed were carried out before the focus was centered upon the procedural policy for the whole year.

2. Description of the problem

The power-station produces as a final product steam of 10 bar and 1.5 bar for the manufacturing process and for heating, as well as electricity for use in various areas within the plant. Complete self generation of the required end-use energies is not possible as the production of heat and electricity are coupled with one another in such a way as to make an exact coordination of the production levels with the requirements data impossible. In practice, this problem of cogeneration of thermal and electric energy is taken into account by first of all using the units to produce heat. After the given volume of process and space heat has been supplied, the amount of electricity is then the balance of the production, a balance which has to be increased through outside supply in order to meet the respective current demand. It is possible that the self generation of electricity in larger amounts would be more profitable, but this would cause pass-out steam, which would have to be disposed of. In view of the resulting costs of the disposal, and of the fact that equipment for recycling the steam as condensed water is not at present available, it is not necessary to discuss such a policy any further.

The analysis is rather more concerned with short-term objectives, and is aimed at the operational policy of the power-station on an hourly basis. All cost estimates are based on peak demand values which are regarded by the management as representative of the respective season and shift. The fluctuations in demand within the periodic delimitations are so small they can be ignored, so the deviations between the peak values and the average values are no longer relevant for the purposes of the analysis.

The generation of the end products process heat (10 bar), space heat

(1.5 bar) and electricity takes place by means of a multi-stage production process which can be exclusively described by means of linear technologies, and which ends in a joint production of all end products at the final stage.

At the first level, completely desalinated water is produced from surface water, mains water and recycled condensate. This can be converted into steam at the next level by means of three oil-fired boilers, K1, K2 and K3 alternatively. K1 generates steam at 64 bar, K2 at 32 bar and K3 at 10 bar. The steam pressure is then throttled before the steam can be consumed in the generation of electricity or directly used in the production area of the firm. The throttle valves E1, E2 and E3 are available for pressure reduction; the generation of electricity can take place using two turbines, T1 or T2, which have different capacities and different efficiencies. The utilization of the turbines causes a cogeneration of electricity and steam, i.e. the amount of electricity produced results at the same time in the generation of steam at 10 bar and 1.5 bar in fixed proportions.

Pollution emission standards prevent the combining of all technical process variations in the power-station here under scrutiny. Although the throttle valves and the turbines are not subject to any restrictions other than those of capacity, there are special regulations which have to be observed when using the boilers. In particular, the use of boilers K2 and K3 is ruled out while boiler K1 is in operation. Boilers K2 and K3 can, however, be operated at the same time.

3. Results

As far as the periodic differentiation of the analysis is concerned, it is worth noting that in 6 out of 9 survey sections the most satisfactory generation of energy in terms of cost was achieved mainly by using boiler K1 and turbine T1. A rigid joint production using T1 leads automatically, nowever, to the situation that the total demand for space heating cannot be satisfied, as the ratio of the energy exploitation does not correspond to the ratio of the demand quantities. In the present case the demand for 10 bar steam cannot be fully satisfied, as surplus production from T1 would lead to the inadmissible pass-out of 1.5 bar steam. Because of this, the demand for 10 bar heat is partially met (sometimes more than 50%) through the throttling of the original 64 bar steam by feeding in desalinated water. This causes, however, a simultaneous reduction in the amount of steam available for the generation of electricity. If the environmental regulations were relaxed so as to make a surplus production of 1.5

bar steam possible, or to enable steam to be eliminated or condensated at reasonable costs, T1 could then definitely be employed economically to full capacity. Under the prevailing circumstances the turbine's maximum capacity of 5,985 kWh can only be used to 44% . The remainder of the required electricity has to be dispatched from outside.

An exact hourly observation leads to the recommendation that, for reasons of economy, the smallest boiler K3 be utilized in the summer months July and August in which the demand for space and process heat is comparatively low. In the early shifts of this months boiler K2 is to be used as a back up for the steam generation, while K1 does not come into service at all. Because K3 can only generate 10 bar steam this policy means that at the same time almost all the necessary electricity has to be supplied from outside. The power-station's activities are reduced to the minimum of self-supply of heating and process heat.

The analysis for the whole year demonstrates, however, through the concrete requirement figures, that the simple addition of the optimum results in each period does not lead to an optimal policy. Merely the fact that for July and August a policy existed of obtaining the total amount of electricity from outside would have a major influence on the contract price for the whole year. This spillovers must be subsequently measured and assessed. Compared with the solution mentioned above a permanent utilization of boiler K1 even in the summer months is successful in reducing the costs of outside electricity, but at the expense of an increase in running costs in those periods. This offset in costs is effective inasmuch as the advantages of a lower electricity contract throughout the whole year outweigh the increase in the power-station's operating costs in the summer months. The savings amount to about 225,000 DM; in view of the annual costs of 15 million DM this means a saving of 1.5%.

A thorough-going recommendation of a strategy for the utilization of the power-station would have consisted ultimately, given a priori knowledge of the demand figures used in the calculations, in electricity generation with turbine T1 being pushed permanently as far as possible and using outside supplies only in cases of technologically caused shortages. This is justified by the fact that, on the one hand, self generation is fundamentally advantageous as compared with outside supply because of the existing oil and electricity prices and the more favourable technological alternatives. On the other hand, in a comparison of these technological possibilities in a narrower sense, 64 bar steam provides the best efficiencies as regards the generation of electricity and heat, an advantage which justifies the insignificant increases in oil and personnel cost as compared with the other alternatives. The amount of water necessary remains irrelevant as regards the utilization of boilers, because the degree of effectiveness is constant for all process variants - one ton of water provides one ton of steam. Instead of desalinated surface or mains water the cheaper condensate, recycled from the manufacturing process, is to be preferred for the generation of steam. Only the residual requirements need then be treated.

4. Subjects of further research

The aforesaid practical study was occasioned by the general, non-specific uncertainty in the firm as regards the economy of its own power-station, which was planned and constructed under completely different environmental conditions. Changes in the price of raw materials, especially fuel oil, and pollution emission standards were at that time either not foreseeable or played a minor part. The economic analysis, therefore, had to take on the function of a control test, which was to provide additional information about the flexibility and stability of management policy as pursued up to the present, and to identify any corrections which might prove necessary. A further aim of the study was to recognize, as exactly as possible, the structure of the multi-stage production problem for processing various energy products, and to use this to propose instruments for a systematic method of assessment and planning, which would generally be of assistance in decision support.

In order to find a practical solution precisely for this second object it is necessary to extend the problem's horizon, starting directly at the chosen definition of the problem at hand. There are first of all fluctuations in demand for the individual types of energy and developments in the resource prices, unforeseeable when the yearly plans were being made, which necessitate a sensitivity analysis of the proposed solution. For example, the breakeven point for electricity supplied from outside as compared with self generation using turbine T1 is roughly 7.22 Pfg/kWh. In comparison, the overall delivery price of electricity at present totals 14.04 Pfg/kWh.

Fluctuations in demand within the plant can be allowed for with greater accuracy by coupling the decisions concerning the energy generation more closely to the production planning in the firm as a whole. WOLF (1987) provides information about this possibility.

As far as long-term strategy for the utilization of the power-station is concerned there are several investment possibilities which present themselves after taking stock of the whole situation, and which can certainly be of service as regards the optimal policy. Firstly, it is worth checking the capacities of future replacement units. In view of the fact, for example, that the average load of the boiler K1 is only about 60% and that its maximum load does not reach 90%, and that boilers K2 and K3, and turbine T2, are mainly only used during breakdowns of K1 and T1, the question can be posed as to whether breakdown probabilities and environmental uncertainties justify organizational slack on such a large scale. Secondly, an increase in investment can also be considered as a means towards long-term optimal policy. For example, policies based on the profitable use of electricity will require the installation of extra condensers to enable the surplus generation of steam in the first place.

References:

[1] Albach, H., Produktionsplanung auf der Grundlage technischer Verbrauchsfunktionen, in: L. Brandt (ed.), Arbeitsgemeinschaft für Forschung des Landes Nordrhein-Westfalen, No. 105, Westdeutscher Verlag GmbH, Köln-Opladen 1962, p. 45-98.

[2] Constantopoulos, P., et al., Decision Models for Electric Load Management by Consumers Facing a Variable Price of Electricity, in: B. Lev (ed.), Energy Models and Studies, North Holland Publishing Company, Amsterdam-New York 1983, p. 273-291.

[3] Fatti, L.P., Optimal Smoothing of Demand for Industrial Gas, in: The Journal of the Operational Research Society, 1983, p. 583-590.

[4] Wolf, J., Industriebetriebliche Energienutzung und Produktionsplanung, Physica-Verlag, Heidelberg-Wien 1987.

DYNAMIC ENERGY PRODUCTION MODEL

Knut RICHTER[1)]

A dynamic one-costumer double fixed cost model of producing and storing energy ist studied. An algorithm will be provided which generates optimal one-peak solutions, i.e. solutions satisfying that production is switched on at most one time.

1. Introduction

A power station is producing and storing energy for one (aggregated) costumer over T periods. The demand of the costumer is assumed to be known and is denoted by d(t), t=1,2,...,T. The main problem is to decide how much energy x(t) has to be produced at every period. Energy can be stored by the producer. The capacity of the energy store is restricted by s(t) units. At one period not more than out units can leave the store and not more than in units can be transfered to the store. Furthermore, the capacity of the producer is limited by c(t) units. The energy available at the store before the planning periods is known by yb and ya units must be at the store at the end of period T. So far supposing that $y(t) = y(t-1) + x(t) - d(t)$ (0) holds for all t the problem is to decide whether there exists a feasible vector $x = (x(1),...,x(T))$ satisfying

$$0 <= y(t) <= s(t), \quad (1)$$

$$0 <= x(t) <= c(t), \quad (2)$$

$$d(t) - out <= x(t) <= d(t) + in, \quad (3)$$

$$y(0) = yb, \; y(t) = ya. \quad (4)$$

1) Technische Universitaet Karl-Marx-Stadt, Sektion Wirtschaftswissenschaften, PSF 964, Karl-Marx-Stadt, 9010

The energy y(t) stored is to be nonnegative if the demand d(t) shall be satisfied.

Two types of cost inputs are included in the model:

(i) Cost cg appears when switching on production, i.e. if

$(1 - \text{sign } x(t-1))*\text{sign } x(t) = 1.$

(ii) Cost cp appears when the production is not equal zero, i.e. if

$\text{sign } x(t) = 1.$

The ojective fuction of the model is, therefore, given by

$$\sum_{t=1}^{T}(cg(1 - \text{sign } x(t-1)*\text{sign } x(t) + cp*\text{sign } x(t) =$$

$$\sum_{t=1}^{T}(cg*(1 - \text{sign } x(t-1)) + cp)*\text{sign } x(t) \quad \text{---> min.} \qquad (5)$$

In many cases production figures x(t) should not be too small. This is the reason to demand $x(t) \in \{0\} \cup [low,up]$, (6)

i.e. if x(t) is not zero then it is expected to exceed the value low. Finally, the production figures of period 0 (and perhaps of T+1) are assumed to be known, i.e. x(0) = xb and x(T+1) = xa. (7)

The restrictions and the objective (0) - (7) provide the model under study. Introducing low(t) = max{d(t) - out,low} and up(t) = min{d(t) + in, c(t), up} the restrictions (2), (3) and (6) can be replaced by

$$x(t) \in (\{0\} \cup [low(t),up(t)]) \cap [d(t) - out,up]. \qquad (8)$$

Model {(0),(1),(4),(5),(7),(8)} is a dynamic two-stage model and a recursion can be developed if the stages y(t) and x(t-1) are fixed for given periods. In other word, two-dimensional dynamic programming can be applied to the problem (Bellmann 1957). The situation to be studied here is that the production shall be switched on at most one time. Then switch-on and switch-off periods and the distribution of the production between these periods are to be determined. A solution of this type is called one-peak solution. The next section will provide an algorithm generating an optimal one-peak solution.

2. Generating optimal one-peak solutions

Let t1 and t2 be the switch-on and switch-off periods, respectively. If SET denotes the set of feasible pairs (t1,t2) and COST(t1 ,t2)= = (t2+1-t1)*cp+(sign(T-t2)*sign xa+(1-(1-sign(t1 -1)*sign xb)))*cg then the problem reduces to

COST(t1,t2) ---> min,

$(t1,t2) \in$ SET. (9)

The algorithm sketched here uses two main procedures

(P1) Generating candidates (t1, t2).

(P2) Testing $(t1,t2) \in$ SET and comparing the costs.

Let the two procedures be regarded in more detail.

Lemma 1: $(t1,t2) \in$ SET implies t1 <= on and t2 >= off where

on = 1 + max $\{t<=T: \sum_{r=1}^{t} d(r) <= yb,\ d(r) <= out\}$ and

off = -1 + min $\{t>=1: \sum_{r=v}^{T} d(r) <= s(v) - ya,\ d(v) <= out,\ v=t,...,T\}$.

Lemma 2: $(t1,t2) \in$ SET implies min <= t2 + 1 - t1 <= max where

min (max) = min (max) $\{t2+1-t1: \sum_{t=t1}^{t2} low(t) <= gd <= \sum_{t=t1}^{t2} up(t)\}$

and $gd = \sum_{t=1}^{T} d(t) + yb - ya$.

Procedure (P1) will systematically generate pairs (t1,t2) such that t1 <= on, t2 >= off and min <= t2 + 1 - t1 <= max.

Let now (t1,t2) be provided by procedure (P1). Then two problems have to be solved. First, $(t1,t2) \in$ SET must be tested. Secondly, COST(t1,t2) is to be compared with the previously determined values. Since there is no problem in comparing the costs let the test be studied. Checking $(t1,t2) \in$ SET means to determine whether the system consisting of restrictions (0),(1) and (8) for t=t1,...,t2 and y(t1-1) = yt: $= yb - \sum_{t=1}^{t1-1} d(t)$, $y(t2) = ya + \sum_{t=t2+1}^{T} d(t)$ is solvable.

This system is denoted by (S).

Theorem: System (S) is solvable if and only if the algorithm TEST finds a feasible solution.

Algorithm TEST:

```
Procedure forw;
begin for k: = t to t2 do
   begin delta: = low[k]-d[k]-y[k]+y[k-1];
   if delta >0 then y(k):=y(k)+delta end;
   if y[t2] >0 then end. {no solution}
end;
begin d[t]: = d[t2] + y[t2]; y[t2]: = 0;
   for t: = t2  to t1 do begin
   y[t-1]: = max(0,y[t] + d[t] -up[t]); forw end;
   if y [t1 - 1] > yt then end.{no solution}
   if y [t1 -1] < yt then begin k: = t; y[k-1]:= yt; forw end;
   for t:=t1 to t2 do if y[t]>s[t] then end. {no solution}
   {(t1,t2) is in SET and x(t)=d(t) + y(t-1)-y(t) is feasible}
end.
```

Example: t1=1, t2=3, yb=0, d(t) = (1,4,2), low(t) = (2,3,2)

up(t) = (2,4,3). Then the algorithm proceeds as in the following

table.

t	y(t)	y(t-1)	k	delta	y(k)
3	0	0	3	0	-
2	0	0	2	0	-
1	0	0	1	1	1
			2	0	0

Table 1: Illustration of the algorithm

The vector y = (1,0,0) will be found by the algorithm and tested if it fulfils the store restriction (1). The corresponding vector x = (2,3,2) can be easily determined. In this case - if s(t) = 1 - the algorithm recognises the solubility of the system and secures that (t1,t2) $\in$ SET.

This statement generalizes the solubility condition of the author (Richter 1982, p.60) for the system with low(t) = 0 for all t.

The full algorithm for generating optimal one-peak solutions has been implemented by F. Baumgartl (TU Karl-Marx-Stadt).

References:

Bellmann, R., Dynamic Programming, Princeton University Press, Princeton (N.J.) 1957

Richter, K., Dynamische Aufgaben der diskreten Optimierung, Akademie-Verlag Berlin 1982

THE GROWTH OF ELECTRIC ENERGY PRODUCTION AS AN EXAMPLE FOR THE APPLICATION OF THE EVOLON MODEL

W. Mende[1] and K.-F. Albrecht[2]

The growth of electric energy production in different countries can be discribed by the evolon model during sufficiently undisturbed periods. The found cooperativity parameters of the model are in agreement with a classification rule, connecting these values with the character of the growth process. The data for some countries show decreasing absolute growth rates during the last decade. This leads to important changes in the growth dynamics of this key indicator of national economy. The evolon model reflects this increasing dominance of the saturation process with high accuracy. On the basis of this model an attempt can be made to project the electric energy production for the next decades. According to that model and in agreement with the data of some countries the increasing dominance of the saturation process characterized by falling absolute growth rates leads to a completly new situation in the growth dynamics in comparison to the situation in the last decades.

The evolon model $\frac{dx}{dt} = \dot{x} = cx^k(B-x)^\ell$, $x(t_0)=x_0$, $\emptyset \leq x(t) \leq B$ /3-5,7/ has proved to be a suitable tool for the description of growth processes in systems characterized by a highly aggregated growth indicator x(t). The classification rule /7/,/1/ connects the values of the cooperativity exponents k and l with the character of the corresponding growth process phase. These values are:

Table 1: Expectations and results for the cooperativity exponents

Nr.	phase of the growth process	country	fit region	expectation k	expectation l	result $k^{+\sigma^+}_{-\sigma^-}$	result $l^{+\sigma^+}_{-\sigma^-}$
1	substitution during initial phase	USA	1902 - 1929 (-1921, -1922)	< 1	/	0.664 + 0.026 - 0.027	/
2	substitution during initial phase	Germany	1900 - 1920	< 1	/	0.689 + 0.019 - 0.041	/
3	reconstruction	USSR	1946 - 1955	< 1	/	0.648 + 0.018 - 0.015	/
4	overfulfilment of the GOELRO-plan	USSR	1925 - 1931	?	/	1.494 + 0.100 - 0.095	/
5	advanced electrification with known principles but unknown final target	USSR	1956 - 1985 - (1981 - 1983)	≈ 1	> k	1.028 + 0.038 - 0.027	2.2 + 0.[illegible] - 0.6
6	advanced electrification as pioneer-process (no known principles no final target)	USA	1925 - 1955	> 1	> k	1.461 + 0.012 - 0.013	?

a) less than one (parabolic growth or saturation) if the process has a final target and follows a plan (Processes of substitution and recon-struction are special cases of this process type)

b) greater than one (hyperbolic growth or saturation) if the process is an unplanned unique evolution process.

----

1) Institute for Geography and Geoecology, Academy of Sciences, Berlin (G.D.R.).

2) Central Institute for Cybernetics and Information Processes, Academy of Sciences, Berlin, (G.D.R.).

The used data for electric energy production are taken from statistical year books /2,6,8,9-12/. The data selection and the method for parameter identification, the mean relative deviations of the data from the model (R) as well as the projection of the evolon model curves for some decades into the future are discussed in /1/. The main results are the following:

For all of the six investigated growth periods the obtained values for the cooperativity exponents are in agreement with the classification rule (table 1).
For the initial phases of electrification dominated by substitution processes, growth cooperativity exponents k have been obtained which lie clearly in the parabolic region (fig.1).

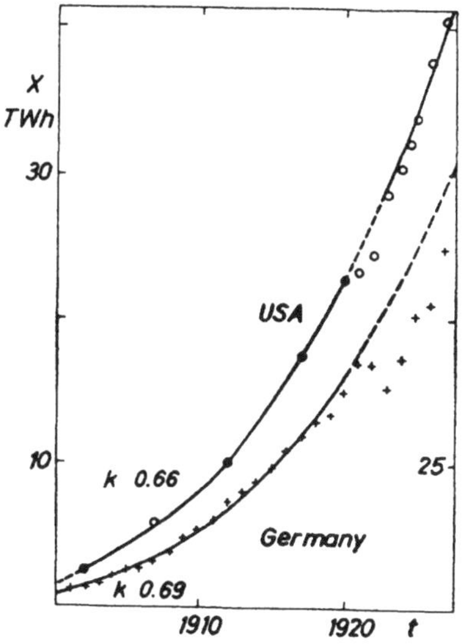

Fig.1 Parabolic growth at the beginning of electrification (substitution).

For the investigation of the undisturbed growth of electric energy production in the advanced electrification period the data from the Soviet Union after World War II seem to be most suitable. The evolon model analysis of these data leads to the following results:

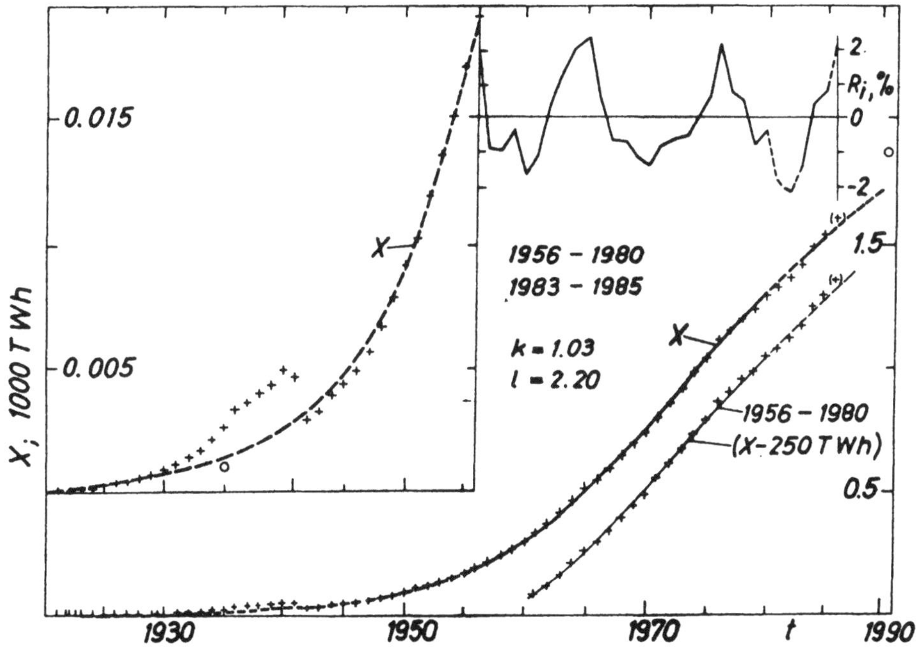

Fig.2 Evolon model curves and data for the electric energy production in the Soviet Union. Fit region 1956-1985 and fit region 1956-1980 (downward shifted by 250 TWh). The relative data deviations from the model values $R_i = (x_i^d - x_i)/x_i$ are illustrated also.

a) For the first ten years after the war when reconstruction was going on parabolic growth comes out (k=Ø.648 , $\tilde{R}$=Ø.83%).

b) For the following ten years (1956-1965) k=1.ØØ9 and R=Ø.81% was obtained. Therefore, as well known, in this period the data can be described well also by the exponential growth model.

c) After 1965 the influence of the saturation process is not further negligible. With additional reasonable assupmtions one gets for the description of the data from 1956 to 1985 the following intervals for the evolon model parameters: 1.Ø<k<1.Ø66; 1.8<l<3.Ø and 255Ø TWh<B<35ØØ TWh. For the best parameter combination (k=1.Ø3, l=2.2 and B=2872 TWh.) the mean relative deviation of the data from the model values is 1.26% (fig.2). The large value for the saturation cooperativity exponent l>1.8 reflects the great complexity of the saturation process.

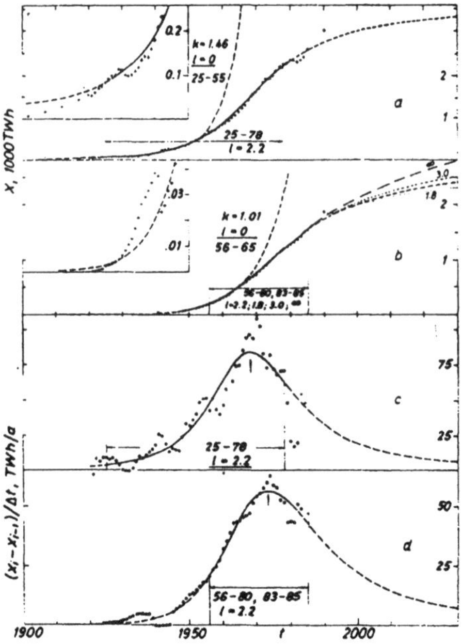

Fig.3 Evolon model curves for the electric energy production:
a) in the USA (fit region 1925-1978) without saturation (l=0) and with saturation (l=2.2 fixed).
b) in the USSR (fit region 1956-1985 without 1981 and 1982, l=1.8; 2.2; 3.0 and l=∞.
c) as fig.4a (l=2.2) but here the annual increase is represented and compared with the corresponding data values (crosses) and with the 5-year moving averages of the data (circles).
d) as fig.4b but here the annual increase is represented and compared with the corresponding data values (crosses) and the mean values over 5 years of the data (circles).

d) For several years the transition from the extensive growth process of the electric energy production to the intensive saturation process has taken place. The saturation began already to dominate. This comes out from the evolon model but can also be seen from the data itself (fig.3d). In 1973 the maximum for the absolute growth rate of the model was reached. The model values for the year 199Ø came out clearly smaller than the target value.

e) Oscillations of the electric energy production data arround the general trend curve have been found. A spectral analysis gives domina-ting oscillation periods at 11, 6 and 4 years (fig.2).

f) The projected model values up to the year 2Ø3Ø depend only weekly on l-values inside the region 1.8<l<3.Ø (fig.3b).

For the investigation of the pioneer process in the advanced electrification period only the data from the USA can be used. Here the developement is more irregular and the determination of the saturation parameters is more difficult. The results obtained are:

a) With the assumption l=2.2 one gets k=1.43. The growth trend for the pioneer process is clearly hyperbolic as it is expected from the classification hypothesis.
b) Although there are great deflections of the data from the general trend one can see that the saturation process for the growth of electric energy production in the USA is going on already (fig. 3a, c).
c) The considerable retrogression of the electric energy production after 198Ø is in contradiction with the monotonous long term trend of efthe evolon model. Therefore on the basis of this model a return to the monotonous rising long term trend has to be expected. The data for 1983 and 1984 and the prognostic values given in the stastisical abstracts of the USA confirm this expectation.

Because saturation is taken into account the evolon model values for 199Ø lie 8-9% lower than prognostic values recently published for the USA and for the USSR.

We would like to express acknowledgement to Professor M. Peschel and Professor J.M Swireschew and his coworkers from the computer centre of the Academy of sciences in Moscow for promoting interest and for stimulating discussions.

References

1. Albrecht K.-F., Mende W. Growth analysis of electric energy production by means of the evolon model, Syst.Anal.Model.Simul.5, (1988)
2. Històrical Statistics of the United States and Statistical Abstracts of the United States
3. Mende W. The evolon model - a model for description, characterisa tion and classification of growth processes, in preparation
4. Mende W., Peschel M. IFAC-Symposium on Control Mechanisms in Bio- and ecosystems, Plenary paper, Leipzig, 1977.
5. Mende W. The evolon Model and its application to Natural processes in Approach to Cooperation and Competition in Dynamic Systems, ed. W. Ebeling and M. Peschel, Akademie Verlag, Berlin 1985, p.261-271
6. Mitchell: European Statistics 175Ø - 197Ø.
7. Peschel M. Mende W. The Predator Prey Model, Akademie Verlag, Berlin 1986.
8. Statistisches Jahrbuch der BRD, Jhg 1984
9. Некрасов А.М., Троицкий А.А. Развитие электро-энергики Союзных республик, Энергоатомиздат, М. 1982.
10. Стеклов В.Ю 60 лет Ленинского плана ГОЭЛРО, Энергия, М., 1980.
11. Народное хозяйство СССР за 60 лет, Статистика, М., 1977.
12. Народное Хозяйство СССР в 1960г 1985г, Статистика, М., 1986.

ANALYSIS AND SIMULATION OF THE DYNAMIC BEHAVIOUR OF SATURATED STEAM TURBINES OF PWR NUCLEAR POWER PLANTS

Gacs, Andras [1)]

ABSTRACT

A digital simulation model of the components of a PWR (pressurized water reactor) type nuclear power plant has been developed in a modular structure in CRIP. This paper deals with the detailed modelling of the transient characteristics of turbines using saturated steam. This model calculates the discontinuity of the expansion line during transients simulating the turbine more exactly than by the conventional models.

KEY WORDS: dynamic simulation, continuous nonlinear systems, modular model structure, weak coupling, stiff character, power plants, secondary circuits, saturated steam turbines.

INTRODUCTION

The analysis of the dynamic behaviour of the secondary circuit of a PWR nuclear power plant type WWER-440 is needed to optimize the transport processes and to fulfil strict safety requirements. There is a demand to construct real time simulator equipments for training plant operators and to prevent serious accidents this way.

A real time digital simulation model has been developed and verified to measurements in CRIP. The basic aspects of the modelling are discussed in [1], the modular structure of component models in [2], the detailed turbine model with other secondary circuit models in [3] and the turbine controller with other subsystems are discussed in [4].

This paper deals with the detailed modelling of the transient characteristics of turbines using saturated steam. The turbine model is a complex structure of continuous nonlinear subsystems in a form of a series of stage submodels. These are described by coupled sets of ordinary differential equations with stiff character, integrated with changing time steps and by different integration methods.

The turbine model is a component in the modular system in which the models represent the main physical processes separated from each other by the weakest possible coupling. The coupling parameters are selected in most cases from measurable ones in order to use real transient data for tuning the models.

STRUCTURAL CONDITIONS

The modules of the system follow the technology (cf. Fig.1). The turbine model has two submodels: a one-flow high-pressure cylinder (HP) and a two-flow low-pressure one (LP). Their connections to other models with the coupling parameters are given next, the scope and approach of these models and their solution are mostly quite non-conventional (e.g.: the direction of the information prevents the time-consuming successive approximation technique needed for the conventional models).

The model of the turbine control valves (CV) is a set of control nozzles with sonic flow of the main saturated steam supplied from the steam generators (SG) through an isolation valve (IS). The open cross-sections of these valves and that of the intercept valve (IV) are governed by the turbine controller and protection (TC) model providing all the important features of the real turbine control of the plant. The steam consumption of the HP is calculated by the HP submodel, the inlet steam pressure is given by the CV.

1) Computerized Reactor Control Dept. of CRIP
(Central Research Institute for Physics)
P.O.Box 49, Budapest, H-1525, Hungary

The moisture separator and reheater unit (MSRH) is connected between the HP and LP to dry and overheat the working steam, the steam parameters and flows are calculated in a similar way (HP gives its exhaust data to and LP gets its inlet data from the MSRH model) and there is no need for any iteration.

The exhaust of the LP is led to the main steam condenser (CO) and the exhaust pressure is given by the CO, the steam flow is calculated by the LP. This method ensures the simple approach of the turbine model (a two-step solution) and that of the condensate- (LPR) and feedwater- (HPR) reheaters (bleed pressures are continuously available from the HP and LP submodels, bleed flows are calculated by the models of these reheaters and by that of the deaerator and feedwater tank). Check valves built in the bleed lines are modelled as well.

The electrical generator model (EG) calculates the speed based on the rotor moment (mechanical power of the turbine). The EG model has its own control submodels (voltage, exitation) and protections.

Besides the mentioned controls, inputs from technological logics (TS), commands from the unit power controller (PC) and the direct commands of the trainee from the "plant operator's desk" of the simulator equipment (OD) are also taken into account.

The parameters of the grid (electricity network) and that of the cooling water (temperature) have direct effect via the EG and the CO models, respectively. The other components of the power plant have indirect coupling effects (e.g.: the reactor through the main steam parameters and the steam consumption of the turbine).

The model system contains two secondary circuits (two turbine lines) according to the technology. The turbine models of these circuits are identical.

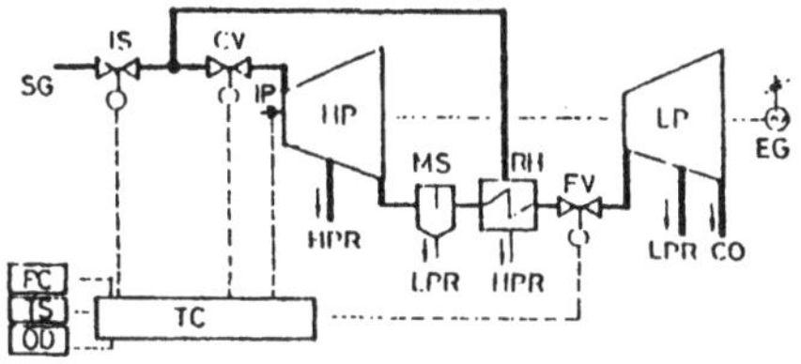

Fig. 1

Technological Chart

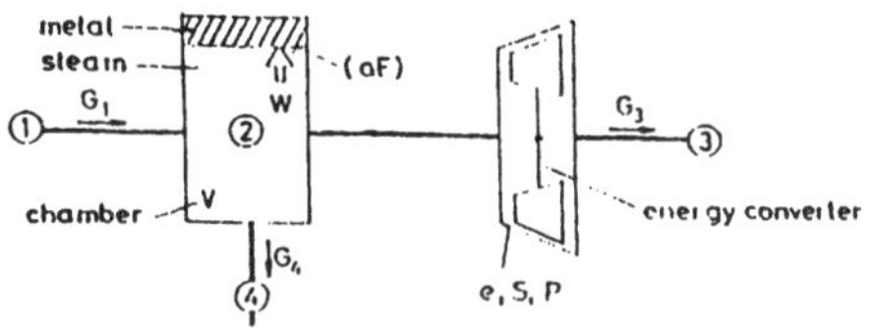

Fig. 2

Scheme of the Model

STEADY STATE CONDITIONS

To describe transient characteristics correctly, it is essential to define proper integral variables. It is presumed that the initial state of simulation is a stationary one so it is calculated by steady state models setting first order time derivatives by zero. Dynamic simulation is started from steady state only. Thus two model-versions developed for each component of the modular model structure. One is for steady state (initial values are calculated) and one is for the dynamics (using the initial values at the first time step and the disturbances later).

The steady state solution is rather detailed to provide accurately balanced initial parameter values. (The dynamic solution keeps these values until the disturbances.)

DYNAMIC SIMULATION

The dynamic turbine models (HP and LP) are composed of point-models of turbine stages. Each of them has two parts (cf. Fig.2): one is a chamber for storing and distributing inlet steam, the other is an energy converter assuming the blading (nozzle- and blade-rows) of the stage as one unit. Tunable parameters are choosen for adjusting the model to the measured real transients.

As for the chamber, the tunable parameters are: its volume (V), the equivalent mass of the wall (m), the surface (F) and the coefficient (a) of heat transfer from the steam to the wall. The simulated parameters are the specific density (r) and enthalpy (h) in the chamber and the wall temperature (t). Mass flows (G), heat flows (W), specific heat (c) and a lot of other values are also calculated.

The corresponding time (T) derivatives are obtained from mass- and heat-balance equations and from the calculation of heat transfer from the steam stored in the chamber to the equvalent mass of the wall. Assuming isothermic conditions outside the cylinder, but taking into account the amount of heat stored in the metal of the stage, the most simplified forms of these differential equations are (indices are according to Fig.2 and "w" means "wall"):

$$dr(2)/dT = [\ G(1) - G(3) - G(4)\ ] / V(2) \tag{1}$$

$$W(w) = a * F * [\ t(2) - t(w)\ ] \tag{2}$$

$$dt(w)/dT = W / m(w) / c(w) \tag{3}$$

$$Q(2) = G(1) * [\ h(1) - h(2)\ ] / V(2) / r(2) \tag{4}$$

$$dh(2)/dT = Q(2) - W(w) \tag{5}$$

Enthalpy and relative density define the other state properties of the steam: entropy (s), pressure (p), temperature (t) and relative steam content (x).

As for the energy converter, assuming that the mass flow of steam through the turbine follows the cone-law for the performance of stages at varying loads, the Stodola-constant (S) is given by the G(3) flow and the p(2), p(3) pressures. The steady state model provides the accurate constant for S, hence G(3) is given by the pressures.

The thermodynamic efficiency (e) and the expansion line vary with the flow. The various losses and their flow-dependencies are tunable parameters. After an adiabatic expansion from p(2) to p(3), the h(A,3) adiabatic specific enthalpy can be calculated from steam properties. Calculating the next, h(3) point of the expansion line after a non-adiabatic expansion, this and the power output of the stage (P) can be defined easily by:

$$h(3) = h(2) - \langle\ e * [\ h(2) - h(A,3)\ ]\ \rangle \tag{6}$$

$$P = G(3) * [\ h(2) - h(3)\ ] \tag{7}$$

The calculation of the CV-HP-MS-RH-IV-LP-CO model line is solved in two steps: first the pressures are given by the steam property functions from CO to CV, then mass flows are calculated from CV to CO. Values of r(2), h(2), t(w) are integrated parameters, and their derivatives are to be calculated in the second step as well.

RESULTS OF THE ANALYSIS

All components of the model structure were independently subjected to a series of tests to verify that the models adequately simulate the corresponding technological units. The coupling of all models made possible to form a dynamic simulation model of the thermodynamic behaviour of a nuclear power plant.

The tunable parameters were fitted to achieve good agreement with the available measurements. Here we can present only one but important example: a change in power output (cf. Fig.3). We show some of the main parameters of the power plant: power of one turbine unit (sum of HP and LP), feedwater temperature and flow (after HPR and before SG), turbine speed and CV position. An increase in power output was simulated, so the speed increased, but the setpoint (PC) changed to close CV, therefore the power decreased; then the speed normalized and CV opened to stabilize power output at a lower level. The HP and LP saturated steam turbine models have the expected responses. (Feedwater parameters follow these changes, but the flow has a delayed response due to the long pipelines. This shows the correct couplings of the models.)

These and other results show acceptable characteristics. The deviations were insignificant and there was no oscillation in any case.

This turbine model demonstrates the discontinuity of the expansion line. In steady state the three time derivatives (cf. the equations (1), (3) and (5) above) are equal to zero, therefore h(2)=h(1) and r(2)=r(1), hence states 1 and 2 (cf. Fig.2) are identical, but are (or may be) different during transients.

Our analysis also proved that the effects of inhomogenous flow and spontaneous condensation can be neglected (a metastable expansion is asssumable) even during transients.

As a result, the dynamic behaviour of saturated steam turbines of PWR nuclear power plants can be simulated by this model more exactly than by conventional models.

Fig. 3 CHANGE IN POWER OUTPUT

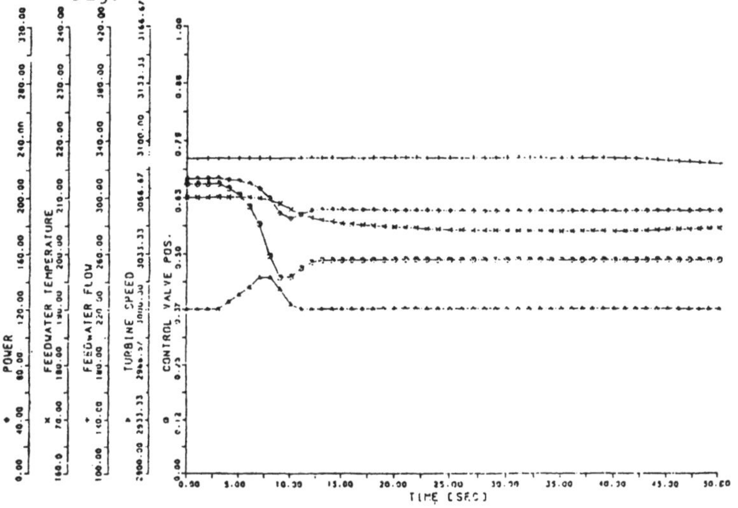

LITERATURE

[1] Vegh, E., Zs. Kiss, A. Gacs: Modelling approaches of a basic principles simulator, Proc. IAEA Course on Nuclear Safety, CRIP, Budapest, 1987

[2] Gacs, A., Zs. Kiss: The dynamic simulation model of the Secondary Circuit, Proc. European Congress on Simulation, Academia, Praha, 1987, Vol.3, 246-250.

[3] Gacs, A., J.S. Janosy, Zs. Kiss: Simulation of the dynamic behaviour of the secondary circuit of a WWER-440 type nuclear power plant - Part 1, KFKI-1987-44/G, CRIP report, Budapest, 1987

[4] Doorenbos, J., A. Gacs, Zs. Kiss: (title v.s.) - Part 2, KFKI-1987-81/G, CRIP report, Budapest, 1987.

Optimal Power Control of a Nuclear Reactor by Means of an Inverse Model

D. Hoppe and J. Götz *)

Summary. The control of nuclear reactor operation during instationary operating states is connected with high demands to process control in order to guarantee nuclear safety every time. Especially during start-up, a dynamic nonlinear process with power changes over several decades, the restriction of the reactor period is to be observed. In this connection problems of time-optimality have to be solved. Based on the point-kinetic equations an inverse reactor model has been derived. It permits to calculate such a control variable, the reactivity ϱ, which produces the wanted period T as the reactor output signal. By means of this inverse model a reactor period controller with a superposed power controller has been developed and realized by a microcomputer. The experiments with the zero-power reactor RAKE show that the controller is able to control both the reactor period and the power.

Keywords. Nuclear reactor, point-kinetic equations, inverse kinetic, nuclear reactor period control, power control, microcomputer application

1. Introduction

This work deals especially with reactor start-up. Starting from a small stationary power level P_0 the reactor must be controlled to a higher level P_1, whereas the power rise - characterized by reactor period T - is restricted. That is a time-optimal problem, and its solution is known. Searched is the solution of the control problem: Estimation of such a reactivity ϱ (as command), which causes the wanted reactor period.

Up to now, it depends on experience and care of the reactor operator to observe the allowed minimum period to be nearly timeoptimal and to avoid overriding of the period restriction. Override would lead first to an alarm and second to an automatic shut-down by the safety system. From literature some results are known /1-6/ concerning conventional or computer based automation of start-up procedure with strong simplified reactor models.

2. Point-kinetic and derivation of the inverse reactor model

In small nuclear reactors, where local effects are absent the process model is given by the point-kinetic equations /7,8/:

$$\dot{n} = \frac{\varrho - \beta}{l} n + \sum_{i=1}^{6} \lambda_i c_i \qquad (1)$$

$$\dot{c}_i = \frac{\beta_i}{l} n - \lambda_i c_i \; ; \; \sum_{i=1}^{6} \beta_i = \beta = 0{,}00641 \qquad (2)$$

where n(t): neutron density (proportional to reactor power)

$c_i(t)$: i-th delayed neutron precursor density

$\varrho(t)$: reactivity (control variable)

λ_i: decay constant of the i-th deleyed neutron precursor; $\lambda_i = 0{,}0124\ s^{-1} \ldots 3\ s^{-1}$

β_i: i-th fraction of delayed neutrons

l: life-time of prompt neutrons ($l = 5 \cdot 10^{-5}$s in our case)

Besides of neutron density n the reciprocal period $\frac{1}{T}$ is a reactor output signal too:

$$\frac{1}{T} = \frac{d}{dt} \ln \frac{n}{n_0} = \frac{\dot{n}}{n} . \qquad (3)$$

Because of the nuclear safety the reactor period is restricted:

*) Zentralinstitut für Kernforschung Rossendorf der Akademie der Wissenschaften der DDR, Postfach 19, Dresden - DDR, 8051, German Democratic Republic

$$\frac{1}{T} \leq \frac{1}{T_o} = \frac{\ln 2}{30\ s} \tag{4}$$

so that the start-up process becomes time optimal, if

$$\frac{1}{T} = \begin{cases} \frac{1}{T_o} & \text{for } 0 \leq t < t_1 \\ 0 & \text{for } t_1 \leq t \end{cases} \tag{5}$$

$$n(t) = \begin{cases} n_o \cdot e^{\frac{t}{T_o}} & \text{for } 0 \leq t < t_1 \\ n_1 & \text{for } t_1 \leq t \end{cases} \tag{6}$$

and

$$\rho(t) = \begin{cases} \sum_{i=1}^{6} \frac{\beta_i}{1+\lambda_i T_o}(1-e^{-(\lambda_i + \frac{1}{T_o})t}) + \frac{1}{T} + \\ + \sum_{i=1}^{6} (\beta_i - \lambda_i l \frac{c_{io}}{n_o}) e^{-(\lambda_i + \frac{1}{T_o})t} & \text{for } 0 \leq t < t_1 \\ \sum_{i=1}^{6} (\beta_i - \lambda_i l \frac{c_i(t_1)}{n_1}) e^{-\lambda_i (t-t_1)} & \text{for } t_1 \leq t. \end{cases} \tag{7}$$

For $c_i(t)$ an according equation can be given.

For the solution of the control problem we want to avoid a control loop, because a control by inverse kinetic seems more suitable. Inversions of eq. (1,2) are well known /7,8/, but the inverse kinetic derived there uses $n(t)$ as an input, whereas we need such a model with $\frac{1}{T}$ as an input. Introducing the new state variables

$$x_i = \beta_i - \lambda_i l \frac{c_i}{n} ; \quad i = 1,2,\ldots,6 \tag{8}$$

from eq. (1), (2) and (3) we get the special reactor kinetic equations (symbol R)

$$\begin{aligned} \dot{x}_i &= -\lambda_i x_i + (\beta_i - x_i)(\rho - \sum_{j=1}^{6} x_j) \\ \frac{1}{T} &= \frac{1}{l}(\rho - \sum_{i=1}^{6} x_i) \end{aligned} \tag{9}$$

and the corresponding inverse reactor kinetic equations (symbol R^{-1})

$$\begin{aligned} \dot{x}_i &= -\lambda_i x_i + (\beta_i - x_i) \frac{1}{T} \\ \rho &= \sum_{i=1}^{6} x_i + \frac{1}{T} \end{aligned} \tag{10}$$

.

By means of integration over scanning time $\Delta t = 0{,}1$ s for $\frac{1}{T}$ = const it leads to the recursion formula

$$x_{i,n+1} = x_{i,n} - (1-e^{-\lambda_i \Delta t})\left[x_{i,n} - (\beta_i - x_{i,n})\frac{1}{\lambda_i T_n}\right] \tag{11}$$

The models R^{-1} and R are nonlinear, therefore is their connection in series only than capable to an identical reactor period transmission, if the corresponding state variables x_i in R and R^{-1} are identical. Therefore the non-measurable reactivity-disturbance z must be compensated. A conventional control loop according to Horowitz /9/ for this reason is providingly suitable, because the disturbances of the measuring channel $\tilde{z}$ are dumped to few. By this reason the system shown in scheme 1 has been developed.

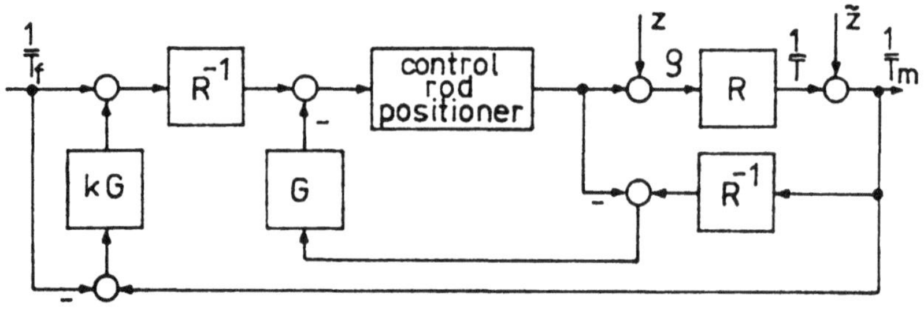

Scheme 1: Principle of start-up control

k: Constant $\lessapprox 1$, T_f: wanted reactor period, T_m: measured reactor period

G : first order lag with small time constant

The presented system may both suppressing state differences nearly and dumping the measuring disturbances strongly. A detailed foundation and explanation will be given in /10/.

3. Realization of start-up control

For the experiments besides of the demands for nuclear safety the technical circumstances of the small zero-power reactor RAKE and that of the used computer had to be respected. This led to the special kind of realization presented in Scheme 2:

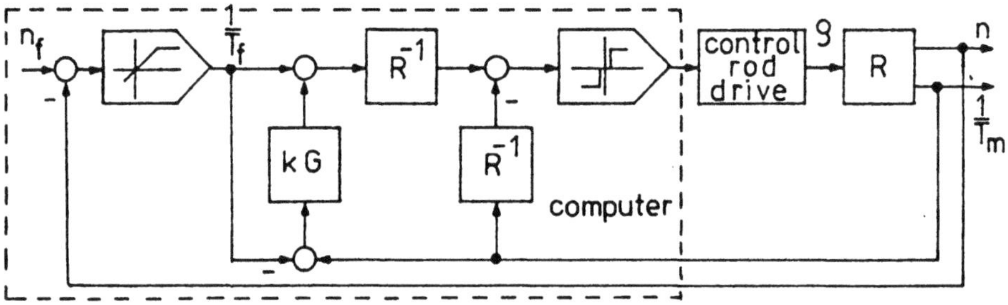

Scheme 2: Realizing of start-up controller by a single board computer

The demand variable $\frac{1}{T_f}$ is generated by a power control loop. The nonlinear characteristic of its controller guarantees the exponential neutron density rise during start-up with $T=T_0$ and the dumping of $\frac{1}{T}$ near the final power level n_1. The whole start-up controller has been realized by using a single board computer Robotron K 1520 M developed for control and surveillance of processes with high safety demands /11/. The application programm will be started cyclically every 0,1 s. It needs 35 ms computing time and a memory of 1,6 kByte.

The results gained by the experiments at the reactor RAKE may be summarized as following:

- the derived kinetic model and its inverse are suitable for describing a nuclear reactor during the start-up process
- based on this new principle of describing a start-up controller for nuclear reactors will be realized on a small single board computer
- the computer based start-up process meet the requirements of time optimal size of reactor power in a good approximation.

Further development should be directed on the permanent installation of such controllers at small reactors to solve the problems of nuclear safety and availability.

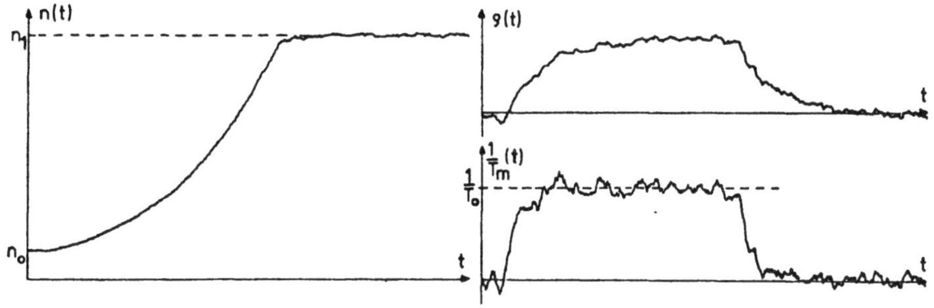

Scheme 3: Results of start-up experiments with the start-up controller at reactor RAKE

Literature

/1/ Takahashi, Y., Takamatsu, S.: Digital start-up control of a research reactor. IEEE Trans. N. S. 12(1965)4, 355-366

/2/ Mohler, R. R., Shen, C. N.: Optimal control of nuclear reactors. Academic Press, New York, London 1970

/3/ Philipp, L. D., Bunch, W. L., Wood, M. R.: Prototype instrument for controlling reactor start-ups. IEEE Trans. N. S. 18(1971)395-402

/4/ Potapenko, P. T.: Sintez sisstemy svjazannych regulatorov dlja upravlenija raspredeleniem moscnosti v jadernom reaktorje. Atomnaja Energija 27(1969)189-193

/5/ Gossanyi, A., Vegh, E.: Computerized reactor power regulation with logarithmic controller. Hungarian Academy of Sciences, Central Research Institute for Physics, Budapest, Report KFKI-1982-85

/6/ Hirata, K. et al.: Computerized automatic start-up control for water reactor nuclear power plants. 9. IFAC-Congress, Budapest 1984, Vol. 1, p. 153-158

/7/ Keepin, G. R.: Physics of nuclear kinetics. Reading, Palo Alto, London 1965

/8/ Ash, M.: Nuclear reactor kinetics. Mc. Graw Hill, New York 1965

/9/ Horowitz, I. M.: Synthesis of feedback systems. Academy Press, New York and London 1963

/10/ Hoppe, D.: Rechnergeführte Anfahrsteuerung eines Kernreaktors. Dissertation TU Dresden, 1987

/11/ Müller, F.: Ein zuverlässiges Mehrprozessorsystem für kerntechnische Automatisierungsaufgaben durch Fehlerprophylaxe, Fehlerdiagnose und Fehlertoleranz. Dissertation, ZfK Rossendorf der AdW, 1987

SIMULATION MODEL AND MICROPROCESSOR SYSTEM FOR DIGITAL CONTROL OF HYDRAULIC SYSTEM IN A HYDROELECTRIC POWER PLANT

Prof. Dr Milan ŠODAN, Dr Seid TEŠNJAK, Ante MARUŠIĆ, M.Sc., Tomislav TOMIŠA, M.Sc.
Faculty of Electrical Engineering, High Voltage and Energy Dept.
University of Zagreb, Yugoslavia

SUMMARY: In this paper mathematical and corresponding simulation model for real hydroelectric power plant hydraulic system (lake, two headrace tunnels, three surge shafts and four pressure pipes) have been presented, together with microprocessor system which controls the observed system, and, if necessary, the generating unit so that the wave propagation does not exceed the, so called, critical level in the three surge shafts regardless of the requirements during the hydroelectric power plant operation. A similar microprocessor system (by the same author) has been in function in a similar hydroelectric power plant and has proved successful so far.

KEY WORDS: hydraulic system, hydro-electric power plant, mathematical model, simulation model, microprocessor system, digital control

1. INTRODUCTION

The lay-out of the hydroelectric power plant depends greatly upon the dynamics of its hydraulic system. While starting and stopping the generating unit the changes in the flow of water may cause serious damages, especially in case, when the water level in surge shafts decreases or when air "enters" into system.

In order to prevent such dangerous situations, it is necessary to estimate dimensions of the surge shafts by using automatic control system for wave propagation.

Comparing the previously mentioned solutions and with the respect to the necessary investment, a smaller surge shafts and wave propagation control system seem to be the solution. The state of the hydraulic system can be successfully supervised and may be influenced according to the specific algorithm due to the development of electrical and non-electrical value converters as well as microprocessors.

2. MATHEMATICAL MODEL

The mathematical model in the investigated hydraulic system comprises the lake, two headrace tunnels, three surge shafts and four pressure pipes. The dynamic relations in headrace tunnels and pipes are expressed by the continuity equation and impuls law /1, 2/:

$$\partial H/\partial t + Q/A * \partial H/\partial x + (a^2/A*g) * \partial Q/\partial x = 0 \qquad (2.1)$$

$$\partial Q/\partial t + Q/A * \partial Q/\partial x + A*g*\partial H/\partial x + (1/2*A*D)*\lambda*|Q|*Q - A*g*\sin\beta = 0 \qquad (2.2)$$

The surge shaft dynamics is generally described by using water volume balance law in it, i.e.:

$$Aw(Z) * dZ/dt = Qi - Qa \qquad (2.3)$$

where Z stands for water level in the surge shafts, Aw(Z) stands for its cross-section as the level function, Qi means the water flow in front of the shaft and Qa means the water flow behind the surge shafts.

The equations (2.1) and (2.2) may be substantially simplified if the following is assumed: a) incompressibility and one dimension water flow, b) headrace tunnels and pipes as elements of concentrated parameters, c) water velocity smaller than wave propagation of hydraulic shock, and d) quadratic dependency between water flow and friction. The appropriate analysis /3, 4/ and presentation for the real hydraulic system may be found in /5/.

3. SIMULATION MODEL

Simulation model and software has been developed on desktop computer HP 9845-B in BASIC program language and implemented in corresponding microprocessor system (chapter 5.) - computer "GEPARD-II"; IBM PC-AT compatible. Mathematical model consists fifteen non-linear differential equations. The problem, in fact, the system of these equations, is solved by numerical method Runge-Kutta of the fourth grade, and it gives exact results (if the steps are small enough), even with the quick time changes in the observed variable. The software consists of the main program and 28 sub-programs. If the ways of starting and stopping the generating units are observed, the following forms may be apllied:

- maneuver by step-change law,
- maneuver by linear law,
- maneuver by exponential law.

The simulation model input values are water flow changes through the turbine and generating unit power changes, respectively. As the output value, any of the fifteen values of the state may be chosen. The generating unit control may be performed automatically in the simulation model according to the programed schedule, or manualy by means of functional keys based on the information display.

4. SIMULATION RESULT ANALYSIS

For the investigated hydraulic system, a large number of starting and stopping of the generating unit has been made in order to investigate almost all possible situations as far as low water level in any of the three shafts may be concerned.

The possibility of low water level in surge shafts as well as air presence in the hydraulic system have been checked by simulation in Fig. 1. By starting two generating units from zero to the rated value in unsuitable moments for the surge shafts, "stirring" of the water level was intended to occur. The generating unit starting and stopping schedule was following:

- at t=0 sec starting the generating unit 1 and 2 from 0 to 100%,
- at t=55 sec starting the generating unit 1 and 2 from 0 to 100%,
- at t=945 sec stopping the generating unit 1, 2, 3 and 4 from the rated value to zero,
- at t=1282 sec starting the generating unit 1 and 2 from 0 to 100%,
- at t=1625 sec starting the generating unit 3 and 4 from 0 to 100%,

From such maneuver forms and the response it is possible to observe that at t = 1687 sec starting from t=0 the critical water level decrease occurs in all the three surge shafts.

5. MICROPROCESSOR SYSTEM FOR SURGE SHAFT SUPERVISION

Basic blok diagram of the developed technical system for surge shaft supervision is presented in Fig. 2. The equipment should meet the following requirements:

- application of reliable devices,
- usage of programable devices (microprocessors with good data processing),
- redundant desing that would increase the availability of the whole system,
- possibility to display the situation and system operation in terms of quality.

Measuring equipment includes signalling necessary to establish and monitoring states in surge shafts (Fig. 2). Water level in surge shafts is constantly monitoring by means of hydrostatic transmitters whose analoguous output signal (4-20 mA) is proportional to the water level in surge shaft. Redundant (double) pressure measuring has been done in common surge shaft only.

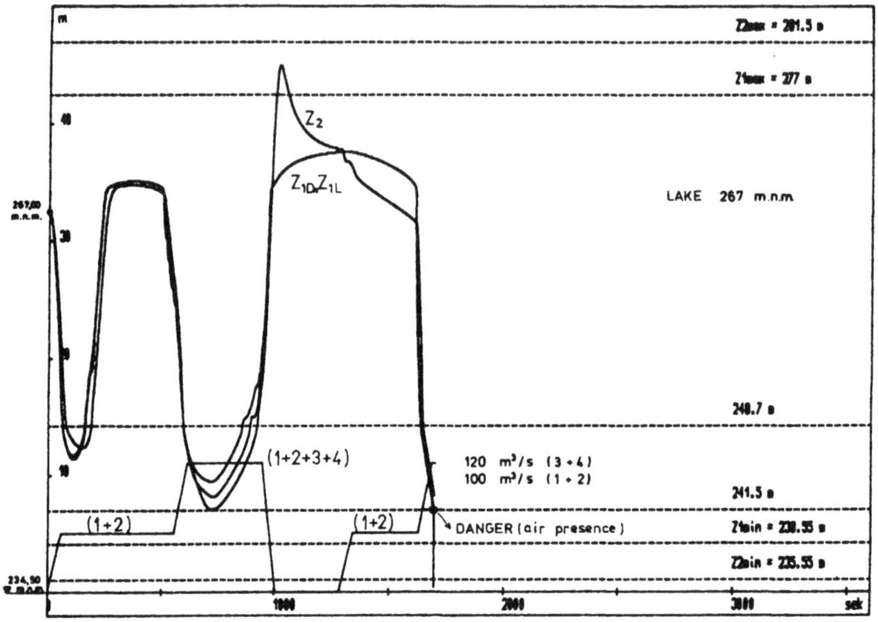

Fig. 1 Simulation of water level oscilations in surge shafts

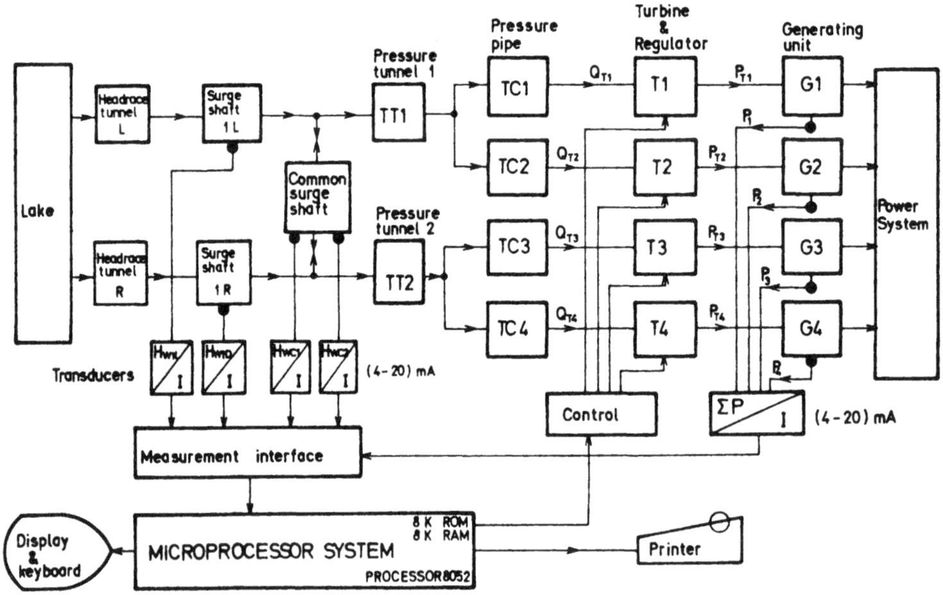

Fig. 2 Microprocessor system structure for surge shaft monitoring and control the HEPP

In order to control the hydroelectric power plant (HEPP) in operation, it is necessary to check a number of signals from the power house (operational state, kind of active power regulation, generators power, etc.). In this case, the generating unit active power change, the overall active power of the HEPP and operational state of the generating unit switchers are controlled.

Blocking logic prevents the change of the relevant value of turbine regulators or the group of regulators when the situation in any surge shaft, for safety reasons, does not permit the increase of water flow through turbines.

The display equipment should enable the staff to supervision the situation, particularly if it is necessary to control the hydroelectric power plant manually. In such a case, display should be in the control room.

Continuous recording of water level changes in surge shafts and overall active power ($\sum$ P) are performed by a four-channel recorder and in the case of a complete breakdown of microprocessor system, it enables manual control of HEPP in order to prevent damage situations caused by low water level in surge shafts.

Recording of the events connected with the microprocessor system operation may be performed by printer which is an integral part of this system.

6. CONCLUSION

Control of HEPP in its transitive states may be successful if time restrictions of different maneuvers with generating unit are introduced. The states mentioned above include those that might cause water level oscilation in the surge shaft beyond permitted limits. From experience it can be seen that in case of pre-damage situations, the staff is not capable to control the HEPP properly since there is uncertainly concerning the right maneuver. For safety reasons, when damage "threatens" HEPP, the only possibility is to put HEPP out of operation until normal conditions are established. In this case, the period for hydraulic system settling is important. Longer standstill of the HEPP essentially disturbs its availability, especially if it is a regulating hydroelectric power plant.

Increassing the surge shaft volume (exsisting HEPP with shafts whose dimensions have been "weakly" estimated) is one of the possible solutions but as far as construction work is concerned, investment would be considerable. More convenient solution would be to automatize the control of HEPP with the existing hydraulic system by means of device which determines the optimal time coordination between maneuvers with generating unit and water level oscilation in the surge shafts.

Algorithm of such device is based on energy redundancy calculation in hydraulic system for water supply and it, in turn, determines the HEPP's power changes at a specific moment. It enables maximal usage of the existing hydraulis system in all operational situations with simultaneous control and prevention of dangerous maneuvers in critical transitive situations. It is possible to solve such dangerous water level oscilation while new plants are designed so that simulation of different operational situations and optimization of surge shaft volume decrease costs without affecting the avaliability of hydroelectric power plant.

REFERENCES

1. Davis C.V.: Handbook of Applied Hydraulics, Mc Graw - Hill Book Co., New York, 1952.
2. Parmakian J.: Waterhammer Analysis, Dover Publications, New York, 1963.
3. Hoppe M.: Die regelung von Systemen mit Alpass - Eigenschaften - dargestellt durch theoretische und experimentelle Untersuchung einer Wasserkraftanlage, Schriftenreihe am Lehrstuhl fur Mess- und Regelungstechnik, Ruhr - Universitat, Bochum, 1981.
4. Hoppe M., Tešnjak S.: Modellbildung und Simulation des dynamischen Verhaltens von Wasserkraftanlagen, Ruhr - Universitat, Bochum, 1983.
5. Sodan M., Tešnjak S., Marušić A., Tomiša T.: Hydraulic system dynamics in high-pressure derivational hydro-electric power plant, European Congress on Simulation, Prague, 1987., Vol. C, 185-189

POWER SYSTEM FREQUENCY CONTROL CHARACTERISTICS AS FUNCTION OF NUCLEAR POWER PLANT PARTICIPATION

Z. Domachowski[1)]

Introduction

When the participation of nuclear power plants in electric power system increases then they have to be expected to take an increasing part in the frequency and power control of the power system. However there are specific regulation problems in nuclear power plant control [2,3]. They cause, among others, a deterioration of power system frequency control characteristics.

In general three types of power plant cooperate in an electric power system. They are fossil fuel, hydraulic, and nuclear ones. Such a power system has been simulated as three cooperating power plants ("three-plant system") . Linear model has been considered. Despite of constructional differences between fossil fuel and nuclear power plants an analogous structure of power plant control system model has been assumed.

Power system frequency control characteristics have been considered under power disturbances. These last ones have been simulated as electric load simultaneous step changes in each turboset.

1. Plant Control System Model

The detailed model structure as well as transfer function parameters of the fossil fuel and nuclear plants, and of the electric power network, are the same as those considered in [2]. The parameters of each steam generator controller have been optimized under fuel flow step change. The criterion of minimum integral of steam pressure error absolute value has been assumed. The parameters of each turboset controller have been optimized under electric power set point value step change. The parameters of steam generator and turboset controllers have been optimized assuming autonomous circuits. The proportional-intergrating controllers have been considered. No limitations have been imposed upon the control system stability margin, and the overshoot of turbine control valve, and the inlet pressure drop.

The hydraulic plant control system model has been adopted according to the IEEE Committee recommendations [4].

The cooperation between three plants may be represented basing on the following equation:

$$T_{im}\frac{d}{dt}f_i + Z_i f + \sum_{\substack{j=1 \\ j\neq i}}^{3} \frac{1}{T_{ij}}\left(\int f_i\,dt - \int f_j\,dt\right) = P_{mi} - P_{di}$$

----

1) Institute of Fluid-Flow Machinery of the Polish Academy of Sciences, Gdańsk, Poland

where f_i denotes turboset frequency, $P_{mi} - P_{di}$ - error between mechanical and electrical power of turboset, T_{im}, T_{ij} - time constants, Z_i - damping torque coefficient.

2. Simulation of Nuclear Power Plant Participation

So called "three-plant system" has been applied as power system model. There have also been considered some cases of so called "two-plant system". The principal power system models taken into account have been shown in Table 1, where H denotes hydraulic plant, F - fossil fuel plant, N - nuclear plant.

Table 1

				% F	% N	% H	vibration period, s	total response time, s
1	F	F		100	0	0	2.5	17
2	F	F	H	66,6	0	33,3	2.7	236
3	F	F	N	66,6	33,3	0	3.0	23.5
4	F	N	H	33,3	33,3	33,3	3.1	225
5	F	N	N	33,3	66,6	0	3.4	36.5
6	N	N	H	0	66,6	33,3	3.5	225
7	N	N		0	100	0	3.8	53

3. Influence of Nuclear Power Plant Participation on Power System Frequency Control Characteristics

First a power system composed of fossil fuel and nuclear plants only has been considered, see 1, 3, 5 and 7 in Table 1. When the participation of nuclear power plants in a power system increases then the power system frequency vibration period increases too. This relationship has been shown in Fig. 1. E.g. when the nuclear power plant participation increases from 33 % up to 66 % then the power system frequency vibration period increases of about 13 %.

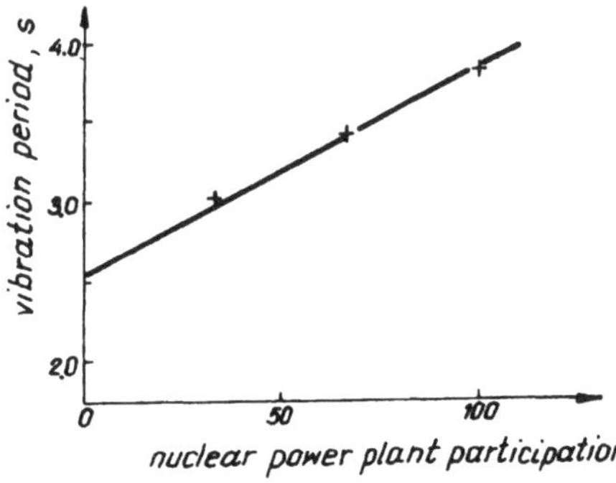

Fig. 1. Power system frequency vibration period as function of nuclear power plant participation

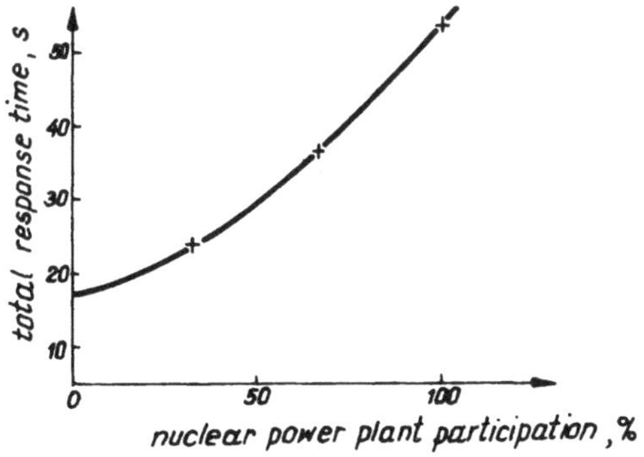

Fig. 2. Power system frequency total response time as function of nuclear power plat participation

The relationship between the total response time of power system frequency and the nuclear power plant participation has been represented in Fig. 2. E.g. when the nuclear power plant participation increases from 33 % up to 66 % then the power system frequency total response time increases of about 60 %. Here the so called total response time means time after which the error between the power system frequency and its steady state value is less than 15 mHz, i.e. less than $3 \cdot 10^{-4}$ of power system frequency nominal value.

When hydraulic power plants are included into the power system frequency control then damping of power system frequency transient process increases considerably. As a result the electric power system frequency transient process becomes aperiodic. Only some short duration vibrations appear in the initial part of that process. However hydraulic power plant participation in the power system frequency control makes several times longer the power system fraquency total response time, compare cases 1 to 2, 3 to 4, 4 to 5, and 6 to 7 in Table 1.

4. Final Remarks

Increasing nuclear power plant participation in an electric power system makes worse power system frequency control characteristics. As a viewpoint the vibration period as well as the total response time of power system frequency have been taken into account. This is resulting from considerablly longer both vibration period and total response time of turboset frequency in a nuclear power plant with so called "driving turbine", and with so called power output feedback [1], than in a fossil fuel plant. Therefore a control system ensuring required damping of transient process should be recommended to a nuclear power plant participating in electric power system frequency control.

5. Literature

[1] Domachowski, Z.: Simulation Studies on Optimum Turboset Feedback in Generating Unit Control System. Syst.Anal.Model.Simul. 4 (1987) 6, 565-569

[2] Domachowski Z. and G.Bugała: Simulation as a Tool of Comparison between Nuclear and Fossil Fuel Plant Control. Simulation in Research and Development, A.Javor (Editor), Elsevier Science Publishers B.V. North Holland 1985, 201-206

[3] Pioger, G. and G .Testud: Etude du comportement dynamique du réseau sur des temps longs. Application au reglage fréquence-puissance. Electricite de France. Bulletin de la direction des études et recherches, série B (1978) 1, 5-59

[4] Dynamic Models for Steam and Hydro Turbines in Power System Studies. IEEE Committee Report Paper T 73 089-0

DESIGN OF A REDUCED MODEL FOR A NUCLEAR-POWER PLANT UNDER EARTH-QUAKE EXCITATION

N. Model; S. Grossman[1]
P. Dineva; L. Hadjikov[2]

1. Introduction

In this paper the nuclear-power plant will be discretized by using the finite element method. We get a system of ordinary differential equations. To reduce the simulation-time the original system will be approximated by a lower order system. For beginning we use in this paper a system of linear ordinary differential equations. In the next paper we extend the investigations also for nonlinear systems. The design of the reduced model will be carried out by using the method of minimizing the equation-error. The quality of the approximations will be examined by comparing the simulation results of the nonreduced model with the reduced model.

2. The model of the nuclear-power plant

In the following figure the structure of a nuclear-power plant under earthquake excitation is illustrated [1].

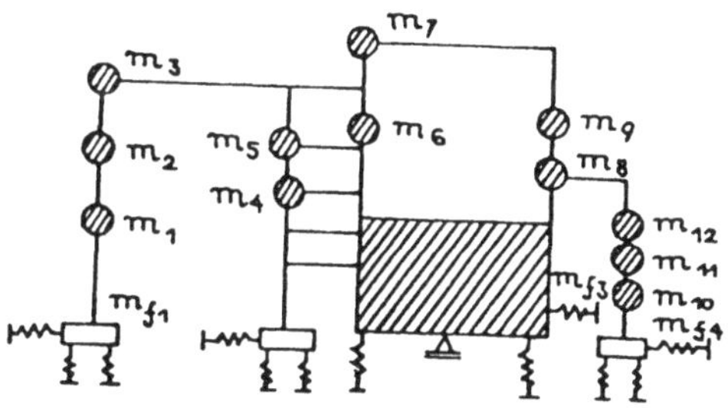

m_i - MASS CONCENTRATION IN THE i KNOT
m_{fj} - MASS OF THE j - FOUNDATION

FIG. 1

[1] Central Institute of Cybernetics and Information Processes, Berlin, GDR
[2] Institute of Mechanics and Biomechanics, Sofia, Bulgaria

The aim is to describe earthquake-motion of the complex structure (for instance power plant) by finite element method and to reduce the order of model. The effects of reduction of the dynamical soil-structure system is studied. The structure is decomposed into elastic elements assembled in separate units where the masses are to be appropriately concentrated. The equation of motion are derived by applying D'Alembert's principle to structure masses, at first the displacements and rotations of the foundations. Then the equations of motions of the foundations are derived analogously by taking into account loads that are excercised by the structure on ground. The equation of motion of the discretized system (system with concentrated parameters) can be written in the form :

$$M \frac{d^2x(t)}{dt^2} + C \frac{dx(t)}{dt} + K\, x(t) = -Mb\frac{d^2x_o(t)}{dt^2}$$

where M,C,K are matrices of mass, damping and stiffness resp.; $\ddot{x}_o$ is the earth acceleration; b is the vector of direction of external loading. With

$$y(t) = \begin{bmatrix} x(t) \\ \dot{x}(t) \end{bmatrix} \qquad x\in R^n \quad y\in R^{2n} \qquad u\in R^n \; ; \; u_i = \ddot{x}_{i0}$$

follows

$$\dot{y}(t) = \begin{bmatrix} 0 & I \\ -M^{-1}K & -M^{-1}C \end{bmatrix} y(t) + B\,u \qquad B = -\begin{bmatrix} b \\ \cdot \\ \cdot \\ b \end{bmatrix}$$

$$\dot{y}(t) = A\,y(t) + B\,u(t) \tag{1}$$

3. Reduction of model

The behavior (stability, dynamics, stationarity) of the 'essential' ordinates of the original model (1) is to approximated by a lower order model (2).

$$\dot{\tilde{x}} = A_r\tilde{x} + B_r u \quad ; \quad \tilde{x}\in R^r \;,\; r<n \tag{2}$$

In relation to the used model the choice of the essential ordinates is to decide. For these the socalled reduction matrix R is used.

$$x_r = Rx \quad ; \quad x_r\in R^r \; ; \; x\in R^n \tag{3}$$

R-(r,n) matrix

The elements of R are zero or one.
The equation-error d(t) has the form

$$d(t) = \dot{x}_r(t) - A_r x_r(t) - B_r u(t) ,$$

with the unknown reduced matrices A and B .
It can be shown [4], that

$$B_r = A_r RA^{-1}B \qquad (4)$$

$$A_r = RASR^T(RSR^T)^{-1} \qquad (5)$$

with

$$S = \int_0^\infty e^{At}(A^{-1}B)G(A^{-1}B)^T e^{A^T t} dt \qquad (6)$$

$$G = \mathrm{diag}(g_1,\ldots\ldots,g_n) \quad g_i - \text{weights}$$

is valid.
If

$$G(t) = Q\, e^{2\alpha t}$$

than the computing of the integral (6) is not necessary.
In this case the matrix S is the solution of the algebraic Liapunov-Matrices-Equation.

$$(A + \alpha I)\, S + S\,(A + \alpha I)^T = -(A^{-1}B)Q(A^{-1}B)^T \qquad (7)$$

where α has to be suited suited choosen.
It is also possible to stabilize a instabil system [1].
The solution S of the equation (7) has the form [3].

$$H = \left[\begin{array}{c|c} A & O \\ \hline Q & -A^T \end{array}\right] \quad ; \quad \sum_{i=0}^{n} a_i \; H^{n-1} = \left[\begin{array}{c|c} S & T \\ \hline U & V \end{array}\right]$$

$$S = V^{-1}U \qquad a_i\text{-coefficients of the characteristic polynom}$$

4. Results

For the described s o i l - s t r u c t u r e - system we have a quadratic (40,40)-systemmatrix A. When we don't into account the foundation displacements and rotations and we suppose that the construction is rigid fixed, then we get a (24,24)-systemmatrix A. The numerical investigations were carried out for both cases and the reduced matrices A and B were computed.

REFERENCES

[1] N.Model;P.Dineva;L.Hadjikov : Optimization of mechanical properties of structure from the point of a aseismic design;Transactions of the 9-th International Conference on Structural Mechanics in Reactor Technology, Lausanne, 17-21 august 1987,Switzerland

[2] O.Foellinger : Reduktion der Systemordnung, Regelungstechnik 1982, Heft 11

[3] P.C.Mueller : Die Berechnung von Liapunov-Funktionen und von quadratischen Regelflaechen fuer lineare, stetige, zeitinvariante Mehrgroessensysteme, Regelungstechnik 1969, Heft 8

[4] E.Eitelberg : Model reduction by minimizing the weighted equation-error, Int. Journal Control, 1981, Vol. 34, No. 6

MATHEMATICAL MODELLING AND SIMULATION TECHNIQUES OF COMPLEX POWER PLANT PIPE NETWORK

Häusler, Frank
Rösel, Gert-Steffen[1]

The analysis of thermohydraulic transients is essential to plant design and safety. Phenomenological description of two phase flow is based on integral balance of mass, energy, and momentum *DELHAYE* [2]. The balance laws can be written for each control volume divided into subvolumes V_1 and V_2.

$$\frac{\partial}{\partial t}(\varrho_k \psi_k^{\mathbb{B}}) + \operatorname{div}(\varrho_k \psi_k^{\mathbb{B}} w_k) + \operatorname{div} J_k^{\mathbb{B}} = 0 \qquad (1)$$

Jump conditions result on the interface A_I.

$$\sum_{k=1}^{2} \left[\dot{m}_k \, \psi_k^{\mathbb{B}} + n_k \, J_k^{\mathbb{B}} \right] = 0 \qquad (2)$$

balance	$\psi_k^{\mathbb{B}}$	$J_k^{\mathbb{B}}$
mass	1	0
momentum	w_k	$-\mathbb{T}_k$
energy	u_k	$q_k - \mathbb{T}_k w_k$

$\mathbb{B} \in$ { mass, momentum, energy }
w - velocity vector
ϱ - density
u - specific total internal energy
q - specific heat flow vector
$\mathbb{T}$ - stress tensor
$\dot{m}$ - mass transfer rate

In view of manifold geometrical situations (space - time position of phase volumes) it is impossible to solve the 10 partial differential equations (1). Appropriate averaging techniques are applied to (1). Using time- and space averaging method in consideration of a fixed pipe wall (1) is rewritten in the following form in one space dimension z.

$$\frac{\partial}{\partial t} \mu_k^{\mathbb{B}} + \frac{\partial}{\partial z}(\mu_k^{\mathbb{B}} w_k) + \frac{\partial}{\partial z} \Phi_k^{\mathbb{B}} = \Gamma_k^{\mathbb{B}} \qquad (3)$$

These 6 partial differential equations result from projection on z axis. Handling products of averages is possible under the assumption that the average of a product can be expressed by the product of averages.

1
Academy of Sciences of the GDR
Institute for Informatics and Computer Technique
Rudower Chaussee 5
GDR Berlin 1199

As further hypotheses are generally admitted :

- longitudinal heat conduction terms and their derivatives are neglegible
- phase viscous stress derivatives and their powers are neglegible
- the pressure is constant over a cross section ($p_1 = p_2$)

balance	$\mu_k^{\mathbb{B}}$	$\Phi_k^{\mathbb{B}}$
mass	$\mathcal{A}_k \varrho_k$	0
momentum	$\mathcal{A}_k \varrho_k w_k$	$\mathcal{A}_k p$
energy	$\mathcal{A}_k \varrho_k u_k$	$\mathcal{A}_k p\ w_k$

$\mathcal{A}$ - area fraction

p - pressure

w - velocity in z-direction

Transfer laws must be known to express $\Gamma_k^{\mathbb{B}}$ at interface and pipe wall. Irrespective of further model reductions and the number of equations transient two phase flow modelling requires the solution of a set of partial differential equations (3). The especial character of momentum balance arises from possible sign changing of velocity w_k. Compared with energy u_k and density ϱ_k (called cell variables) the velocity is named a flux variable.

The finite difference equations in space are written in terms of the staggered difference scheme on Eulerian mesh. State variables, such as pressure and internal energy, are obtained at the centre of the mesh cell, and flow variables are obtained at cell boundaries. With this configuration the donor-cell scheme is expressed. In cell balances the divergence term is described as transport process of cell variables quantities depending on flux sign.

$$\frac{\partial}{\partial z} (\mu_k^{\mathbb{B}}\ w_k)\Big|_{z=c} = \frac{1}{z_c} \begin{cases} \mu_c w_r - \mu_l w_l & \text{if } w_r \geq 0 \text{ and } w_l \geq 0 \\ \mu_r w_r - \mu_l w_l & \text{if } w_r < 0 \text{ and } w_l \geq 0 \\ \mu_r w_r - \mu_c w_l & \text{if } w_r < 0 \text{ and } w_l < 0 \\ \mu_c w_r - \mu_c w_l & \text{if } w_r \geq 0 \text{ and } w_l < 0 \end{cases} \qquad (4)$$

In right hand terms are omitted phase denominator k and balance type $\mathbb{B}$. Subscripts on right hand side denote :

c - cell centre, r - right cell or flux, l - left cell or flux.

Slight modifications are used to apply (4) to branching cells declared below. In flux balances the divergence term is expressed by central differencing.

The structural description is closely connected with the discretisation of space. Technical description elements are links (length of tube with a constant geometry), branchings, and fittings (valves e.g.). Each link is limited by two fixed nodes. These nodes present a branching point or a boundary point or a connection point. Links can be subdivided into cells depending on space discretisation. Following types of nodes are used :

a connection node	-	a flux variable described by momentum balance (flux balance)
a connection and boundary node with fitting	-	a flux value described by steady state flow relation.

The distribution node connected with 3 (without limitation of universality) links can not be described by a momentum balance in z-direction. A branching cell limited by 3 nodes is introduced to handle the mass and energy storage behaviour. The flux direction is a priori determined at distribution cell, if the neighbourhood of two such cells is excluded. The limiting nodes are described by a normal momentum balance. Following types of cells are used :

a connection and distribution cell	-	a cell variable described by mass and energy balance (cell balance)
a fictive cell	-	a cell value (boundary condition).

Reducing the memory requirements the structural description is stored in form of neighbour lists. The number of cell N_c or flux N_f balances is determined by the number of links N_L, number of nodes N_N, distribution nodes N_D, boundary nodes N_B, and number of fittings N_F.

$$N_C = N_L - 2 N_D \quad ; \quad N_f = N_N - N_B - N_D - N_F$$

Usual numbering of nodes is insufficient for accessing to different types and appropriate evaluations. Renumbering and ordering in consideration of different cell and node types permit efficient access of data elements in conjunction with a clearly arranged notation of various equations. The computational effort with respect to the solution of linear equations are not regarded on this level.

Uncompletly known boundary conditions can processed with the knowledge of time series of cell transients. This form of dependence has to be registered in sparsity scheme. Reconstruction of unknown parameters is based on adaption of these transients.

On solving linear systems under the assumption of their sparsity and structural symmetry the *minimum degree algorithm* GEORGE [3] is minimizing computational operations. The sparsity depends on neighbourhood of integration variables. Each Jacobian of the complete system based on time stepping of equation (3) is arranged in view of sparsity by the following types of dependence :

cell balance - cell variable	cell balance - flux variable
flux balance - cell variable	flux balance - flux variable.

A complete system matrix can be always composed by above mentioned relations independent of further model reductions.

The obtained initial value problem is solved fully implicit by backward differentiation formula. Time stepping is performed on local error estimation. The efficency of numerical algorithm depends essentially on the performance of the Newton step in a modified way.

The Jacobian updating takes place by analysing the solution behaviour and the convergence.

The software package *INROL* [1] is dealing with a instationary condensing steam flow including computation of flow- and pressure forces on piping, wall heat conduction, and air blowdown by start up.

Literature

[1] *BUSSE, U.* et al.:
Berechnung thermofluiddynamischer Vorgänge in Rohrleitungen
Preprint 88.03,
AdW der DDR, Institut für Informatik und Rechentechnik 1988.

[2] *DELHAYE, J. M.; GIOT, M.; RIETHMULLER, M. L.*:
Thermohydraulics of Two-Phase Systems for Industrial Design and Nuclear Engineering, New York 1981.

[3] *GEORGE, A.; W-H LIU, J.*:
Computer Solution of Large Sparse Positive Definite Systems
New Jersey 1981.

STRUCTURAL ASPECTS OF THE (NATURAL AND ARTIFICIAL) SEMANTIC MEMORY

(P. Érdi: Central Research Institute for Physics, Hungarian Academy of Sciences, H-1525 Budapest, P.O.B.49, Hungary)

ABSTRACT

The idea of semantic memory has proved to be useful both for interpreting psychological experimental data and as a key term of a computational technique for knowledge representation and information retrieval. The problem of the complexity of the structure of semantic networks and some qualitative aspects of kinetic models of the ontogenetic formation of semantic memory are discussed.

1. Introduction

Semantic memory is one of the most influential notion of both cognitive psychology (for a recent review see Chang 1986), and of that field of artificial intelligence researches which deals with 'knowledge representation' (for a selected readings see Brachman and Levesque 1985).

Conceptual models of semantic memory are based on the assumption that knowledge structure stored in this type of memory system can be identified with network of nodes representing concepts joined by (in the overwhelming majority of the cases binary) links expressing relations among them.

In general terms, the theory of networks of large number of nodes constituting very complex structures offers, at least in principle, a common model for the treatment of natural and synthetic information processing system. While one direction of the researches, motivated by statistical physics adopts elementary units (e.g. neurons or computing elements) as nodes (Choi and Huberman 1983, Hopfield 1982), another one, sometimes labeled as "neoconnectionist" school defines networks with intermediate level computational - cognitive elements. These elements (i.e. "schemas", adopting the slightly vague term) are the building blocks of many network models switched to this neoconnectionist school of the 'parallel distributed processing' (McClelland and Rumelhart 1986).

Network of semantic memory can be physically realized by the Connection Machine (Hillis 1986). One processor corresponds to one concept, and the main question to be answered is, whether what kind of topological arrangement can be considered as optimal one is some sense? Designers of multi-computer networks also deal with the problem of the topological arrangements of the elements (Uhr 1984). It is a general question whether how to give different definitions for the complexity of graphs, and how to switch these notions to the storing and recalling capabilities of the semantic memory.

It is interesting to notice that chemical and physical analogies proved to be useful for analysis of the formation and the properties of semantic memory (Masui and Lumsden 1985, Huberman and Hogg 1987). Analogously to chemical polymers, tree-like, branching and cross-linked semantic memory could occur. Therefore kinetic models of polymerization can be useful for describing semantic memory formation. An important qualitative property of these kinetic models is the exhibition of phase transition-like phenomenon. In one phase the size of the cluster of nodes are bounded, in the other one most of the nodes are connected, constituting a single giant 'molecule'. The transition occur if the ratio of number of links / number of nodes is greater than a critical value.

2. Semantic memory: some concepts and models

It is more or less accepted the existence of difference between episodic and semantic memory. While the former preserves information about spatiotemporal relations of events having autobiographic character, the latter contains general conceptual information ('knowledge').

In the early semantic memory models (Quillian 1968, Collins and Quillian 1969) nodes corresponding to word concepts were arranged hierarchically. Furthermore, the cognitive economy principle assured that properties are stored at the highest level only. Hierarchical arrangement implies the existence of unique distance between nodes. Somewhat more is also true: not only metricity, but a stronger property, namely ultra-metricity holds for such kind of structures. The scope and limits of ultrametric structure of semantic memory has been analyzed earlier (Érdi 1988).

Though the hierarchical - network model has had obvious merit in interpreting artificial and natural language comprehension, a number of alternative models have been elaborated. The spreading activation theory (Collins and Loftus 1975) answers the question whether how activation spreads after priming (i.e. activating by seeing, hearing, etc.) a node. Activation of a node implies the activation of adjacent nodes, and so on. Dynamic behaviour of the activation vector A(t) is described by the equation

$$A(t+1):=E(t)+MA(t) \quad (1)$$

where E(t) specifies the priming, and M is defined by the connectivity matrix R, and the relaxation constant g, and the amplification constant a, as follows:

$$M:=(1-g)I+aR \quad (2)$$

where I is the identity matrix.

Experiments with this model supported the view, that more connected, than tree graphs are more natural objects of knowledge representation. Of course, for structures represented by such kind of graphs the notion of distance between nodes can not be defined, though the applicability of the concept of neighbourhoodness is still preserved. The space of word concepts seems to be topological, and at most locally can be (ultra) metric.

3. Complexity of graphs

A number of complexity measures can be given for characterizing the different structural aspects of a graph.

A complexity measure defined for trees has been given by Huberman and Hogg (1986) expressing the diversity of subtrees. The D(T) measure can be constructed recursivly for the values for each of the subtrees by the formula

$$D(T) := (2^k - 1) \prod_{j=1}^{k} (D(T_1)), \qquad (3),$$

taking into account the n-ary interactions among k distinct subtrees. If all the subtrees are isomorphic, the measure has minimal value.

Another complexity measure is associated to the notion of density (Uhr 1984). The possible densest graph can be assigned to a given degree, and diameter. (The degree is the number of links to each node, while the diameter is the longest shortest path between any pair of nodes). The upper bound of the nodes in a graph of a given degree and diameter is:

$$N_{(d,k)} = \frac{d(d-1)^k - 2}{d-2} \qquad (4)$$

A densest graph can be constructed from a degree N tree (its radius - i.e. the distance from leaf to root - is equal to the half of the diameter) by adding new links to each leaf node. The diameter can be reduced, while its value is equal its radius.

A third graph - theoretic complexity measure is related to the cyclomatic number to be calculated. As it is well known from elementary graph theory (e.g. Berge 1973), the v(G) cyclomatic number of a graph B with n vertices, e edges, and p connected components is

$$v(G) := e - n + p \qquad (5)$$

In a strongly connected graph, the cyclomatic number is equal to the maximum number of linearly independent circles. Since inserting a new edge in G increases v(G) - by unity - , it can be applied as a complexity measure. This notion has been applied to measure software complexity (McCabe 1976, Hall and Preiser 1984, Stokłosa 1986).

4. Kinetic approach to semantic memory development

The skeleton of a stochastic model is given to describe the ontogenetic development of the network of semantic memory. Connections between nodes are changing randomly. From mathematical point of view the evolution of a random graph is studied. The investigation of the random graphs started with the seminal paper of Erdős and Rényi (1960).

The basis conceptual assumptions are:

(i) semantic memory is represented by the network of elementary 'units',

(ii) the unit A_r can change its internal state by the 'isomerization reaction' $A_r \underset{\kappa_{sr}}{\overset{\kappa_{rs}}{\rightleftharpoons}} A_s$

(iii) the units can associate with each other, as it is described by the "reaction" $A_r + A_s \underset{\lambda_{-rsu}}{\overset{\lambda_{rsu}}{\rightleftharpoons}} A_u$

(iv) the velocity of the structure formation can be given by the combinatorial mass action kinetic rule of the continuous time discrete state space stochastic models of chemical reactions (Whittle 1986, Érdi and Tóth, in press).

The stochastic model can be given in terms of the generating function F(s,t):

$$\partial F(z,t)/\partial t = \sum_r \sum_s \kappa_{rs}(z_s - z_r)\frac{\partial F}{\partial z_r} + \sum_{r<s}\sum \sum_u (z_u - z_r z_s) \left[\frac{\lambda_{rsu}}{V} \frac{\partial^2 F}{\partial z_r \partial z_s} - \lambda_{-rsu} \frac{\partial F}{\partial z_u} \right] \qquad (6)$$

There are two questions referring to the qualitative properties of the model, which are relevant both in the terminology of polymer kinetics and that of semantic memory development.

(1) What are the conditions that only trees are generated?

(2) What are the conditions of the occurence of 'critical' i.e. phase transition-like behaviour?

Instead of detailed answers only some hints can be given here.

(1) A particular class of stochastic processes, namely Galton - Watson, or branching processes, which can be considered as a special class of stochastic models of chemical reactions, lead to tree structures. The main statistical properties of the trees (i.e of the branching) can be given (e.g. Kliemann 1987).

(2) If the average branching ratio is less than a critical value (particularly to trees it is just 1), the size of clusters is finite, but for greater values most of the units are connected. This phenomenon - gelation in terms of polymerization - might be interpreted in terms of semantic memory, too!

References: can be requested from the author

ESTIMATION OF DRUG NEED FROM THE HEALTH PROJECTION - THE CASE OF INSULIN

František Hauser, Michal Anděl[1)]

1. Introduction

The most important factors influencing drug need are disease prevalence and employed therapy. Our ability to predict future drug need is in most cases limited by the lack of adequate information on both these factors. One of the notable more favourable cases is diabetes though even here we are still far from the desired situation. A permanent rise in diabetes prevalence in most countries attracted much attention in recent years and epidemiological data about diabetes are gathered with a rapid pace. Even when these data are still rather imperfect, they can - together with the data about the demographic composition and characteristics of the studied population - form the necessary basis for building an epidemiological simulation model for the projection of the diabetic population. The second basic factor affecting the drug need is the employed therapy. Insulin therapy is indicated in a relatively well defined portion of diabetic patients and in Czechoslovakia the numbers of these patients are monitored by routine medical statistics.

2. Material and methods

Evolution of the diabetic population has been estimated by means of a medico-demographic model. Population projection normally used in demography was adapted to include medical aspects of the population structure. Population members were classed not only according to sex and age but also according to their health state. Three categories of health state were distinguished: non-diabetics, type 1 diabetics and type 2 diabetics. The distinction between type 1 and type 2 diabetes is in aggreement with the recommendations of the World Health Organization |1|.

The population projection is calculated year by year from the age-specific incidences, mortalities and fertilities in each health state category. This means that the model requires to specify the age composition of each health state category in the initial year of the projection and for every following year of the projection age-specific mortalities and fertilities for each category and age-sex specific incidences of type 1 and type 2 diabetes |2|.

[1)]František Hauser, Institute of Social Medicine and Health Services Organization, Vítězného února 54, CS-121 39 Praha 2
Michal Anděl, Institute of Clinical and Experimental Medicine, Vídeňská 800, CS-146 22 Praha 4

These data have been obtained for the Czech Socialist Republic (CSR) from statistical and health care yearbooks and from recent domestic and foreign epidemiological studies. Some input data have been derived by indirect standardization from the corresponding data for the German Democratic Republic (GDR) where better and more detailed statistical data about diabetics are available and where environmental, social and genetical conditions are similar to those in the CSR.

From the definition of type 1 diabetes as insulin dependent diabetes it follows that all type 1 patients have to be treated regularly by insulin. By contrast only a small fraction of type 2 patients require insulin therapy. This fraction changes with age from about 15% in young adults to 3% in very old patients. Summing up the patients using insulin therapy and multiplying the sum by the average daily dose, we arrive at the total insulin need.

However, with some type 2 diabetics it is possible to choose between insulin therapy and other types of therapy. For several reasons the use of insulin therapy in the CSR is on the increase. On the other hand recent years have witnessed a certain rationalization of insulin therapy and the average daily dose has decreased. The influence of these two developments on the future total insulin need was the subject of this study.

In the following projections it was assumed that the demographic characteristics such as fertility, mortality, perinatal mortality, etc. remain unchanged on the level of the early 1980´s. Incidences, too, are assumed to remain on the high level attained in the early 1980´s.

3. Results

The above stated assumptions define the base-line projection variant. Let us sum up its main features. The rise in prevalence is slowing down asymptotically. However, the increasing prevalence is superimposed over the growth of older age groups and this results in large increases in numbers of elderly type 2 diabetics. Numbers of type 1 diabetics on the other hand correspond to the demographic composition and the prevalence remains the same. Changes in the numbers of type 1 diabetics thus depend on the numbers of newborns which with unchanged fertilities depend on numbers of women in the reproductive age.

Our analysis of future insulin need is sumarized in Fig. 1. Related to the year 1980 (100%), the figure shows the real insulin need in the past years (crosses - x). The prognosis is based on the facts about insulin therapy in the CSR and percentages of insulin treated diabetics in the CSR and GDR. These percentages are in the GDR slightly higher and as the development in the CSR follows with a delay of about 4 to 5 years the development in the GDR, we assummed that between 1985 and 1990 CSR could gradually reach the current situation in the GDR (curves I and II). If we assume a mere conservation of the current proportion of insulin treated diabetics, we obtain curves III and IV.

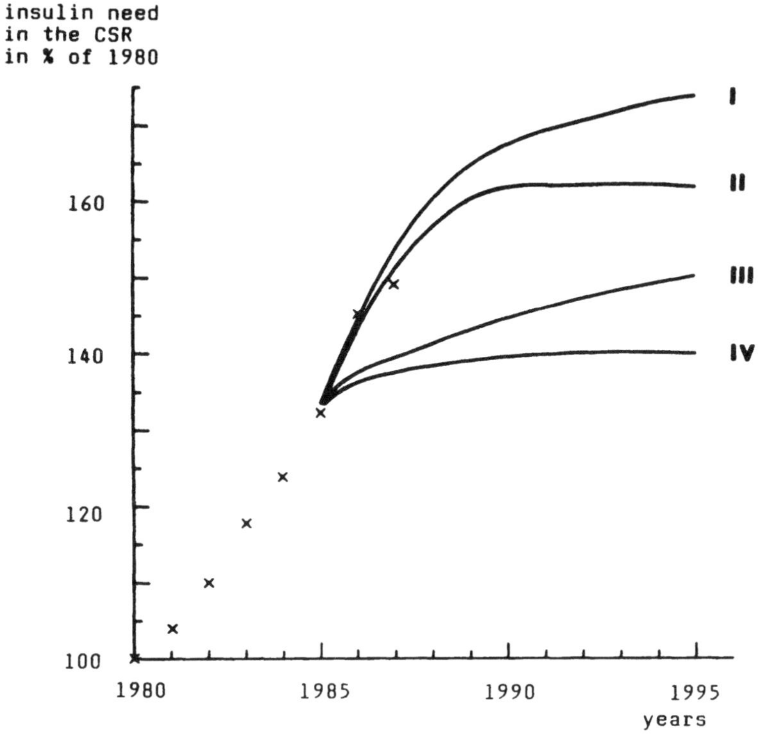

Fig. 1. Four variants of future insulin need in the CSR. See the text for detailed explanation.

The second varied parameter is the average daily dose of insulin which declined in recent years. Thus, the curves I and III represent continuation of the current practice while the curves II and IV result under the assumption that the average daily dose of insulin would be decreased by 7% within the next 10 years. The latter case can also be viewed as more physiological and consequently more desirable from the viewpoint of insulin therapy.

4. Discussion

Anticipation of future development always includes both the analysis of sure past data and uncertain possible development variants which are extensions of combinations of various assumptions. From the viewpoint of our case we arrive at four extreme situations: a higher percentage of insulin treated diabetics with a higher (curve I) and a lower (curve II) average daily dose of insulin and a lower percentage of insulin treated diabetics with a higher (curve III) and a lower (curve IV) average daily dose of insulin. All these possibilities were considered but it is probable that the last one would be ideal and the first one the least desirable. The daily medical practice and thus also the insulin dosage is more and more influenced by improved medical knowledge about pathophysiology of type 1 and type 2 diabetes and it can be assumed that this will lead to the insulin dosage approaching as closely as possible the physiological one. Hence the univocal desirability of the variant IV. On the other hand many physicians use older approaches which they learned at school including older principles of insulin therapy and that is the reason why the most desirable development will not materialize. In fact the preliminary data for the year 1987 indicate that the variant II is the more probable one.

5. Conclusion

Although limited in precision and predictive ability by insufficient quality of input epidemiological data the presented approach demonstrates what means in practice the old proposition that health care needs should be determined according to the health state of the population. In addition this approach indicates how medico-demographic simulation models can make currently available data more valuable and how they can be used in estimating future health care needs.

Literature

|1| Diabetes Mellitus, WHO Technical Report Series 727, WHO, Geneva, 1985

|2| Hauser, F., Anděl, M.: Projection of Type 1 and Type 2 diabetes prevalence, in: Advances in Biomedical Measurement, E.R. Carson (ed.), Plenum Press, London, in press

NUMERICAL SIMULATION OF PULSATILE FLOW IN ARTERIES

Thomas Pohl #

I. INTRODUCTION

Since many years arterial blood flow has been systematically investigated, - both by experimental and theoretical methods. The contraction and relaxation of the left ventricle initiates a non-linear transient flow at the root of the aorta which results in complex wave patterns in the aorta. The waves are propagated throughout the arterial tree and reflected at major bifurcations, stenosis and other pathological phenomenas. So the characterization of pulsatile wave patterns is of particular interest to physiologists since irregularities in human circulation may be stated by analysing the time-dependent graphs of pressure and flow. Some of the related questions concerning the pulse wave propagation and the pulse wave reflection in arterial vessels may be answered by numerical simulations. Therefore a mathematical theory, based on an unsteady one-dimensional model of an incompressible fluid flow in a tapered,elastic tube is presented. The aortic valve is incorparated by special boundary conditions. The elasticity of the wall is described by an arctangent model cited from Langewouters et al. /3/. Wall friction is also taken into account as well as diffusion of fluid through the arterial wall. The hyperbolic model equations are discretizised in analogy to the implicit Beam - Warming finite differences method /8/ and numerically solved.

II. MODEL EQUATIONS

The numerical simulation requires a model which is able to describe the wave propagation and reflection. Following proposals of Gerrard & Taylor /1/ and Kurz /2/ the equation of continuity is

$$\frac{\partial A}{\partial t} + \frac{\partial(uA)}{\partial x} = -\Psi(x) \tag{1}$$

where x= space variable along the vessel, t= time, A= internal cross-sectional area, u= cross-sectional mean velocity and Ψ describes the flux through the wall. Replacing the derivatives of A in time and space by

$$\frac{\partial A}{\partial t} = \left.\frac{\partial A}{\partial p}\right|_x \frac{\partial p}{\partial t}, \quad \frac{\partial A}{\partial x} = \left.\frac{\partial A}{\partial p}\right|_x \frac{\partial p}{\partial x} + \left.\frac{\partial A}{\partial x}\right|_p$$

where p= pressure, and defining a local wave speed

$$c = \sqrt{\frac{A}{\varrho} \left.\frac{\partial p}{\partial A}\right|_x} \tag{2}$$

(ϱ= 1060 kg/m^3= density) the continuity equation can be rewritten

$$\frac{\partial p}{\partial t} + \frac{\partial(up)}{\partial x} + (\varrho c^2 - p)\frac{\partial u}{\partial x} = -\frac{\varrho}{A} c^2 \left\{ u \left.\frac{\partial A}{\partial x}\right|_p - \Psi(x) \right\} \tag{3}$$

The first term to the right of the equal sign arises from the taper of the aorta, the second one from the loss of mass by diffusion through the wall. The differential force balance equation is

$$\frac{\partial(uA)}{\partial t} + \frac{\partial(u^2 A)}{\partial x} + \frac{A}{\varrho}\frac{\partial p}{\partial x} + 2\frac{\partial A}{\partial x}\left(\frac{\partial u}{\partial x}\right)^2\left(\frac{dx}{2}\right)^2 = -\pi D \frac{\tau_w}{\varrho}\frac{u}{|u|} \tag{4}$$

The skin friction between the fluid and the wall is represented by the term including τ_w. The term is completed by the factor u/|u| for consideration of the flow direction. Neglecting the term of higher order and simplifying by the continuity equation (1) the momentum equation (3) becomes:

$$\frac{\partial u}{\partial t} + u\frac{\partial u}{\partial x} + \frac{1}{\varrho}\frac{\partial p}{\partial x} = \frac{u}{A}\left(\Psi(x) - \frac{1}{\varrho |u|}\tau_w \sqrt{4\pi A} \right) \tag{5}$$

Abtlg. Theoretische Biologie, Botanisches Institut der Universität, Kirschallee 1, 5300 BONN 1, FRG

The system of equations is completed by an elasticity model representing the wall displacement in dependence of the static pressure of the flow. In this work the model of Langewouters et al. is choosen because the presented arctangent model is based on an extended study human thoracic and abdominal aortas. A preliminary study has shown the human physiology is described most sufficiently by Langewouters et al. in comparison with some other models. Especially the calculated mean wave speed is close to physiological data (cp./6/). The here mentioned equation is

$$A(p) = A_m\left\{\frac{1}{2} + \frac{1}{\pi}\arctan\left(\frac{p - p_0}{p_1}\right)\right\} \tag{6}$$

To consider the taper of the vessel the equation (7) is extended for the purpose of this paper in replacing the constants by linear dependences in space:

$$A(x,p) = A_m(x)\left\{\frac{1}{2} + \frac{1}{\pi}\arctan\left(\frac{p - p_0(x)}{p_1(x)}\right)\right\} \tag{7}$$

$$A_m(x) = A_{mt} - (A_{mt} - A_{ma})\frac{x}{L}$$

$$p_0(x) = p_{0t} - (p_{0t} - p_{0a})\frac{x}{L}$$

$$p_1(x) = p_{1t} - (p_{1t} - p_{1a})\frac{x}{L}$$

The corresponding data are taken from /3/ as follows: A_{mt} = 4.61 cm**2, A_{ma} = 2.06 cm**2, p_{0t} = 38.5 mmHg, p_{0a} = 9.0 mmHg, p_{1t} = 45.7 mmHg, p_{1a} = 29.1 mmHg and the aortic length is choosen L= 50 cm. The local wave speed can be calculated in respect to equations (2) and (7) from

$$c(x,p) = \left\{\frac{p_1(x)}{\varsigma}\left(\frac{\pi}{2} + \arctan\left(\frac{p - p_0(x)}{p_1(x)}\right)\right)\left(1 + \left(\frac{p - p_0(x)}{p_1(x)}\right)^2\right)\right\}^{1/2} \tag{8}$$

II. RESULTS

The simulation of the pulsatile flow in the aorta requires the integration of the eqs. (2, 5, 7) in two separated elastic tubes interconnected by special boundary conditions modelling the aortic valve. In difference to the above model equations the flux ψ and the wall friction τ_w are neglected. The differential equations are numerically solved by the implicit method of Warming & Beam. The scheme is first-order-time and second-order-space accurate (FTCS). A detailed description for also more-dimensional systems is to be found in /8/, respectively in Roache /5/. The boundary conditions are provided by the method of characteristics/4/ of hyperbolic systems. The characteristic directions are given by

$$\left.\frac{dx}{dt}\right|_{1,2} = (u \pm c) \ (9), \text{ corresponding to } \quad du \pm \frac{1}{\varsigma c}dp = 0 \tag{10}$$

which allows the determination of the flow variables at the boundaries. The first part (length l=5cm) serves to get a directed flow onto the valve initiated by a periodic pressure function for each oscillation:

$$p_V(t) = \begin{cases} p_m - p_{am}\cos\left(2\pi\frac{t}{\Delta T}\right) & \text{for } \hat{t} \le t \le \Delta T + \hat{t} \\ p_m - p_{am} & \text{for } t < \hat{t},\ t > \Delta T + \hat{t} \end{cases} \tag{11}$$

The period length is defined by f= 68/min and the other constants are ΔT= .4sec, $\hat{t}$= .1sec, p_m = 65mmHg and p_{am}= 55mmHg. At the right (x=l) of the small vessel and at the left of the aorta the boundary conditions are specified by the model of the aortic valve. The valve opens when the pressure gradient at the valve position becomes positive. Afterwards it does not change its state until the "closing" condition becomes true, which means that there is a resulting flow in negative x - direction (u< -25 cm/sec). The closure of the valve does not change until the pressure gradient becomes positive again. As long as the valve stays open pressure and velocity right at the valve position are calculated from the mean values of the variables at the boundaries determined by eqs. (9) and (10). During the "closed" period of the valve the velocity is put u= 0. and the pressure may be determined by characteristics, - both at the edges of the vessels and at the valve position itself. The right of the aorta finishes with a strong resistance and the flow variables are calculated from the Bernoulli eq./7/ in correspondence to eqs.

(9) and (10). The resistance is choosen to get always a static pressure of p_{per} = 65mmHg in the peripheric regime. The equation the velocity is calculated from is explicitly given by

$$\tilde{u} = -\frac{c}{M} (\pm) \sqrt{\left(\frac{c}{M}\right)^2 + \frac{2}{M}\left\{ \frac{p - p_{per}}{\rho} + cu \right\}} \qquad (12)$$

where M denotes the resistance

$$M = \zeta + \frac{1}{m^2} - 1$$

with

$$\zeta = \left(\frac{1}{m\mu} - 1\right)^2, \quad m = 0.36, \ \mu = 0.6$$

The pressure results from equation (10).

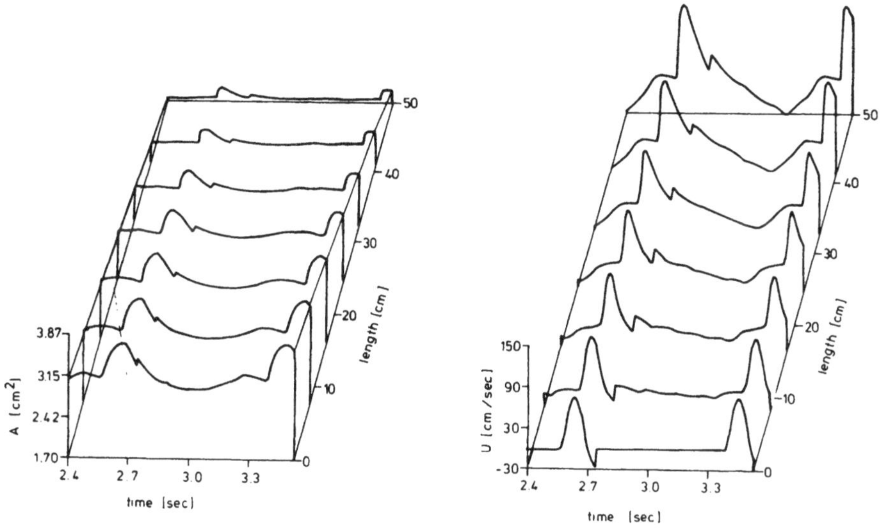

Fig.(1): Cross-sectional area A (extended by pulse wave) and velocity u versus time t and aortic length x

The calculations are started with initial datas

$$p\ (t=0,\ 0 \leqslant x \leqslant L+1) = 65\ \text{mmHg}$$

and $$u\ (t=0,\ 0 \leqslant x \leqslant L+1) = 0.\ \text{cm/sec}.$$

Fig.(1) presents the calculated graphs of cross-sectional area A and velocity u occuring in the aorta($0<x<L$) versus time and space in three-dimensional diagrams. Following the direction to the background one can see at the left and at the right border that the originally linear-tapered vessel is slightly bend by the pressure waves. In front of the figure the situation right behind the valve position is shown. Remarkable is the sharp incision caused by the closure of the aortic valve,as it does occur in human physiology too. This incision decreases while the wave is propagated through this artery with a wave speed close to physiological data. In the diagram u(t,x)it becomes obvious that the heart valve closes when the required back flow results. Since the valve keeps closed the velocity u equals zero.

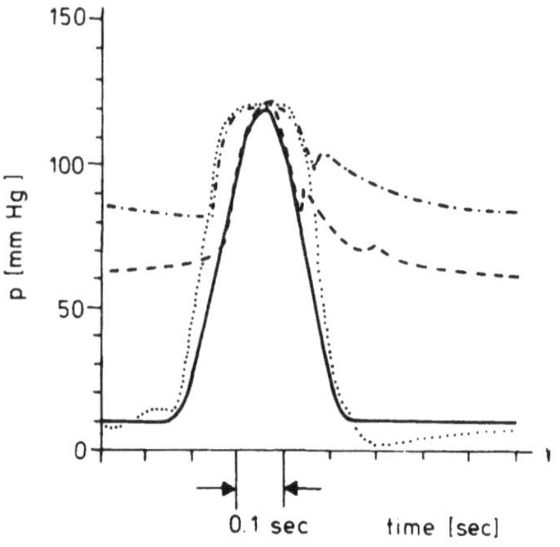

Fig.(2): Comparison between calculation and experiments/6/.
aortic pressure: _._._._.(exp.); -----(calc.)
ventricular pressure ....(exp.) ——— (calc.)

Fig.(2) compares experiments /6/ and the here presented calculations. The fat line and the dashed line mark the calculation of pressure development in the ventricle (determined by equation (11)) and directly at the root of the aorta (calculated by (2,5,7)). The aortic pressure follows the ventricular pressure determined by the differential equation system precisely as it does in nature, even when the aortic pressure is too low. As a conclusion it may be stated close agreement with physiology can be reached when non-linear terms of the model equations and pulse wave propagation and pulse wave reflection are included into calculations.

ACKNOWLEDGMENT -- The author acknowledges the assistance of F.Langner (Aachen) in performing the computer simulations. It is also a pleasure to thank W.Limberg and R.Opitz (Aachen) for some helpful remarks during the course of the work and H.Geithmann (Bonn) for preparing the script.

REFERENCES

/1/ Gerrard, J.H.; Taylor, L.A.: Mathematical Model Representing Blood Flow in Arteries; Medical and Biological Eng. + Computing,15,1977,611
/2/ Kurz, W.: Nichtstationäre Strömung in Modellen arterieller Gefäße, Dissertation, RWTH Aachen, 1980
/3/ Langewouters, G.J.; Wesseling, K.H.; Goedhard, W.J.A.: The Static Elastic Properties of 45 Human Thoracic and 20 Abdominal Aortas In Vitro and the Parameters of a New Model; J.of Biomechanics,17,1984,6,425
/4/ Lorenz, J.: Über die Wellenausbreitung in viskoelastischen Leitungen und ihre gasdynamische Analogie; Dissertation, RWTH Aachen, 1973
/5/ Roache, P.J.: Computational Fluid Dynamics; Hermosa Publishers, Albuquerque, 1985
/6/ Siegenthaler, W.: Klinische Pathophysiologie; Thieme Verlag, Stuttgart, 1982
/7/ Truckenbrodt, E.: Fluidmechanik I, II; Springer Verlag, Berlin, 1980
/8/ Warming, R.F.; Beam, R.M.: On the Construction and Application of Implicit Factored Schemes for Conservation Laws (Manuscript); SIAM-AMS Proceedings, 11, 1977

"BALANCE" SEGMENTS IN MULTICOMPARTMENTAL MODELS

Milan Kotva[1)]

1. Introduction

Among the very important components of the General Method of Multicompartmental Models (GMMM - /1/) are so called segments. A segment is defined as a non-compartmental model approximation of the multicompartmental subsystem behaviour. One of the many possible types of segments are "balance" segments.

2. Basic principles of "balance" segments

"Balance" segments were originally created for simulating the kinematics of (bio)chemical processes for which only equilibrium of dissociation constants but not rate constants are known. For this reason "balance" segments are constructed on the assumption of the immediate equilibrium setting and on the principle of "components balance". It was this principle which gave the name to this type of segments.

Not having rate constants it is impossible to create the kinematic multicompartmental scheme of the process under study. Nevertheless, a kinetic scheme can be constructed. Each compartment represents the quantity of molecules of one of the compounds or ions taking part in the given process. In the next step, the basic components have to be defined. Naturally, most of the ions taking part in the process represent such a basic components, the substrata of more complex compounds can be the other ones.

As an example, basic (bicarbonate) part of the plasma buffer system is shown in the compartmental representation in Fig. 1. Basic components of this system are H^+, OH^- and CO_2.

According to the graphic model from Fig. 1 three balance equations can be written down:

for H^+ $$XC1 = X1 + X4 + X5 \quad (1)$$
for OH^- $$XC2 = X2 + X4 + X5 + X6 \quad (2)$$
for CO_2 $$XC3 = X3 + X5 + X6 \quad (3)$$

Besides the balance equations, two dissociation and one equilibrium equations for steady state hold:

$$X1 \,.\, X2 = K1 \,.\, X4 \,.\, V \quad (4)$$
$$X1 \,.\, X6 = K2 \,.\, X5 \,.\, V \quad (5)$$
$$X3 \,.\, X4 = K3 \,.\, X5 \,.\, V \quad (6)$$

[1)]Institute for Social Medicine and Organization of Health Services
Vítězného února 54
121 39 Praha 2
Czechoslovakia

where Xi - quantity of compound or ions in grammemolecules
Ki - dissociation or equilibrium coefficient
V - volume of the solution in litres

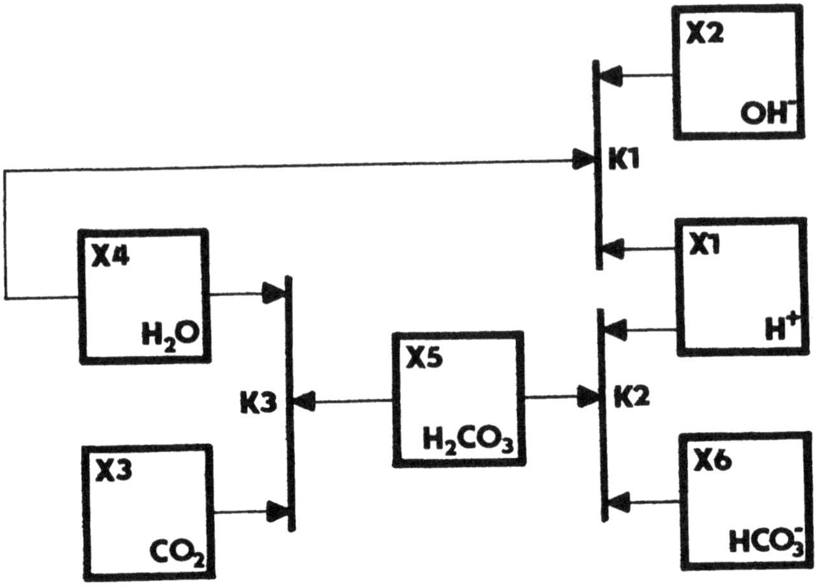

Fig. 1. Kinetic multicompartmental scheme of bicarbonate part of the plasma buffer system.

Under the assumption of the immediate equilibrium (dissociation) setting, the above written equations can be used directly for computation of all Xi in time t. Knowledge of total quantities of basic components (XCi) and of values of coefficients Ki is the conditio sine qua non for such a computation. When any XCi changes its value by Δ XCi in the course of the time interval Δ t, the new values of Xi in time t+ Δ t can be calculated from equations (1) to (6) using XCi+ Δ XCi on the left hand sides of equations (1), (2) and (3). Repeated computation of Xi is also necessary when values of Ki change because of the change of solution temperature etc.

From the view point of mass exchange between the multicompartmental subsystem (approximated by the "balance" segment) and its environment it follows that:

$$\Delta XC1 = \pm \Delta X1 \pm \Delta X4 \pm \Delta X5 \quad (7)$$

$$\Delta XC2 = \pm \Delta X2 \pm \Delta X4 \pm \Delta X5 \pm \Delta X6 \quad (8)$$

$$\Delta XC3 = \pm \Delta X3 \pm \Delta X5 \pm \Delta X6 \quad (9)$$

where Δ Xi - quantity of compound or ions which entered (+) or left out (-) the subsystem in the course of the time interval Δ t.

It is obvious, that entry or exit of some more complex compound (i.e. H_2CO_3) changes more XCi than only one. In every case, mass exchange requires to calculate a new value of volume V, too.

3. Simulation model of the "balance" segment

The resulting set of (generally nonlinear) algebraic equations described above can be solved in an iterative way, or by the use of some other method, demanding many computational steps. However, one very old method from the analog computers application spectrum seems to be more advantageous (few years ago, this method has been rediscovered again - for digital computers).

The basic principle of this method is an incorrect - from the mathematical point of view - trick, when the first derivatives of unknown variables over time are added in suitable order to each one of the given set of algebraic equations. This means that the set of algebraic equations is transformed to a set of first order differential equations. If the solution of such a set of differential equations is stable, the first derivatives will in steady state be equal to zero and the solution will be equal to the solution of the original set of algebraic equations. In our example it means:

$$\frac{dX1}{dt} = XC1 - X1 - X4 - X5 \qquad (10)$$

$$\frac{dX2}{dt} = XC2 - X2 - X4 - X5 - X6 \qquad (11)$$

$$\frac{dX3}{dt} = XC3 - X3 - X5 - X6 \qquad (12)$$

where

$$X4 = X1 \,.\, X2 \,/(K1.V) \qquad (13)$$

$$X5 = X1 \,.\, X2 \,.\, X3 \,/(K1.K3.V^2) \qquad (14)$$

$$X6 = X2 \,.\, X3 \,.\, K2 \,/(K1.K3.V) \qquad (15)$$

As a matter of fact, this trick replaces the assumption of the immediate equilibrium setting by the assumption of continuous transition between two equilibrium states. It is necessary to take notice of the fact, that the time of the substituting set of differential equations is not the time of process under study. However, when Δt used in context with equations (7), (8) and (9) comes near to zero, these equations can be rewritten to the following form:

$$XC1 = \pm F(X1) \pm F(X4) \pm F(X5) \qquad (16)$$

$$XC2 = \pm F(X2) \pm F(X4) \pm F(X5) \pm F(X6) \qquad (17)$$

$$XC3 = \pm F(X3) \pm F(X5) \pm F(X6) \qquad (18)$$

where F(Xi) - input (+) or output (-) flow of Xi.

It is obvious now, that the set of equations (10) to (18) approximatively represents the original biochemical process on the kinematic level. Let us say that the latter set of equations is the approximative functional model of the multicompartmental subsystem shown in Fig. 1. under the assumption, that t is the time of the process passing in this subsystem. Naturally, either the original assumption of the immediate equilibrium setting or the assumption of continuous transition between two equilibrium states do not hold for this model representation.

The above defined "balance" segment has to be used in simulation program not only as a part of the simulation model, but also in the initial conditions setting procedure. It is necessary to find the initial values of Xi from the initial values of XCi. Naturally, the equations (16), (17) and (18) do not play any role in this procedure. On the other hand, a special procedure for check-up on the equilibrium setting is obligatory.

4. Conclusion

As it is mentioned above, "balance" segments were originally created for simulating the kinematics of (bio)chemical processes. However, it turned out that this type of segments can be used as well in multicompartmental representation of other processes, e.g. demographic, epidemiological etc. Hence, in addition to "transport" and "cohort" segments (see e.g. /2/, /3/) the "balance" segments become a usefull component of the GMMM, too.

Literature:

/1/ Kotva, M.: General Method of Multicompartmental Models. Proc. of European Congress on Simulation, ACADEMIA, 1987, Prague.

/2/ Hauser, F., Kotva, M.: Simulation of Continuous Transport Processes with Consideration of Transit Times Distribution. Proc. of 9. IMACS Congress, 1979, Sorrento, Italy

/3/ Hauser, F., Kotva, M.: A "Cohort" Model for the Simulation of Medico-Demographic Processes. Proc. of European Congress on Simulation, ACADEMIA, 1987, Prague.

DYNAMOEXPERT: PC-AIDED SIMULATION AND IDENTIFICATION TOOLS FOR BIOMEDICAL APPLICATIONS

V.V.SHAKIN, A.A.KARASEV, V.V.PRZYJALKOWSKI

1. Bio-medical goals

The proposed system DYNAMOEXPERT is a set of powerfool and flexible tools for development and usage of mathematical models for various biomedical applications. Main concepts of mathematical modelling (simulation) are becoming being accepted now as a common research instrument in many areas including various bio-medical applications. Complexity of the whole human organism and its funcional (homeostatical and adaptational) sub-systems demands thorough development of problem-oriented data storing, processing and interptetation methods. At the same time, available instrumental software does not seem to be satisfactory enough to supply regular everlasting support for large-scale long-term complex bio-medical research projects [1].

2. Need for automation

The proposed system DYNAMOEXPERT was elaborated step by step as a logical consequence of understanding of the fact that aforementioned procedures usually involved in biomedical research should be automatized. The system development was stimulated with an idea to supply mathematical and software support for research projects in the field of space life sciences. It was resently launched research program MEDILAB [2] which was conjugated with multiple inhomogeneous experimental data storing and retrieving to provide their complex exhaustive exploration and processing, mathematical simulation of the human organism along with its homeostatic and adaptive features, model identification, interpretation, formulating newer hypotheses, modifying current aims of the study, designing relevant experiments to reach goal of the whole research project.

DYNAMOEXPERT is aiming dynamic simulation and identification of various biomedical processes and pathological conditions. It is capable of creating, re-viewing and estimating some scenaries which are generated by experts in medicine and physiology.

The main destination of the system is to run investigation more efficiently due to expert knowledge more closely linked with mathematical models through graphical tools and by means of generating

Computer Center, USSR Academy of Sciences
40 Vavilov str., 117967, Moscow, USSR

formal knowledge base when viewing the sequence of events after user has supplied disturbances, which are to be numerical model equivalents to some pathological conditions of biomedical objects.

3. Many separate tools already exist

There exists a rather large amount of program facilities intended to handle the most of the tasks listed above.

Almost every particular procedure has sometimes several relevant commercial software tools to meet many needs for automation. Just some additional applied programs would be supplemented to treat arising problems. It is quite natural wish to aviod re-creating available tools from the very beginning. The new problem however arises from the fact that any one of the program facilities is unsufficient to solve the great deal of actually encountered tasks - due to complexity and inherited informality of the latters. The development of one unversal program would be a hard and rather doubtful activity. More realistic, in our convintion, is to provide user ability to work with any of existing program tools and to switch from one to another when difficulties occur with the method involved. This approach forms the basic idea of DYNAMOEXPERT. The point is that user is not obliged to arrange interface and/or data transfer between any two existing parts of applied software.

The stated way of system operation implies the existence of some universal form of data handling and storage - a kind of Database Management System (DBMS) in fact. Also, there should be involved some kind of Executive able to handle overall process of switching from one system to another, data being transferred if neccesary. Visualization tools newcoming with personal computers (PC) are also very important to provide friendly interface between PC and researcher or, which is even more critical, between PC and practical physician.

4. DBMS as a basis

Effectiveness of such complicated and multi-functional system as DYNAMOEXPERT would not be achieved without use of underlying powerfull universal DBMS. DYNAMOEXPERT implements, as a matter, a network-type DBMS called COMPASS [3].

Using COMPASS is anticipated due to the following benefits:

- centralized storage of overall-system information without duplication and hence without danger of inconsistency of the data stored;

- data integrity supported by means of DBMS;

- shared access to the same data by different sub-systems;

- data access control in view of usage by different sub-systems and users;

- long-time storage of the complex-structured data;

- ability of data handling on the basis of having alternative variants along with ability of fixing up any desirable variant while rejecting others.

The data manipulation language (DML) used in COMPASS is more powerful than the DML of so-called relational database systems. This DML along with the graphical data access allowes DYNAMOEXPERT to be a friendly tool not only for computer novices, but also for such experts who are just able of formulating their task logically.

The DBMS COMPASS makes it possible to store and manipule the following kinds of data:

- different versions of the models of human organism, including the information on subdivision of every model onto various sub-models;

- end-user-defined relations between the attributes prescribed to each model;

- various sets of parameters for each model;

- previously obtained empirical information concerning different model variables and relations, that makes possible for end-user to involve this information in the model computation process and in comparative analysis of the model functioning as well;

- auxiliary information on functioning of each model including data about users and their rights, relations of the models to real human beings observed, etc.;

- necessary data for high-level user-computer dialogue support: "help" data, graphical dialogue arrangement data, etc.

5. DYNAMOEXPERT now

In its backgrond DYNAMOEXPERT has the IBM PC-compatible computer software, MS-DOS operating system being included.

Here follows a brief list of the tools available with their options of interest for explored biomedical applications:

- graphical package with fast access to data from another media;

- information base for data, models, scenaries, and acts;

- numerical integrator to investigate dynamic processes with graphical representation of results and with option to fit them to experimentally observed data;

- supervisor to provide interface between other tools and user and to supply him with prompting messages and help;

- statistical package;

- identification package able of numerical fitting and model parameters estimation and validation [4].

The set of tools is currently modified to be extended to meet needs of biomedical applications.

6. Literature

[1] Balci, O., and R. E. Mance: Simulation model development environments: a research prototype. J Opl Res. Soc. Vol.38 (1987) 2, 753 - 763.

[2] Gazenko, O.G., A.I. Grigoriev, E.A. Ilyin, and Kholin,S.F.: MEDILAB - project of medical laboratory in space. In: Proc. 3rd European Symp. on Life Sciences in Space, ESA-ESTEC Publ. Nordwijk (Netherland) 1987, 275-278.

[3] Filippov, V.I.: Data base management system COMPASS. - In: Problems of Data Base Management Systems, Univ. of Turku. Turku (Finland) 1980.

[4] Walter, E.: Identifiability of State Space Models. Springer-Verlag, Berlin 1982.

A NEW MODEL FOR CURRENT FLOW IN DENDRITES OF ARBITRARY GEOMETRY

A. Schierwagen[1] and J. Francu[2]

1. Introduction

Current flow within dendrites or axons of nerve cells has usually been represented in the framework of the core conductor concept /2/. Starting with its standard assumptions, a new model of current flow in branching dendrites is derived. The method is a variant of the continuum approach employed in many different fields of research /1/. In the present case, the fine structure of the branching dendrites is replaced by an equivalent, locally homogeneous electrical medium described by averaged electrical quantities. For the mean potential and the mean current density in this medium partial differential equations of parabolic type are derived which represent natural generalizations of the simple case of a leaky cable.

2. Problem formulation

Suppose the nerve cell body is in the origin of a system of polar coordinates r, θ, φ. Let each dendrite be contained in a space angle θ. All quantities involved will be considered in dependence on distance r. Assuming extracellular current flow as negligible, the polar angle φ becomes a parameter and can be omitted.
Each dendrite is topologically of tree type, composed of cylindrical segments which join at branch points. The tree as a whole, then, can be replaced by an electrically equivalent medium which averaged quantities are assigned to, depending on distance r: axial resistance R(r), membrane capacity C(r) and membrane conductance G(r).
These parameters depend in a simple way on the "microscopic" structural and electrical parameters of the dendrites (mean segment diameter d(r), specific electrical parameters as axial resistance r_a, capactity c and conductance g). For example, the "macroscopic" value of axial resistance is given by

$$R(r) = 4 \cdot r_a/(n(r) \cdot \pi \cdot d(r)^2),$$

with n(r) the dendritic branch number and taking all dendritic segments at distance r electrically as parallel.

If we consider a volume element ϑ at time t within the angle θ restricted by surfaces S_1, S_2, the application of Kirchhoff's law yields

1) Karl Marx University Leipzig, Institute for Physiology, Liebigstr.27, DDR-7010 Leipzig
2) Purkyne University Brno, Chair for Mathematical Analysis, Janáčkovo nám. 2a, 66295 Brno, CSSR

$$I_1 + I_2 + I_C + I_G = 0 \tag{1}$$

where $-I_1$ and I_2 are the axial currents crossing the surfaces S_1, S_2 at distances r_1, r_2; I_C and I_G denote the summed capacitive and lost currents in $\mathcal{V}$.

The currents in eqn. (1) can be expressed as follows:

$$\int_{\Theta} \int_{r_1}^{r_2} \left(\frac{\partial I}{\partial r} + C(r)\,\frac{\partial V}{\partial t} + G(r)\cdot V(r,t)\right) dr\, d\Theta = 0. \tag{2}$$

Dividing this equation by $|\Theta| \cdot |r_2 - r_1|$ and passing to the limit $|\Theta| \longrightarrow 0$, $|r_2 - r_1| \longrightarrow 0$, the first basic equation is obtained:

$$\frac{\partial I}{\partial r} + C(r)\cdot\frac{\partial V}{\partial t} + G(r)\cdot V = 0. \tag{3}$$

From Ohm's law one can easily derive

$$V(r_2,t) - V(r_1,t) = \int_{r_1}^{r_2} \frac{\partial V}{\partial r}\, dr = -\int_{r_1}^{r_2} R(r)\cdot I(r,t)\, dr,$$

which gives the second basic equation:

$$\frac{\partial V}{\partial r} + R(r)\cdot I = 0. \tag{4}$$

Eqns. (3) and (4) are Kelvin's well-known RC cable equations with space-dependend coefficients.

3. Analysis of the model equations

Combining eqns. (3) and (4), we find a second order equation in V:

$$\frac{\partial V}{\partial t} - \frac{1}{C(r)}\,\frac{\partial}{\partial r}\left(\frac{1}{R(r)}\,\frac{\partial V}{\partial r}\right) + \frac{G(r)}{C(r)}\cdot V = 0. \tag{5}$$

Applying the transformations

$$x(r) = \int_0^r (C(s)R(s))^{1/2}\, ds,$$

$$V(x,t) = \exp\left(-1/2\int \gamma(x)dx\right. \cdot v(x,t), \tag{6}$$

eqn. (5) gives rise to

$$\frac{\partial v}{\partial t} - \frac{\partial^2 v}{\partial x^2} + \delta\cdot v = 0, \tag{7}$$

where $\delta = \beta + \gamma^2/4 + \gamma'/2$, $\beta = G/C$ and $\gamma = (C'R - C'R)/(2CR)$.

Assertions on solutions of eqn. (7) depend on whether δ is a constant. If so, then we have the cable equation with constant coefficients for which a well-developed theory exists. Its many specific solutions can be applied, therefore, to the continuum equation (5) through the transformations (6).

The condition δ = const. implies a generalized "3/2 power law" /2/ to hold, i.e.,

$$n(r)\cdot d(r)^{3/2} = d_o^{3/2} \cdot \exp(\gamma_1 \cdot x)\,(1 + 2C_2\gamma_1 \exp(-\gamma_1 x))^2, \qquad (8)$$

which results from the solution of a Riccati type differential equation as explicated in detail elsewhere /3/.

4. Conclusion

The aim of this study was to expand the range of application for analytical neuron models. Employing a combination of standard core conductor theory with the continuum approach, the 1-dimensional cable equations with space-dependend coefficients have been derived.

Analytical solutions can be calculated in straightforward manner from those of linear cable theory with constant coefficients, provided a generalized branching condition holds. This condition is the most general one derived so far allowing to model a much wider class of neuronal dendritic trees than before.

5. References

/1/ Bachmat, Y., and Bear, J.: Macroscopic modelling of transport phenomena in porous media. 1: The continuum approach. Transp. Por. Media 1 (1986), 213-240.

/2/ Schierwagen, A.: Distributed parameter systems as models of current flow in nerve cells with branched dendrites: Inverse and forward problems. Syst. Anal. Model. Simul. 5 (1988), in press.

/3/ Schierwagen, A., and Francu, J.: A continuum description of subthreshold membrane potential in branching dendrites. J. Theoret. Neurobiol., submitted.

SIMULATION SYSTEM SONCHES BASED TOXICOKINETIC MODELLING AS A TOOL IN BIOLOGICAL MONITORING

Wünscher[1), G.; V. Wenzel[2); M. Flechsig[2); E. Matthäus[2)

ABSTRACT: The use of an interactive simulation system (SONCHES) for toxicokinetic modelling will be described.
The usual physiological based description of the toxicokinetics of inhaled gases/vapors (partition coefficient concept) is completed by an approach for urinary excretion of metabolites, water, creatinine. Various simulation experiments (scenarios: certain inhaled chemicals, different species - human, rat, variation in body build, physical exercise, and exposure pattern) are stored in a result library.
The simulation system make a tool of the user's model bank not only because of the facilities for post-processing of these results. Despite of the progress which has been made, further research is needed in the validation of our models.

KEYWORDS: occupational exposure, biological monitoring, physiological based toxicokinetic models, simulation, data collection

INTRODUCTION

Toxicokinetic modelling is a powerful tool in designing of experiments, the interpretation of data extrapolation from animal to man, extrapolation from one exposure regimen to another, and the evaluation of the effect of a variety of physiological, and pathological changes on the dwelling of toxic chemicals in the body. Industrial hygienist's activities in this field - especially biological monitoring (BM) and risk assessment - can be improved if toxicokinetic techniques are employed. Above all it is the rationale for setting exposure limits. The development of biological monitoring of exposure to chemicals has been hindered by the lack of understanding of the factors affecting the toxicokinetics. An important conceptual advantage of toxicokinetic models, widely used nowadays, is the idea of combining internal toxicant transport system and anatomic-physiological surrounding. The model basic structure of various species (humans, rats) under varying conditions (physical activity, body build etc.) is the

[1] *Chemical Industry Health Center of GDR, DDR-4220 Leuna, PSF 31*

[2] *Central Institute of Cybernetics and Information Processes, Academy of Sciences of the GDR, DDR-1086 Berlin, Kurstr. 33*

same. Only the corresponding parameters concerning physiological conditions and distribution ratios (so called partition coefficients) would need to be modified in the respective simulation experiment. These parameters and further assumtions are based on experimental data, on values reported in literature or on empirical considerations. In this way toxicokinetic simulation models have the capability to supervise the gathering and interpretation of the whole data at hand in the field under study - provided that a flexible user friendly simulation system and adequate resources are available. The requirements of a industrial hygienist's tool can be met with by the implemented toxicokinetic models (as a model base) especially if handled under support of simulation system SONCHES. This tool has been designed to unite within one on-line-connected system the three main capabilities for serving simulation activities:

- an object-related language supporting model building, not necessary asking for predefined explicite equations, but forming model structures rather from empiric and algorithmic details;
- a comfortable simulation environment containing e.g. model and data libraries, scenario tailoring services for single run, multi run and model debugging, components for model change, model self-documentation etc. ;
- a dialogue-based calculus for result processing and visualization from table up to graphic forms.

SIMULATION MODEL

Essentially our approach follows the ideas of FISHEROVA-BERGEROVA and co-workers /1/. There are seven basic compartiments which represent the anatomic-physiological structur of lung, skin, liver, fattissue, muscles, and other parts of body .
The amount of a vapor (i.e. organic solvent) taken up by various body tissues depends on the partition coefficients, which can be determined in vitro. As absorbed vapors are chiefly applicated by inhalation and transported via the blood stream alveolar ventilation and blood flow (perfusion rate) is an other determinant. Since the toxicity of most industrial chemicals is related to the average tissue concentration over a given period of time we have regard a compartiment "toxicodynamics". As a rule it has little effect on toxicokinetics, but is required to conduct quantitative risk assessment.
Complementary our model has a substructur concerning the urogenital system. Not blood, for all practicle reasons, urine is the only easily accessible BM sampling site (indicator) in man

(and in the intact animal). Therefore the model incorporates a compartiment "bladder" with processes of urine excretion (inclusive water, metabolites, creatinine). Water and creatinine are of particular interest to adjust values of urinary biological indicators. In biological monitoring of rapidly metabolized substances such as solvents, spot samples taken at the end of the work shift should provide information on the exposure during that day of work. It is common practice to adjust the values determined from spot samples to creatinine. Simulation experiments deals with this issue, in order to assess whether creatinine adjusted parameters meet the BM requirements. A sub-model for urinary excretion is difficult to derive. For that reason we use empirical functions (excreted water/creatinine time series, time of urinary output) observed in experiments. These tables are linked to the model and automatically included whenever a simulation run is started. Performing different simulation experiments linkage can be altered to any other function table stored in table library curently connected with the system.
Skin absorption as an essential source of errors in the evaluation of BM data is also considered.
Physiological parameters for the model were determined by measurement or were founded in biological tables or in reference books. The partition coefficients of many solvents used in industry were published in various papers (/1/,/2/,/4/).

SIMULATION STUDIES

As is suggested by the preceding model description, the role of driving forces for the dynamical behaviour mapped by this model is played by physiological or environmental factors, that can alter in particular respiratory and cardiovascular functions, producing thereby a significant effect on the toxicokinetics.
Defining different scenarios means in first order to vary these driving variables and to study sensitivity of such changes. Generally, simulation experiments begin with the definition of scenarios (verbal-descriptive form attributed to concrete numerical values of factors/parameters). In view of these factors the term "scenario" stands for a set of these mutually interrelated values depicting the changes in such categories as

- physiological surrounding (human, animal)
- exposure pattern (occupational exposure, chamber experiment)
- physical activity (resting, working conditions)
- body build (total body fat)
- skin absorption (compared with inhalation)
- BM indicator medium (blood, urine, exhaled air).

It.has been shown the scenario concept can support the communication of workers are concerned with modelling. The model was tested under various conditions. Experimental data used in the testing were descended from experimental exposure (benzene: /2/, and own data). The experimental data and the simulation data matched well if the comparisons based on mean values obtained from a group of volunteers exposed under well controlled conditions (physical activity!). Such a verified model is ready to be used for simulation of working conditions. The occupational conditions simulated using the physiological parameters of the workers and the workweek exposure schedule for input. The model used for setting of bioequivalents for an unusual shift schedul (/4/). The simulation studies with respect to further chemicals are going on.

LITERATURE :

/1/ *V. Fisherova-Bergerova:*
Toxicokinetics of organic solvents
Scand. J. Work Environ. Health 11 (1985) Suppl 1, 7-21

/2/ *A. Sato; T. Nakajima:*
Pharmacokinetics of organic solvent vapors in relation to their toxicity
Scand. J. Work Environ. Health 13 (1987) 81-93

/3/ *W. A. Ritschel; P. S. Bauerjee:*
Physiological Pharmacokinetic models: Principles, Applications, Limitations and Outlook
Meth. and Find Exptl. Clin. Pharmacol. 8 (1986) 10, 603-614

/4/ *M. E. Andersen; M. G. MacNaughton; H. J. Clewell; D. J. Paustenbach:*
Adjusting Exposure Limits for Long and Short Exposure Periods Using a Physiological Pharmacokinetic Model
Am. Ind. Hyg. Assoc. J. 48 (1987) 4, 335-343

/5/ *V. Wenzel; E. Matthäus; M. Flechsig:*
SONCHES-Nutzerhandbuch, Dokumentation f)r das Simulationssystem 1985

/6/ *V. Wenzel; K. Bellmann; E. Matthäus; M. Flechsig:*
SONCHES- An Interactive Simulation Language for Design, Validation and Usage of Ecological Models
In System Analysis and Simulation 1985,
Proceedings of the 2. Int. Symp., Berlin 1985, 129-132

/7/ *E. Matthäus; M. Flechsig; K. Bellmann; V. Wenzel:*
Use of Task Facilities of the Simulator SONCHES for Agroecosystem-Model Investigation and Control
In System Analysis and Simulation 1985,
Proceedings of the 2. Int. Symp., Berlin 1985, 133-140

PARAMETER ESTIMATION OF THE BLOOD PRESSURE CONTROL SYSTEM IN DOGS

Christian Fritzsch[1)]

Abstract: Experiments were carried out with trained conscious dogs and anaesthetized mongrels to identify the short-time regulation of blood pressure. The readings were used to establish parametric models in the form of discrete transfer functions, and the parameters were estimated by recursive least squares method and weighted recursive generalized least squares method.

Key words: Short-time blood pressure regulation, closed loop identification, chronic animal experiments, parameter estimation

Introduction: Short-time regulation in the closed blood pressure loop of dogs was identified within the framework of experimental system identification /1/. For an unbiased estimation of the parameters occurring in a closed loop, an additional test signal must be provided which sufficiently and continuously stimulates the circulatory system. Extra disturbance took the form of binary stress sequences using a tilting table and lower body negative pressure (LBNP).
Points of interest included the quality of the estimated partial models, the reproducibility of the estimated parameters of experiments executed under identical conditions, and parameter variation when changes occurred in circulatory regulation.

Animal experiments: Readings were obtained from dogs anaesthetized with alpha-chloralose, and from conscious dogs which had catheters implanted. In all experiments, pressure was recorded via catheter in the aorta, in the A. carotis communis at the level of the A. thyreiodea sup. and in the Vena cava sup. immediately before the right atrium. The momentary heart rate resulted from the RR intervals of the ECG. In the tilting table stress and LBNP experiments we followed first order designs of experiments by PLACKETT and BURMAN /2/.

Model and parameter estimation method

The partial systems of the blood pressure regulating loop to be identified were regarded as stochastically disturbed systems whose deterministic portion over time is described by a linear difference equation and, after z-transformation, by a discrets transfer function:

$$G(z) = \frac{b_1 z^{-1} + b_2 z^{-2}}{1 + a_1 z^{-1} + a_2 z^{-2}} z^{-d} \qquad d = \frac{\text{dead time}}{\text{sampling time}}$$

The a and b parameters were estimated by means of weighted recursive generalized least squares method (k-weighting: $c^2 = 0{,}999$).
The model dead time was determined by variation, and the identified parameter models were selected with the aid of the residual sum and the coefficient of determination.

1) Karl-Marx-University, Clinic for Paediatric Medicine, DDR - 7050 Leipzig, Oststraße 21/25

Results: Three different test series were carried out for identification - 1. Tilting table experiments with anaesthetized dogs, 2. Experiments with trained conscious dogs under tilting table stress, 3. LBNP experiments with trained conscious dogs.

A simplified blood pressure control system is shown in Fig. 1, which makes allowance for the data obtained in the animal experiment.

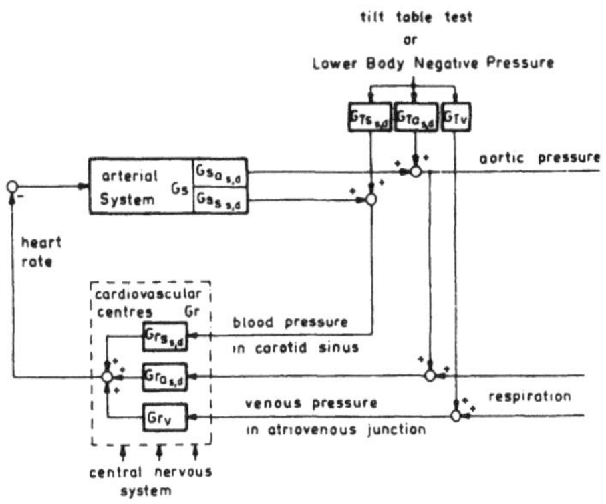

Fig. 1. Simplified blood pressure control system making allowance for data obtained in animal experiment

The tilting table experiments were evaluated manually at intervals of a second and a half second. One-second sampling led to a discrepancy between an excellently performed animal experiment and the model quality of subsequent estimation of parameter suggested a filtration of measured data, and a non-linear model was selected for comparison /3/.

Fig. 2 shows that half-second sampling of the momentary readings alone clearly improved the coefficients of determination for all partial models.

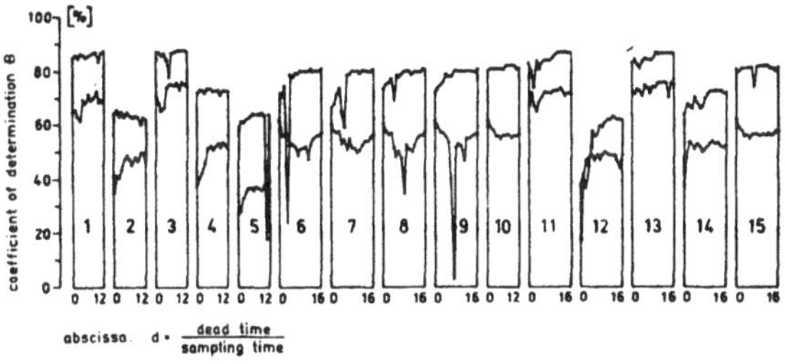

Fig. 2. Coefficients of determination from a tilting table experiment, the lower values resulting from parameter estimation only for every 2nd reading (one-second intervals), for content of models see Fig. 3

If the physiological circulatory conditions are to be distinguished from pathophysiological ones, then the parameter estimation results from experiments repeated under identical conditions must reoccor within a scatter range to be identified. The results from tests repeated on the same day and at intervals of days/weeks were therefore of interest. While there is little difference and standard deviation in the a-parameter of partial models from repeated experiments with anaesthetized dogs, the standard deviation is comparatively large in some of the b-parameters. The a-parameters are described as non-significant if they fulfil the 3σ rule.

They are normally non-significant but become significant in 53 % of all partial models under the effect of angiotensin II, and dead times become longer.

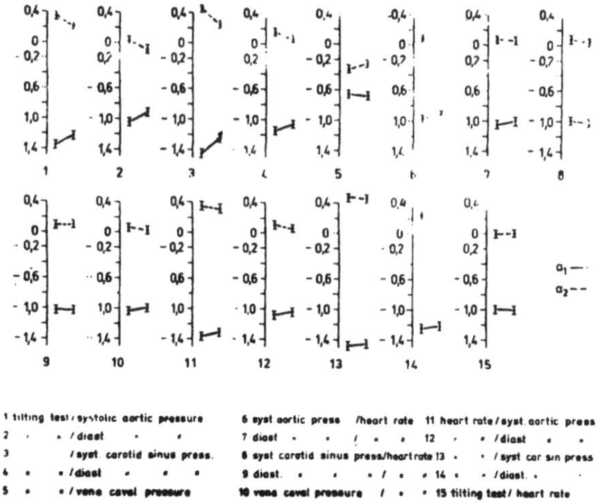

Fig. 3. a-parameters with standard deviation of partial models (1 - 15) from two successive tilting table experiments with conscious dogs

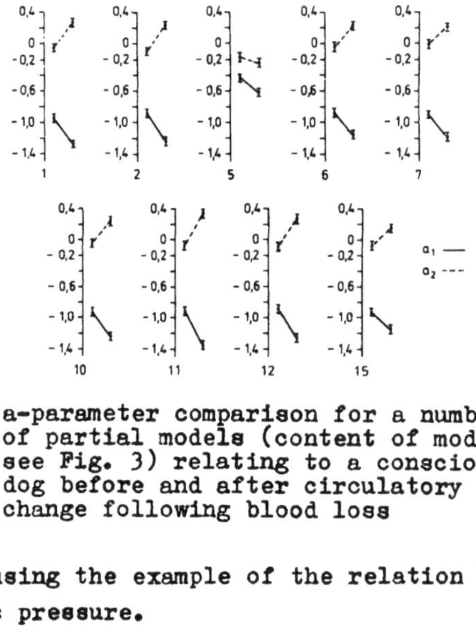

Fig. 4. a-parameter comparison for a number of partial models (content of models see Fig. 3) relating to a conscious dog before and after circulatory change following blood loss

Successive tilting table experiments with conscious dogs normally lead to non-significant a-parameter as can be seen in Fig. 3 (where 3 is an exception). In contrast to this, circulatory changes after blood losses are reflected in the significance of all a-parameters (Fig. 4). Parameter estimations from LBNP tests with conscious animals gave the best results. Here again successive experiments are characterized by non-significant a-parameters.

Depending on the experimental animal, significances may occor in one or two partial models (Fig. 5). Short infusions of angiotensin II significantly change the a-parameters compared with the preceding test (Fig. 6). Over a period 14 - 21 days most but not all a-parameters of the partial models are in the 3 σ range, as is shown in Fig. 7 using the example of the relation between heart rate and aortic systolic pressure.

References

/1/ Fritzsch, Ch.: Chronische und akute Tierexperimente zur Identifikation der Blutdruckregulation unter orthostatischen Belastungen. Promotion-B-Schrift, Karl-Marx-Universität Leipzig, Bereich Medizin 1986

/2/ Plackett, R. L., and J. P. Burman: The design of optimum multifactorial experiments. Biometrica 33 (1947) 305 - 325

/3/ Fritzsch, Ch., and J. Beyer: Verbesserte Systemparameterschätzung der Blutdruckregulation bei anästhesierten Hunden. Biomed. Biochim. Acta 44 (1985) 1729 - 1734

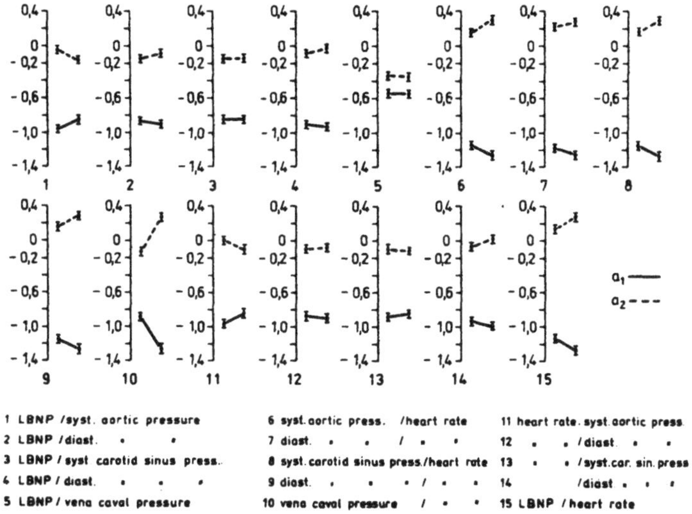

Fig. 5. a-parameters with standard deviation from two successive LBNP experiments relating to a conscious dog

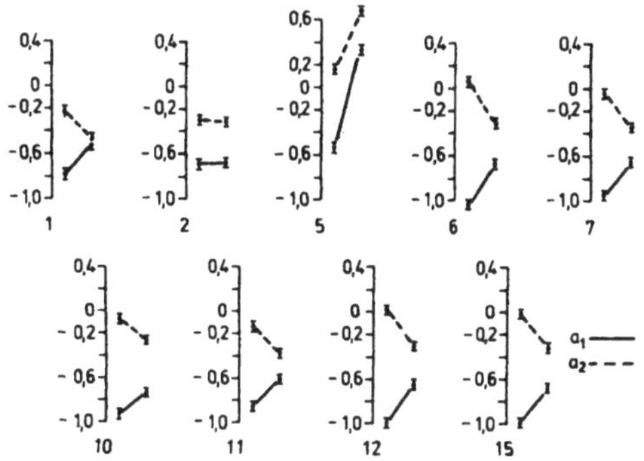

Fig. 6. a-parameter comparison for a number of partial models (model no. as in Fig. 5) relating to a conscious dog under the influence of angiotensin II

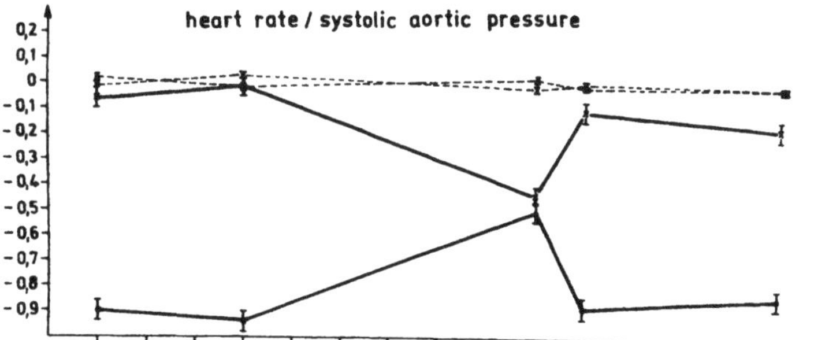

Fig. 7. Time behaviour of parameters with standard deviation of a partial model over 15 days

COMPUTER-GRAPHIC FOR ROTATIONAL AND MIRROR SYMMETRIC FRACTALS WITH m COLOURS AND A DETERMINED BASIC DESIGN

H.Schilling)1

1. Introduction

The production of fractals 1 is controllable under geometric or aestetic points of view. In the sequence of conformal mappings

$$z_{j+1} = w(z_j), \quad (j=1,2,\ldots) \tag{1}$$

the mapping function for a picture is unique represented by K-1 fixed points z_{Fk} and L poles z_{Pl} :

$$w(z) = z + z \frac{\prod_{k=2}^{K} (z-z_{Fk})}{\prod_{l=1}^{L} (z-z_{Pl})} \tag{2}$$

In this representation the source of the coordinate system is the fixed point $z_{F1} = 0$.

2. Parameters for Controlling the Number of Colours

A fixed point is attractor and produces areas of a Julia-set for

$$\left| \frac{dw(z_F)}{dz} \right| < 1 \tag{3}$$

[2]. For K > L the infinite point is attractor. Thus in this case a determined number of m colours limits m-1 absolute values of derivations in a fixed point to the interval from zero to one. These derivations $w'(z_F)$ may be chosen as parameters to control the appearing attractors and the number of colours.
Defining the system of real parameters (0 ... 1)

)1 Central Institute of Cybernetics, Academy of Sciences of the GDR, Kurstr. 33, Berlin 1086, DDR

$B_1, \ldots, B_{m-1} \; ; \; \beta_1, \ldots, \beta_{m-1}$ (4a)
with the property
$B_\mu \exp(2\pi i \beta_\mu) = w'(z_{F\mu})$, (4b)
a nonlinear system of m-1 equations

$$B_\mu \exp(2\pi i \beta_\mu) = 1 + z_{A\mu} \frac{\prod_{j=1, j \neq \mu} (z_{A\mu} - z_{Aj})}{\prod_{l=1}^{L} (z_{A\mu} - z_{Pl})} \tag{5}$$

($\mu = 1, \ldots, m-1$) results.

3. Pictures with Rotational Symmetry

C_n -symmetry is guaranteed by a mapping function w(z) with the period $2\pi/n$ for the dependence on the phase variable $\varphi =$ arc z. Thus rotational symmetry with the centre z_R is produced by an operator of kind

$$w = w([z-z_R]^n) \; . \tag{6}$$

In the following the symmetry centre is defined by $z_R = 0$ and therefore the source of the coordinate system is not a fixed point : $z_{F1} \neq 0$. On this understanding C_n -symmetry with m colours is produced by a mapping function of kind

$$w_{mn} = \frac{\sum_{k=0}^{m-1} a_{kn} z^{kn}}{\prod_{l=1}^{r} (z^n - z_{Pl}^n)} \tag{7}$$

The system of m-1 nonlinear equations from (4)

$$B_\mu \exp(2\pi i \beta_\mu) = w'(z_{A\mu}) \quad (\mu = 1, \ldots, m-1) \tag{8}$$

is completed by the linear system for the coefficients a_{kn}

$$z_{A\mu} = \frac{\sum_{k=0}^{m-1} a_{kn} z_{A\mu}^{kn}}{\prod_{l=0}^{r} (z_{A\mu}^{n} - z_{Pl}^{n})} \tag{9}$$

following for the attractors $z_{A\mu}$ from (7) .

4. Mirror Symmetry

Defining w=u+iv, z=x+iy then from the Cauchy-Riemann-equations follows: If axis of symmetry is defined by the x-axis, mirror symmetry is produced by an operator w(z), which real part u(x,y) depends on the imaginary part y as an even function and which imaginary part u(x,y) depends on y as an odd one. Thus mirror symmetry with m colours is produced by a mapping function of kind

$$w_{mv} = \frac{\sum_{k=0}^{m-1} A_k z^k}{\prod_{l=1}^{r} (z - z_{Pl})(z - z_{Pl}^*)} \tag{10}$$

z^* is the conjugated complex value of z. The coefficients A_k are real values.

C_{nv}-symmetry with m colours is realized by

$$w_{mnv} = \frac{\sum_{k=1}^{(2m-1)n} A_k z_{A\mu}^{kn}}{\prod_{l=1}^{r} (z_{A\mu} - z_{Pl})(z_{A\mu} - z_{Pl}^*)} \tag{11}$$

if A_k is a real coefficient.

5. Basic Design

The attractors define a basic design of different coloured points. A basic design of equal coloured points is to be realized by the poles z_{pl} . Pictures with C_{nv}-symmetry and a basic design of 2nr equal coloured points are determined by r poles z_{pl} in (7) or(11).

6. Multicriterial Decision for the Nonlinear System

In consequence of these demands a system of nonlinear equations with complex coefficients results. It designates the space of the attractors, fixed points and poles. Defining them as control values of the multicritereal decision, they determine exactly the mapping function w and the ranges of convergence and divergence. Equations to calculate these ranges are to be found in [2] .
Goal functions in the multicriterial problem are the deviations from given parameters B_μ and β_μ in (4), a_{kn} in (7) or A_k in (11). For the solution and discussion the program system REH [3] is used, which is developed at ZKI/AdW.

7. References

[1] Peitgen, H.-O., P.H. Richter : The Beauty of Fractals. Springer-Verlag Berlin, Heidelberg, New York, Tokyo 1986.

[2] Schilling, H. : Computer-Graphik fuer Fraktale mit C - Symmetrie aus m Farben . Wissenschaftliche Zeitschrift der TH Karl-Marx-Stadt (in press) .

[3] Straubel,R., A. Wittmuess : Das Programmsystem REH zur rechnergestuetzten Entscheidungshilfe . ZKI-Informationen AdW DDR 2/1986.

SIMULATION ENVIRONMENTS

L. Dekker [1]

ABSTRACT

In the paper it is discussed that in systems simulation nowadays a simulator is becoming a synonym for a computer provided with a simulation environment, i.e. a programming environment dedicated to simulation. A simulation environment is defined as a programming environment of a computer, that is dedicated to systems simulation and that takes care for a flexible and intelligent interfacing between a user (i.e. the experimenter) and the system to be experimentally studied. The tasks a simulator has to support are depicted by a discussion about simulation studies in Research & Development. It is made clear that a simulation environment has to perform extra tasks in comparison to a normal programming environment of a computer: description, modelling, experimentation, knowledge handling, reporting. Several of these tasks ask for powerful and fast computers. An outline of a simulation environment is given. Implications with respect to hardware and software are discussed. It is argued that special-purpose simulators will be constructed as a general-purpose simulator provided with a special-purpose simulation environment rather than as specially hardware-designed simulators.

1. INTRODUCTION

For computers it has become normal practice that both the hardware and the system software are provided with special flexible tools as an interactive interfacing between user and user program. It is a custom to denote these special tools as the programming environment of a computer.
In many fields of application nowadays computers are frequently used for system simulation in experimental studies in R&D projects. For this reason it is important to develop special-purpose programming environments for simulation: so-called 'simulation environments'.
A simulation environment can be defined as a programming environment of a computer, that is dedicated to system simulation and that takes care for a flexible interfacing between a user (i.e. the experimenter) and the system to be experimentally studied, both in the modelling and execution stage.
A simulation environment has to perform extra tasks in comparison to a normal programming environment of a computer. Several of these tasks can be very time–consuming like the execution of 'system model validation' experiments. As a consequence the software and hardware tools must fulfil severe requirements.
In the past hybrid computers were successfully applied for simulation of dynamic systems, not only because they were faster than digital computers at that time but also because the simulation environment of a hybrid computer was superior to that of a digital computer.

Activities in the simulation of a system are:
- system modelling
- model validation
- definition of experiments
- experiment modelling
- execution of experiments
- observation of experimental results
- analysis of experimental results
- storing of analysis results

Other important activities in a simulation study are:
- description of the study, defining:
  general knowledge about the system,
  the system to be studied,
  the aims of the study,
- reporting, documentation

In the recent past only the execution of simulation experiments and the observation of simulation results have become automatized. But at present it has already become quite normal that all activities in the simulation of a system can more or less be performed in a computer-assisted way. However it is not yet common practice that all activities in a simulation study can efficiently be done in a computer-assisted way. Like also in general in experimentation it is essential in a simulation study that a user/experimenter is involved in all these activities.

In this paper a computer provided with a simulation environment will be called 'simulator'. For comparison we quote two definitions given a small decade ago in the papers [1], [2], respectively:

"A future simulator has to contain a powerful data-processing system able to perform four major tasks:
- performing a simulation run, i.e. computation modelling, implementation and execution of both validation and simulation experiments
- utilizing and updating the data base SCE [2]
- performing background activities, e.g. for decision-making purposes
- supporting the previous tasks and the experimenter as well by implemented simulation methodology knowledge."

"A simulator is a data processing system programmed to represent a conceptual model of a system under study."

[1] Dept. of Mathematics & Informatics, Dept. of Applied Physics, University of Technology, Delft, The Netherlands
2 SCE = model base of triplets (s,c,e), where 's', 'c', 'e' stand for 'system', 'cosystem' and 'experiment', respectively.

The differences between the definition in this paper and the quoted ones illustrate the changes in the state-of-the-art with respect to both hardware and software. One decade ago the hardware costs of computers were still the dominating factor and the software tools were still (relatively) under-developed. Nowadays hardware has become (very) cheap. Because of this fact personal computers (PC's) are now worldwide extensively used. For that reason software tools have been developed that are more powerful and more user-friendly. The availability of cheap and powerful PC's has caused that many PC's have been bought by nonexpert users. So from the user's side there was a great need nonexpert-friendly software tools. This is the reason why, after the introduction in the first MacIntosh PC's, menu-based software tools have become popular. But moreover these many nonexpert clients have made it profitable for software houses to develop a great variety of high-quality software tools.
As to simulation methodology there has not made enough progress in the last decade. The actual state-of-the-art in simulation methodology hampers a further development of simulation environments.
In this respect it is worthwhile to quote still two other statements from [2]:

> "Recent advances in hardware have removed hardware as a limiting factor in data processing, hereafter limitations in the design of simulators will arise from the methodology rather than from the hardware."

> "A few decades hence system simulation will become a saturated art limited by human capabilities."

These statements transposed to the present time say:

> Limitations in the design of simulators arise from the methodology rather than from the hardware. Within roughly a decade system simulation will be a saturated art limited by human capabilities.

2. EXPERIMENTATION BY SIMULATION

First the activities in a simulation study will briefly be elucidated.

Aim(s) of the study. The aim 'A' of a simulation study can for example be to contribute to the process of bringing an invention to a commercial success (innovation). But also the aim can simply be a better understanding of the system behaviour or a modification of a system in order to improve its behaviour. Before other activities can start there is quite often a need to extend available qualitative and quantitative knowledge about a system by applying knowledge search and/or experimentation on the system.

Knowledge. Knowledge 'K' is an integrated description of facts 'F', relations 'R' between these facts and laws 'L' where-upon these relations are based.

System knowledge. The knowledge 'Ky' about a system 'sy' that belongs to some class of systems will consist of general knowledge about that type of systems, given facts about 'sy' and given relations between these facts.

System modelling. In an experimental study by simulation one has to transform the description of 'sy' into a system model 'sm'. A precise formulation of the system model 'sm' is necessary to allow a close one-to-one implementation of the system model through a computational model 'csm' as a system 'csy' on a computer.

Experiments. The given system knowledge 'Ky' is often not enough to fulfil the aims of the simulation study. Then experiments must be done to complete the system knowledge as far as necessary to achieve the aims of the simulation study.

Experimentation. Experimentation on a system 'sy' is done to study the static and dynamic system behaviour in order to extend the knowledge 'Ky' about the system. System experimentation requires that the system is or can be made sufficiently controllable and observable. For this reason it may be necessary to make special provisions to obtain enough controllability and observability. This can be realized as follows:

- inputs are provided to make the system controllable,
- input signals are applied to control the system,
- outputs are provided to make the response of the system observable,
- input signals and output signals are measured and are made observable.

Analysis of the experimentation results will result in an improved description of the system behaviour or in other words to an extended system knowledge 'Ky'.

Experiment modelling. In a simulation study it is evident that system modelling has to be applied to the system 'sy'. In case of experimentation on the system 'sy' input variables are provided and output variables are observed by means of special 'control' and 'measuring' equipment. Hence in experimentation there is often no need for experiment modelling. But in simulation it is also necessary that system modelling is applied to the experiment(s), because on a computer both the generation of the input variables and the observation of the output variables must be implemented as a (hardware and/or software) system.
A better understanding of the system behaviour will contribute to the qualitative and quantitative knowledge about a system. In experimentation the experimenter always has to do with two systems: the actual system 'sy' and a 'measuring and control' system denoted as cosystem.: 'cy'. In case of a simulation study system modelling has to be applied to both the system 'sy' and the cosystem 'cy'.
Simulation of a system 'sy' can be defined to be experimenting on a substitute of the system, where the substitute is a computer implementation 'sy_c' of a model 'sm' of the system 'sy'. This computer implementation must be a valid analogy of the actual system 'sy' with respect to the aims of the experimental study.
Both in experimentation and in simulation system modelling plays an important role. In particular in simulation

it is of great importance to make an appropriate choice of the model of the actual system (with respect to the aims of the study). To determe the validity of a system model 'sm' there may be required extra 'validation' experiments on the system 'sy'.
Different aspects of the aims may lead to different types of experiments. Each type of experiments requires a specific 'measuring and control' of the system. A system model has to be a good fit of the system in relation to the objectives of a type of experiments. Hence, in an experimental study it may be necessary to use several different system models.

3. SIMULATION IN RESEARCH & DEVELOPMENT

Simulation has already been applied over a long period in Research & Development (R&D) and has become an important tool in experimental studies of R&D problems. Before discussing the simulation appearing in R&D more in detail [3],[4], it is worthwhile to analyze the different steps in a R&D process. The first objective of R&D is to support the ongoing business by keeping up-to-date both products and processes. A main task is the 'transformation' of complicated problems into a simpler form, but with preservation as much as possible of the important qualitative and quantitative characteristics. In simulation the transformation must result in problem formulations that are adapted to the available experimentation and analysis hardware/software tools.
An R&D problem can be specified as follows:
- G = {knowledge K & nonintegrated facts F and relations R} about the R&D problem
- U = unknown knowledge about the R&D problem
- A = aim(s) of the R&D problem

- G_A = part of G relevant to A: i.e. result of filtering of G by A [1]
- U_A = unknown knowledge [2] relevant to A: i.e. filtration of U by A

In a R&D process one can distinguish:
- Formulation of the R&D problem as a triplet (G, U, A);
- Structuring of the R&D problem formulation by 'filtering' G, U by A to obtain filtrates G_A, U_A relevant within the framework of A. (This filtering may require reduction of large amounts of information);
- Experimentation on the real system in order to enlarge G_A;
- Modelling of the triplet (G_A, U_A, A) into a system formulation by defining (I, P, O) with:
  - I, O = I(nput), O(utput) information,
  - P = P(rocessing) operator;
- Computational modelling;
- Simulation.

Let us consider now the case of innovation, i.e. the process of bringing an (existing or expected) invention to a commercial success by the way of R&D (and Marketing).
Figure 3,1 represents the successive steps in the R&D process:

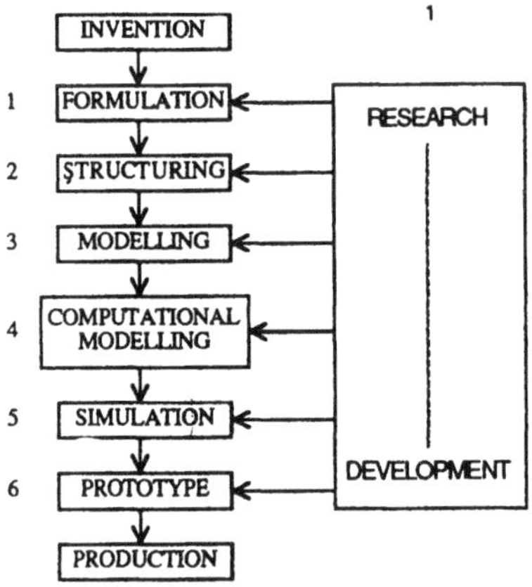

Figure 3,1.
Innovation by Research & Development

1. In the 'formulation' stage the R&D team has to formulate the R&D problem 'F' by describing what is given/known 'G', and what is uncertain/unknown 'U' and by prescribing the aims 'A'.
2. On base of 'A' and 'G' the R&D team has to answer questions with respect to 'U' for further structuring of the R&D problem. 'Structuring' means that it is determined in some way within the frame of 'A' what part of 'G' is relevant , and also how and to what extent 'U' is of importance. This 'filtering' of 'G' and 'U' by 'A' results in 'filtrates' 'G_A' and 'U_A', resp. .
3. In the 'modelling' stage the triplet (U_A, G_A, A) is modelled into a system formulation (I, P, O), where 'I, P, O' stand for the 'input information', the 'processing operator', the 'output information' of a system 'sy'.
4. In the 'computational modelling' stage the obtained system formulation (I, P, O) is transformed into a computational system that can be implemented on a computer.
5. In the 'simulation' stage experimentation by simulation on the computational system is done to validate, improve and extend 'G_A'.
6. The R&D team will also be involved in the 'prototype' stage with respect to the design of the prototype, and the design and validation of the prototype experiments to be done. (These prototype experiments will largely be done by means of simulation).

1 Filtering of G by A means in metaphorical sense associative processing of G and A with A as the reference.
2 'unknown' here means that part of knowledge relevant to A what should and might be found by extrapolation of the laws in K to the nonintegrated facts F.

Nowadays R&D is done in a strongly computer-assisted way. In R&D normally much numeric information processing is necessary. But also an important part of the data processing in R&D concerns nonnumeric information processing, like 'intelligent' searching in and improving of information bases, containing numeric as well as nonnumeric knowledge.

Data processing activities in the above R&D steps are:

1. The problem formulation 'F' is stored in an information base, where 'F' must be described in precise, (not necessarily accurate), structured relations. This information base must be interactively accessible, for example because additional information, obtained during the R&D process will often influence the problem formulation 'F'.
2 The 'filtering' of 'G' by 'A' requires powerful information processing, capable to reduce (interactively) in a 'human-intelligence' way a large amount of information.

3,4 For convenience let us restrict ourselves to those (parts of) R&D problems, where the modelling results in a dynamic system. Then 'I' and 'O' are described by trajectory variables. For this reason in the modelling stage it makes sense to formulate a dynamic cosystem 'csy' fitting in as much as possible with 'U_A', in order to transform trajectory variations into, parametric variations.
Both the modelling and the computational modelling are done more and more in a computer-assisted way. This modelling can still considerably be improved by means of knowledge processing. A model base can be helpful to organize the modelling on the base of 'modelling by exception'.

5,6 In R&D mostly many simulation runs have to be done on complicated systems. Consequently experimentation by simulation often requires powerful information processing. In the modelling stages and the simulation stage it will be necessary to perform research/experimentation to verify, improve and extend the information 'G_A'.

In all R&D steps an important part of the data processing concerns handling of knowledge. For this reason in R&D knowledge processing will become important. There is the need to automate R&D activities also with respect to exploration, transformation, modification and generation of knowledge [1]:

<u>Knowledge exploration.</u> In knowledge exploration the aim 'A' is to excerpt specific knowledge 'U' about some subject(s) from given knowledge 'G' by applying knowledge search, analysis and synthesis methods:
- G = knowledge & knowledge search, analysis and synthesis methods
- U = specific knowledge with respect to some subject(s)
- A = search of this specific knowledge

<u>Generation of knowledge.</u> In knowledge generation the aim 'A' is to generate unknown knowledge 'U' about some subject(s) from some given facts and relations (between these facts) 'G' by applying knowledge generation methods:
- G = facts and relations about some subjects & knowledge generation methods
- U = unknown knowledge [2] with respect to these subject(s)
- A = generation of unknown knowledge in order to:
  exploit it with respect to some aim(s),
  explore it later on

The storage capacity of computer memories has made it possible to store large amounts of knowledge for later computer–assisted retrieval. Distributed parallel–accessible background memories can make retrieval of stored knowledge fast enough. In principle also processing of knowledge can be made fast enough by the way of parallel processing. Unfortunately the methodology necessary for adequate complete automation of knowledge processing is still largely missing. But yet already now an important speedup can be obtained by parallelization of the pre–processing.

As a general rule, in analogy to the laws in system theory, state equations of knowledge processing are proposed as follows:

$K_{om} = K_o - dK_o$

$K_n = K_o + dK_n + dK_{om}$

where:

K_o = old state of knowledge

dK_o = part of old knowledge to be modified by a 'knowledge generation' action

K_{om} = old state of knowledge minus the part to be modified

dK_{om} = dK_o after modification

dK_n = additionally generated new knowledge

K_n = new state of knowledge

There exists a finite interaction range between pieces of 'knowledge'. Therefore knowledge processing has a hierarchical structure. The finite interaction range guarantees that parallellelization of knowledge processing can often be applied succesfully at a low level of aggregation [5]. Many pre-processing activities can be automatized in a nonartificial way and can largely be parallelized with a large speed-up and a good efficiency. In case of human-intelligence activities like inference analysis human interaction is yet indispensable.

---

[1] In the figures 5,1 and 5,2 all activities in knowledge handling are denoted as 'knowledging'.

[2] 'unknown' here has a vague meaning; it may be that it is not clear what is missing

4. SIMULATORS

The state-of-the-art in computer hardware and software has also changed the situation as to the design and construction of simulators. In the past the limited speed of convential digital computers have forced to construct faster computers for simulation of dynamic systems: the analog/hybrid computers.
There is a diminishing need for special-purpose simulators based on computers specially constructed. The reason is that also computers designed for special purposes have to be constructed by means of standard VLSI modules.These VLSI modules are both the fastest, most powerful and cheapest components that can be applied. Therefore as to power and speed it becomes more and more difficult to design and construct a computer for special purposes, that can compete with a general-purpose computer. The construction of specially hardware-designed special-purpose computers may only make sense if it concerns a (very) large field of applications. In other cases the development and construction costs, as to both the hardware configuration and the software tools, will make special`purpose computers too expensive.

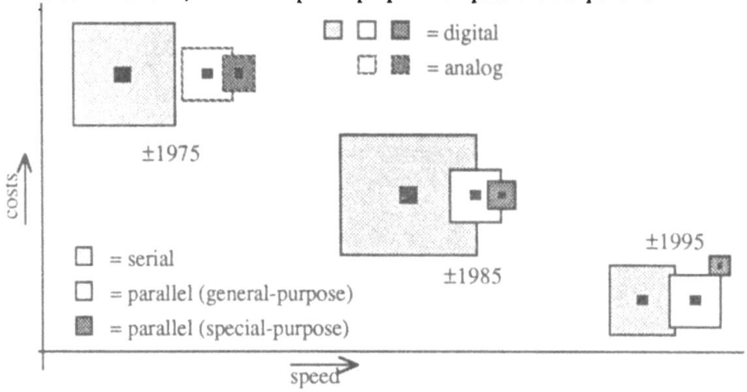

Figure 4,1.Speed, hardware costs and impact of computers on R&D in the recent past, at present and in the near future.

In systems simulation in the last two decades both serial and parallel computers have been used. Both general-purpose parallel computers and special-purpose parallel computers have been applied. Until now these special-purpose parallel computers have always been specially designed as to both hardware and software.
Figure 4,1 represents the speed, the hardware costs and the impact of serial and parallel computers on R&D over a period of two decades from ± 1975 up to ± 1995. The surface sizes of the three categories: serial computers, general-purpose and special-purpose parallel computers indicate their relative importance, where the category 'general-purpose parallel computers' has been taken as the reference. In hardware costs in the period from ±1975 until ±1985 there has not been a significant difference. But already now it has become clear that in the next decade the development going on will result in a significant change for the category special-purpose parallel computers. Figure 4,1 reflects this development in predicting that over the decade ±1985 until ±1995 the special-purpose parallel computers will become relatively more expensive and therefore less important. On the other hand it is likely that the R&D efforts in parallel computation will result in a general breakthrough of data processing into the direction of parallel processing. In figure 4,1 this is reflected in the relative decrease in importance of serial computers.
Both in simulation and even more in parallel computation the state-of-the-art in methodology hampers a faster progress. Successful parallel computers of to-day like in particular supercomputers and to a less extent also transputer systems belong to the category of special-purpose parallel computers. Because the speeding up of supercomputers is based on pipelined processing of the basic numerical operations : multiplication and additon, the supercomputers are merely applicable for large problems in area's like distributed-parameter systems, where 'numerical linear algebra' problems play a dominant role. A transputer system is a MIMD parallel computer, where the speeding up is based on simultaneous operation of 'transputer' processors. Transputer systems are based on a '4 neighbour' processor interconnect and therefore easily extendable to a large size. But transputer systems can merely be applied for problems that fit the restricted processor interconnect like in signal processing and again in distributed-parameter systems.
As to the development of general-purpose parallel computers there are no technical problems left that cannot be solved in a reasonable time [6]. Nevertheless it is rather difficult to predict how much time it will take before a parallel computer model will be available that is completely defined ,validated and widely accepted. But in the meantime both commercial parallel computers like the Hypercube, and university-built parallel computers like the DPP [7],[8], will contribute to the applicability of general-purpose parallel computers.
Figure 4,2 represents the quality of the programming environment, the software costs per computer and the impact of serial and parallel computers on R&D over the same period. Again the surface sizes of the categories: serial computers, general-purpose and (specially hardware/software designed) special-purpose parallel computers indicate their relative importance with the category 'general-purpose parallel computers' as the reference. As to the software costs per computer there has not been a significant difference in the period from ±1975 until ±1985. But figure 4,2 indicates the trend going on already now that over the decade ±1985 until ±1995 the special-purpose parallel computers will become also expensive with regard to the software costs.

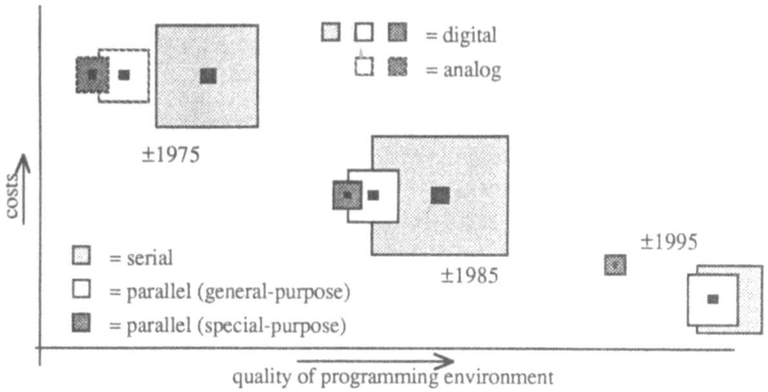

Figure 4,2. Programming environment quality, software costs and impact of computers on R&D in the recent past, at present and in the near future.

Because it is likely that the cost-performance as to both speed, hardware and software of specially hardware/software-designed parallel computers will go down considerably, it appears that these special-purpose computers will become of less importance in comparison to general-purpose parallel computers.

5. SIMULATION ENVIRONMENTS

Figure 5,1 represents the global structure of a future simulator. As defined before a future simulator will consist of a (general-purpose) computer with its programming environment provided a simulation environment as an extension of the programming environment. A simulation environment is composed of mainly software tools, but there will also be needed hardware tools. Main tasks of a simulation environment are:

- description of simulation studies,
- modelling purposes,
- experimentation purposes,
- knowledge processing [9],
- reporting about simulation studies.

As to extra hardware tools as an example it can be mentioned that in case of simulation of lumped parameter systems for observation during run-time a real-time graphical display is required.

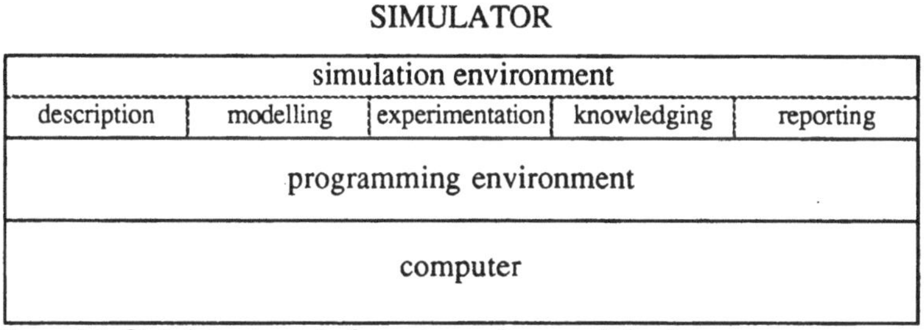

Figure 5,1. Global structure of a future (general-purpose) simulator.

Simulation languages have played an important role in the develpoment of systems simulation on serial digital computers, so far mainly as a convenient description tool. But R&D achievements with respect to methodology-based interactive systems modelling and implementation, a. o. [10], [11]], are beginning now to produce practical results. Simulation languages are even on the way to become also important tools for interactive methodology-supported modelling and experimentation on PC's [12].
Special-purpose simulators will be constructed as a general-purpose simulator provided with a special-purpose simulation environment rather than as specially hardware-designed simulators. By providing simulation environments of general-purpose simulators with facilities to dedicate the activities with respect to modelling, experimentation and 'knowledging' to special fields of application special-purpose simulators can be realized in a 'custom-made by ready-made' way. Figure 5,2. represents the global structure of a future special-purpose simulator.

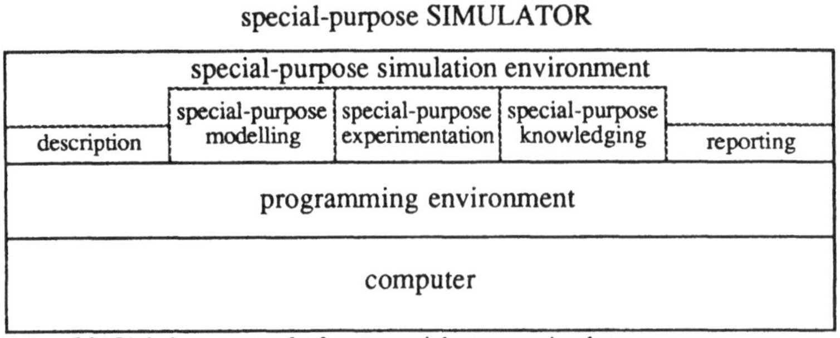

Figure 5,2. Global structure of a future special-purpose simulator.

REFERENCES

[1] Dekker, L., E.J.H. Kerckhoffs, G.C. Vansteenkiste and J.C. Zuidervaart, "Outline of a future Parallel Simulator", Proc.

[2] Dekker, L., "A view on the simulation of systems", Simulation Today, 58, pp 129-132, SIMULATION,30, 3, 1978.

[3] Dekker, L., "Parallel Simulation in Research and Development", Proc. Simulation in Research and Development, IMACS-symp., Eger, 1984.

[4] Dekker, L., "Role of MIMD Supercomputers in Computational Research & Development", Pau, Journée 'Inter-Entreprise Multi-Processeurs', 21 nov., 1985.

[5] Dekker, L., J.C.Zuidervaart, "Some ideas about parallelization of knowledge processing in Research-&-Develpoment", Proc. 2nd European Simulation Multiconference, 1988.

[6] Dekker, L., "Methodology-based parallel digital processors", Proc. Systems, Analysis and Simulation, Part I pp 38-50, Akademie Verlag, 1985.

[7] Dekker, L.and E.E.E. Frietman, W. Smit, J.C. Zuidervaart, "OPTICAL LINK in the DELFT PARALLEL PROCESSOR an example of MOMI–connection in MIMD–supercomputers", Proc. FRONTIERS IN COMPUTING, North-Holland,

[8] Dekker, L., "Dynamic processor clustering in a parallel computer with fully interconnected processors", , Proc. 2nd European Simulation Multiconference, 1988.

[9] E.J.H.Kerckhoffs, G.C Vansteenkiste, B.P. Zeigler (Eds.), "AI applied to Simulation", SCS, Simulation Series (Vol.18, nr.1), ISSN-07359276, 1986.

[10] Spriet, J.A., G.C. Vansteenkiste Eds., "Computer-aided Modelling and Simulation", pp 273-336, Academic Press, 1982.

[11] Dekker, L., M. Elzas, "Methodology-based interactive systems modelling and implementation", Proc. SCS Computer Conf., 1984.

[12] Elzas, M.S., "Desktop Modelling Environments: an idea and its future", Proc. SCS Computer Conf.,1988.

Foundations of a general process theory

G. Wunsch [1)]

The concept of a process is often used in the various branches of mathematics, physics and system theory. Roughly speaking a process means a collection of time functions that in some sence belong together. Examples of processes are the set of all solutions of first order differential equations or sets of strings over an arbitrary set.

The most general definition is the following [4].

A process Ξ on T with values in X is a subset of the set X^T of all maps $\xi : T \to X$, $\xi(t) = x$ from a partially ordered set T into X.

T is called the time set and X the phase space of Ξ.

It can be shown that the following theorem is true [1].

Theorem 1: Ξ is a process on T with values in X if and only if there is a consistent family

$$\alpha := (A_{\underline{t}})_{\underline{t} \in \underline{T}} \quad , \quad \underline{T} := \bigcup_{n \in \mathbb{N}} T^n \tag{1}$$

of phase relations

$$A_{\underline{t}} \subset X^{[\underline{t}]} \quad , \left([\underline{t}] \text{ length of } \underline{t} \right) \tag{2}$$

such that

$$A_{\underline{t}} = \bigcup_{\xi \in \Xi} \{ (\xi(t_1), \xi(t_2), \ldots, \xi(t_n))\} \tag{3}$$

This theorem is an analogy to the fundamental theorem of stochastic processes due to Kolmogorov. The consistent conditions as we have stated before are also defined analogously to the assumptions in Kolmogorov's theorem.

The most important an fundamental process classes must be

1) Technische Universitaet Dresden

defined in the language of logic , e.g. $\Xi \subset X^T$ is a deterministic process, iff

$$\bigwedge_t \bigwedge_\xi \bigwedge_{\xi'} \xi \in \Xi \wedge \xi' \in \Xi \wedge \xi | T^t = \xi' | T^t \Rightarrow \xi | T_t = \xi' | T_t \qquad (4)$$

Here $\xi | T^t$ and $\xi | T_t$ denote the process realisation restricted to the domain $t' \leq t$ and $t' > t$, respectively (t', $t \in T$). Similarly we can define the class of irreversible processes, the class of processes with finite memory and other classes[3].

A process significant in physics and engineering is the Markov process. The concept which is given here is fundamental in the whole process theory with which we are concerned. We shall formulate a definition general enough to include all possible cases.

A process Ξ is said to be a Markov process iff

$$\bigwedge_t \bigwedge_\xi \bigwedge_{\xi'} \xi \in \Xi \wedge \xi' \in \Xi \wedge \xi(t) = \xi'(t) \Rightarrow \xi \overset{t}{\circ} \xi' \in \Xi \qquad (5a)$$

$\overset{t}{\circ}$ denotes the so-called concatenation of time functions:

$$(\xi \overset{t}{\circ} \xi')(t') = \begin{cases} \xi(t'), & t' \leq t \\ \xi'(t'), & t' > t \end{cases} \qquad (5b)$$

We remark that neither the general process nor the Markov process must satisfies the deterministic condition (4). As we have already shown in [2] the following theorem can be proved.

Theorem 2: X^T is a Markov process if and only if there is a consistent family

$$\varphi := (F_{t_1,t_2})_{(t_1,t_2) \in T^2} \qquad (6)$$

of partial functions

$$F_{t_1,t_2} \Leftrightarrow F(t_1,t_2,\cdot) : X \to \mathcal{P}(X) \quad , F(t_1,t_2,x) \subset X \qquad (7)$$

on X with values in the power set $\mathcal{P}(X)$ such that

$$x_2 \in F(t_1,t_2,x_1) \Leftrightarrow (x_1,x_2) \in \bigcup_{\xi \in \Xi} \{(\xi(t_1), \xi(t_2))\} \qquad (8)$$

The consistent condiction are satisfied iff the following equations holds

$$a) \quad \bigcup_{x_1 \in D_{F(t_1,t_2,\cdot)}} F(t_1,t_2,x_1) = D_{F(t_2,t_3,\cdot)} \tag{9a}$$

$$b) \quad F(t_2,t_3,\cdot) \circ F(t_1,t_2,\cdot) = F(t_1,t_3,\cdot) \tag{9b}$$

$\circ$ denotes the composition of functions and D_F the domain of F. It is abvious that (9b) is a generalisation of the well known theorem due to Chapman and Kolmogorov as far as we write for (9b)

$$F(t_1,t_3,x_1) = \bigcup_{x_2 \in F(t_1,t_2,x_1)} F(t_2,t_3,x_2) \tag{9c}$$

We call therefore $F(t_1, t_2, \cdot)$ the transition operator of the Markov process Ξ .

The family φ together with the binary associative operation $\circ$ can be considered as a partial algebra (with the identity $F(t,t,\cdot)$). This algebraic structure we shall call a partial two-parametric Markov semigroup. With it we can prove the

Theorem 3: A process (X,T,Ξ) is a Markov process iff there is a two-parametric Markov semigroup $(X, T, \varphi, \circ)$ so that

$$\xi \in \Xi \Leftrightarrow \bigwedge_{(t_1,t_2)\in T^2} \xi(t_2) \in F(t_1,t_2,\xi(t_1)) \tag{10}$$

Now we are in a position to show that all processes of physics, especially the conventional dynamics and system theory are special cases of the universal Markov processes defined above. For example we consider:

a) Deterministic Markov processes:

$$\xi \in \Xi \Leftrightarrow \bigwedge_{(t_1,t_2)\in T^2} \xi(t_2) = F(t_1,t_2,\xi(t_1)) \tag{11}$$

For time invariance we have especially $(F(t_1, t_2,\cdot) = F(0,t_2-t_1,\cdot))$

$$\xi \in \Xi \Leftrightarrow \bigwedge_{\substack{t\in T \\ \tau \geq 0}} \xi(t+\tau) = F(\tau,\xi(t)) \qquad (T=\mathbb{R}) \tag{12}$$

b) In the mathematical theory of dynamic systems always is

assumed that the phase space X is a C^k manifolds of the dimension n (in a simple case hold $X = \mathbb{R}^n$). F by (12) is a bijection and F and F^{-1} are C^r differentiable. (F is a so-called C^r diffeomorphism).

In this relative simple case we have by (12)

$$\frac{d}{dt} x_i = f_i(t, x_1, x_2, \ldots, x_n) \quad , \quad \begin{matrix} i = 1, 2, \ldots, n \\ (T = \mathbb{R}) \end{matrix} \tag{13a}$$

Another and significant describtion of this "smooth" class of Markov processes is

$$\frac{\partial F_i(t, x_1, \ldots, x_n)}{\partial t} = \sum_{j=1}^{n} \frac{\partial F_i}{\partial x_j} \cdot f_j(t, x_1, \ldots, x_n) \tag{13b}$$

A interesting question to ask is: What is the meanig of our basic process-theoretical equations (13) in view of the fundamental equations of mechanics? For example, we examine the Hamilton equation.

If we assume $n = 2 \cdot m$ and reformutate the variables so that $(x_1, \ldots, x_{2m}) = (q_1 \ldots q_m, P_1, \ldots P_m)$. Further we set in (13a)

$$f = (f_1, f_2, \ldots, f_{2m}) = \left(\frac{\partial}{\partial P_1}, \ldots, \frac{\partial}{\partial P_m}, -\frac{\partial}{\partial q_1}, \ldots, -\frac{\partial}{\partial q_m}\right) H := \text{grad}^* H \tag{14}$$

Then we obtain by (13a) the Hamilton equations

$$\dot{x}_i = [H, x_i] \tag{15}$$

$[\cdot\cdot]$ denotes the Poisson bracket. We see, that from our point ov view the processes of classical mechanics are particular cases of all realistic Markov processes.

As a further example we take into consideration the statistic mechanics. A detail consideration we shall give in [3] , here only we write down the results.

Let $A(\xi_1(t), \ldots, \xi_n(t), t)$ be a "mass-preserving" functional of the Markov process Ξ defined by (13b) .

That means, we assume the condiction

$$\frac{d}{dt} \int_V A(\underline{x}, t) \, d\underline{x} = 0 \quad , \quad \begin{matrix} \underline{x} = (x_1, \ldots, x_n) \\ V \subset X \ , \ X = \mathbb{R}^n \end{matrix} \tag{16}$$

Then we can show that

$$i\frac{\partial A}{\partial t} = L(A) \qquad L = -div(\cdot if) \qquad (i=\sqrt{-1}) \tag{17}$$

In this expression L is a Hermitian operator, and therefore (17) is said to be as a generalisation of the Liouville theorem.

Assume now according to (14) that div f = 0 and f = grad*H. In this "volumen-preserving" restriction we get the common Liouville theorem

$$\frac{\partial A}{\partial t} = -[H,A] \tag{18}$$

of statistic mechanics if we interpret the quantity A with the particle density.

Analogous to the above consideration it can be shown, that also the fundamental equations of quantum mechanics are derivable from the basic equations (13) if the functional A is interpreted as density function of a random variable.

If we summarize our results it is worth of special emphasis that the fundamental processes in (technical) science are the Markov processes to which correspond the defined two-parametric Markov semigroup. Especially the smooth Markov processes with finite dimension defined by (13) stands in connection with the fundamental equations of mechanics.

References

/1/ Wunsch, G.: Verallgemeinerung des Fundamentalsatzes von Kolmogoroff. Wiss. Z. TU Dresden (in vorb.)

/2/ Wunsch, G.: Universelle Markov-Prozesse. Monatshefte für Mathematik. Institut für Mathematik der Universität Wien, Austria (in Vorb.)

/3/ Lewin, A.; Schwarz, W.; Wunsch, G.: Stochastische Signale und Systeme in der Übertragungs- und Steuerungstechnik. Akademie-Verlag, Berlin (in Vorb.)

/4/ Pichler, F.: **Mathematische Systemtheorie, Verlag Walter de Gruyter, Berlin 1975**

FAST AND ACCURATE ARITHMETIC FOR VECTOR PROCESSORS

Ulrich Kulisch[1]

In addition to the four elementary arithmetic operations, more advanced electronic computers such as vector and parallel computers often provide a number of compound operations as additional elementary operations. If pipelined compound operations like "multiply and add", "accumulate", and "multiply and accumulate" contribute essentially to the high speed of the system. Accuracy requirements lead to very similar operations. We identify a set of operations which meet both requirements: high speed and accuracy. After a brief discussion of implementation techniques for the simpler of these operations we present two methods and circuits which allow a fast and correct computation of the more complicated of these operations: "accumulate" and "multiply and accumulate". The first method computes sums and dot products by making use of a matrix-shaped and pipelined arrangement of adders which cover the full floating-point range. The second method requires some local memory on the arithmetic unit. It permits a drastic reduction in the number of adders required. Both methods can also be used to build a fast arithmetic unit for microcomputers in VLSI technology.

[1]Institut für Angewandte Mathematik, Universität Karlsruhe, Kaiserstr. 12, D-7500 Karlsruhe, West Germany

COMPUTER-AIDED SYSTEMS ANALYSIS: METHODS AND TOOLS

A.Sydow, K.Bellmann, E.Matthaeus, P.Rudolph, R.Straubel, A.Wittmuess[1]

Abstract

A survey about software tools for modelling, qualitative analysis, model aggregation and multi-criteria decision making at the ZKI is given. These software tools are developed in connection with industrial projects and combined in a unique framework for all components and a method bank for the steps of the analysis of complex dynamical systems up to the decision support for the control of these systems.

Keywords: Systems analysis, modelling, qualitative analysis, model aggregation and reduction, modelling and simulation support systems, multi-criteria decisions support (DSS).

1. Conception

The analysis of complex systems starts with the development of a model and ends with a control strategy. To support all steps of a complex systems analysis by the computer one needs software tools and links between them on a mathematical level c oncerning unique model classes and data transfer via model banks.

At the Central Institute of Cybernetics and Information Processes software packages for modelling (SONCHES) and decision making for the control of complex processes (REH) in connection with project research for the industry have been developed.

SONCHES as an interactive software tool includes besides a process oriented language for model design in its simulation environment components for model simulation and result processing, model change and documentation. The system helps the modeller to concentrate mainly on the modelling and model handling process.

To support model validation and decision on control strategies software for qualitative analysis for special classes of systems (stability analysis, bifurcation points, chaotic behavior etc.) has been worked out. At the same time the results of qualitative analysis can be used for aggregation and reduction of complicated models for decisions (for example the determination of different areas of the control space to simplify

[1] Zentralinstitut fuer Kybernetic und Informationsprozesse,
DDR-1086 Berlin, Kurstrasse 33

the model by linear or bilinear equations etc).

Model aggregation and reduction or decomposition are necessary to handle with the model in a more efficient and transparent way, i.e. for example to shorten the time for simulation runs for optimization purposes or to reduce model complexity in respect to special control purposes for PC use. For large scale systems different methods have been implemented (exact and modal reduction, equation error method, singular perturbation). For appropriate nonlinear systems model aggregation by GMDH as well as Wiener- and Hammerstein-models have been provided.

To support multi-criteria decision making for the control of complex systems software was developed which includes the method of rank order of objective functions. On the base of this rank order the procedure finds a Pareto-optimal solution for the control. The menue for the dialog offers an userfriendly service on three levels:test level, checking feasable solution, determining efficient solutions).

All this software tools have an unique graphic environment. A linkage between these software elements via model banks has been provided.

The present and futural research work for computer-aided systems analysis at ZKI is directed to an extention of the method bank, to include knowledge based components for using the full software and for decision making and to improve computing time by parallel processing for the simulation runs (special versions).

These software elements were developed in connection with solving control problems in water management, environmental protection (reconstruction of power stations), agricultural production and robotics.

All together about fifty industrial or economical tasks were treated with these software tools.

A rough survey about software for CA Systems Analysis at the ZKI

Tasks of Systems Analysis	Software
1) Model design and simulation - Parameter estimation - Sensivity analysis - Monte Carlo analysis	SONCHES
2) Computer Analysis of Nonlinear Dynamical Systems - Stability analysis of deterministic and stochastic systems - Evaluation of cycles and bifurcation points - Analysis of chaotic behaviour	CANDYS

3) Model Aggregation
   - Methods for linear systems (exact and modal aggregation, singular and regular perturbation)
   - Wiener and Hammerstein models
   - Group Method of Data Handling (GMDH)

4) Design of control strategies REH
   - Paremeter optimization by evolutionary procedures
   - Multi-criteria decision support (rank order of objective functions, relaxation)

2. CA Modelling and Simulation (SONCHES)

Computer aided modelling is aimed to a more and more effective modelling and simulation process using predefined object types, structures and relations to automatize as much as possible simulation and the data and model object handling for validation and result presentation. Methodologically this means to find out general and recurrent substructures (generic objects) with their common behaviour and interconnections. We understand this as object-oriented modelling within a definite set of be haviour attributes guiding the interactions. This allows well-ordered decomposition of large systems by reusable system elements.
Modelling within this restricted set reduces arbitraryness and generality in the application but enlarges the posibilities for the software designer to incorporate much more AI support facilities like, e.g., model semantics checks based on the procedural implementation of the laws and rules of the supported scientific branch, and enables the user to communicate with the modelling and simulation software within its accustomed technical terms. Simulation environment is nowadays a fixed, but different developed part within simulation software, including communication language elements, data base, model base, sophisticated result storage and management abilities, and graphic visualization. Branch-specific software finds therefore an international growing attractivity.
To determine the appropriate simulation tools the question is no longer: "In which mathematical form the model can be represented ?", but rather: "What kind of problem on the base of which knowledge is to be solved ?".
The simulation system SONCHES developed at ZKI is dedicated to problem solving processes mainly in ecology (matter, energy, informational flows between compartments, autochthone compartment dynamics and compartment interactions). It supports not only building a model in analogy to a programme, but implies also model debugging, model change (parameters, processes and structures), proving of hypotheses, incremental/decremen-

tal modelling, data management, comprehensive table function use and the performance of sophisticated simulation experiments with high degree of automation (behaviour analysis, sensitivity analysis, error analysis, Monte-Carlo simulation, optimization). Macro variables (vectors) can be defined by subcompartmentation with a process description valid for all subcompartments (e.g. for the mapping of age class structures) and macro processes can be defined by process refering. More complex simulation experiments can be carried out using macro parameters, in the sense that simulative case studies are the most important tool for model validation.

SONCHES was implemented in FORTRAN 77 and it is running on 16-bit or 32-bit minicomputers (under RSX and VMS, resp.) and on IBM-AT compatible microcomputers (under DOS 3.2).

3. Computer Analysis of Continuous Dynamical Systems (CANDYS)

To understand the dynamic behaviour of real processes an analysis of the computer model has to be done. The analysis is necessary for validation of the model as well as for preparation of control strategies. Systems of deterministic or stochastic differential equations with nonlinear dynamics and free parameters form a wide class of models for stochastically influenced real dynamical systems. The analysis of systems is concentrated on the following problems:

(i) A quantitative analysis for model validation.

This means that, firstly, a system of differential equations (DE) or stochastic differential equations (SDE) has to be developed on the basis of data. Secondly, parameter estimation has to be done under the assumption that the system proceeds in the neighbourhood of stability points.

(ii) A qualitative analysis for a full understanding of long range behaviour of different models by their qualitative properties.

This means that limit points and cycles including stability behaviour have to be computed. The dependence of the stability on parameters has to be analyzed (bifurcation diagramm).

(iii) A stochastic analysis to understand the stochastic influences.

This means that stochastic simulation itself has to be provided as well as statements about the probability distribution of the system (significance band for realizations, the distribution of events and of its first passage times) have to be made.

The program system CANDYS is a software support to realize the above problems of systems analysis:

(1) Modelling of real systems by DE and SDE within a sufficiently wide for practical purposes class of process types (DE and SDE systems with chain structures and a given class of probability distributions).

Estimation of a large number of nonlinear parameters.
(2) A full analysis of the qualitative behaviour of deterministic dynamical systems (limit cycles, computing of a graf of qualitative events).
(3) Simulation of Markov process paths and computation of statistical and probabilistic statements of stochastic influences.

An application of the developed software is the analysis of an energetically open, stochastically disturbed system describing water quality processes in flowing water systems. The model consists of a system of nonlinear DE or SDE with about 50 free parameters for which a model validation to different real systems (parameter estimation), an analysis of stability behaviour and of the influence of stochastic inputs is done.

4. CA Decision Support (REH)

The program-system REH supports the elaboration of control strategies for complex time-dependent decision processes. It provides in a user-friendly way proposals for the decision-maker. These proposals represent compromises between contradictory objectives, on which the decision process is based.
For instance, for a production process such objectives can be:
(1) High increase in production, (2) High quality of products, (3) Small production costs, (4) High safety in risc situation. The aim must be to find a compromise between these nonlinear co-operative objectives.
The program-system REH offers for the user some procedures for achievements of the above mentioned aim. The use of a special procedures of REH depends on the ability of the user to formulate his subjective preferences concerning the different objectives. For example, if the user is able to specify a rank order of importance of the objectives, he can use the implemented procedure "sequential rank order optimization".
The implemented procedures are based on the mathematical elements of multiobjective optimization. If the user wants, he gets only so-called efficient (Pareto-optimal) control strategies as proposals for decision making. The program-system REH yields efficient (Pareto-optimal) control strategies also in the case of non-connected control regions and /or non-concave/non-convex objective functions.
In order to apply the program-system it is needed to formulate an algorithm, which describes the dependence of the objective functions on the control variables. For the programming this algorithm interactive proprocedures exist which make this task easier.
The program-system REH is characterized by four levels.
1. Level: Preparation
This level presents the user the possibility to manipulate external

scenario data (parameters).

2. Level: Learning

This level offers the user the possibility to learn the behaviour of his process model by different types of simulation games.

3. Level: Test of admissibility

The user can formulate target values for the objective functions. The program-system tries to find a control strategy which realizes the chosen target values.

4. Level: Pareto-optimality

There are three different procedures for elaboration of efficient (Pareto-optimal) control strategies at this level.

Two procedures allow the user to articulate in an easy way his subjective preferences regarding the objectives in order to obtain as result efficient (Pareto-optimal) control strategies.

The third procedure supplys the user with a set of efficient (Pareto-optimal) control strategies.

PROGRAM SURVEY

The program-system consists of about 150 routines. It was tried to built a robust program. Each input of the user is examined for admissibility, if needed the program-system gives to the user hints for correction of false input. All activities of the user are noted in the background into a session protocol. The program-system is able to solve problems with 10 objective functions and 150 control variables.

APPLICATIONS

Up to now the program-system REH has been used for about 30 concrete static or dynamic multiobjective decision problems. Some examples are listed in the following:

- design of brake facilities of trucks
- three-dimensional control of a robot arm
- optimal control of a chemical process
- allocation of purification plants on a river
- reduction of SO_2-deposition at regional distributed receptors
- macro-economic planning.

5. Applications (PEMU)

Besides of lots of applications in different areas of industry, economy and research work the very last application in a new field is described in the following. The user friendly scenario optimization model PEMU is imbeded in the G.D.R. Environment Protection Projêct. PEMU is a national simulation and optimization model for computer aided design of control strategies of emission and site specific immission of air pollutants. PEMU has been designed to be useful especially for decision makers involved in SO2 emission control in forestry planning. PEMU is of modular construction. It consists of the submodel I m m i s s i o n and I m p a c t submodels for forest decline and productivity (pine, in preparation spruce) and it is open for further impact submodels.

In PEMU software for CA Systems Analysis are implemented:

- a branch specific simulation system SONCHES for building, validation and using ecological models,
- a multicriterial optimization system REH by which the user is flexiblely supported in preparation decision alternatives.

Possible are:

- scenario techniques and multi-criteria optimization under economical restrictions and restrictions of foresty
- the pine forest pollutant impact model yields predictions of the dynamics of foliage damage and the consequences of growth and timber mass dynamics/ha and timber harvesting increase and decrease of pine forest stands of different age classes, stem density, degree of soil fertility and moisture, all of them avaible by measurements being possible in foresty and by corresponding authorities.
- PEMU considers both the direct forest impact (air pathway: necrosity, encymatic internal bioindication) and the soil pathway taking explicitely into account the degree of acidicy of immission, and with this the site specific compensation ability of the forest ecosystem by alkaline dust and industrial and agricultural borne nitrogenium immission.
- PEMU is implemented on 32 bit technique and has been developed at ZKI since 1983 by an interdisciplinary research team.

References

see the papers of: K.-F. Albrecht, K. Bellmann, R. Boettner, U. Feudel, M. Flechsig, R. Funke, W. Jansen, B. Koch, T. Lange, P. Lasch, E. Matthaeus, N. Model, P. Rathke, P. Rudolph, R. Straubel, A. Sydow, V. Wenzel and A. Wittmuess.

A FLEXIBLE SYSTEM OF PRECEDENCE COEFFICIENT AGGREGATION AND CONSENSUS MEASUREMENT

Jan W. Owsiński, Sławomir Zadrożny

Systems Research Institute
Polish Academy of Sciences
Newelska 6, 01 - 447 Warszawa, Poland

The paper presents a practical method for a software meant for the in-session processing of precedence data obtained from m judges and concerning n items. Precedence data of a judge can be given in a variety of ways - from a unique ordering to $\frac{1}{2}\cdot n(n-1)$ fuzzy precedence indices. First, the principles of processing are presented, then problems and suggested solutions. The latter refer to two notions: "expert agreement" as to the final precedence coefficients on the one hand and "resulting ordering" of the items on the other.

1. INTRODUCTION

In the development of computer-based data analysis tools for decision support in multi-person, multi-item situations it is necessary to find proper means for acquiring, processing, aggregating and assessing data reflecting various opinions. This paper highlights somewhat specific, but sufficiently general situation in which m judges (experts, voters...) express opinion as to the precedence (comparison of value, of importance, of temporal sequence,...) of n items (options, policies, candidates,...).

Within that context two fundamental problems appear: that of approximating the consensus among experts, and of determining the resulting ordering. When all the data obtained from the judges are expressed in the form of orderings then the formulation of Marcotorchino and Michaud (1979) allows one to obtain the solution of both these problems, whether by application of the original method mentioned or a heuristic approach of Owsiński and Zadrożny (1986). Thus, there exists in the space of orderings such one that minimizes aggregate distance from the orderings given by judges.

Experts may, however, be unwilling or incapable of giving information through orderings. The requirement of providing orderings can also be treated as advantageous, since thereby judges are forced to provide more information. Still, in many cases this requirement cannot be kept. Thus, precedence coefficients may have values greater than 0 and smaller than 1, indicating doubts of experts or actual non-uniqueness of precedence. When this broadened space of data is also accepted as solution space then the two distinct problems mentioned arise. It is obvious that within this broadened space solutions can be found that are simultaneously near to all of the opinions expressed than any of the orderings is.

2. BASIC NOTIONS

n - is the number of items ("options", "candidates",...) considered, where n=card I, $i,j,l \in I$,

m - is the number of judges ("experts", "voters",...) providing assessments of precedence of items, with m=card K, $k \in K$,

d^k_{ij} - degree of precedence of item "i" before item "j" as defined by judge "k" when considering "i" and "j" apart from other items, with $d^k_{ij} \in [0,1]$ where

$$d^k_{ij}=\begin{cases}1, & \text{when i strictly precedes j,}\\ 0, & \text{when j strictly precedes i,}\end{cases} \quad \text{with } \{d^k_{ij}\}_{ij} = D^k \text{ (when}$$

$d^k_{ij} \in (0,1)$ then precedence coefficients shall be referred to as "fuzzy" while for $d^k_{ij} \in \{0,1\}$ - as "crisp"),

$o^k(i)$ - serial number of item i in the ordering defined by precedences d^k_{ij}, whenever determination of such ordering is feasible and wished, the ordering itself being denoted O^k,

I_α^{kf} - subsets of I, numbered α, $\alpha \in \{1,...,a_k\}$, defined by the judge k, for which precedence data are given in different ways, with f denoting the type of data definition, and

$$I = \bigcup_{\alpha=1}^{a_k} I_\alpha^{kf} \tag{1}$$

d_{ij} - aggregate precedence indices, i.e. variables forming the solutions to aggregation problems, with $\{d_{ij}\}_{i,j}=D$,

Q - proper objective functions.

3. PURPOSE OF THE SYSTEM

The purposes of the system are as follows: to perform a number of input management functions, to aggregate precedence indices determined for the particular judges, to calculate a number of other, related indices, such as e.g. "degree of consensus", and to perform the session management functions, all of these in an interactive environment of a session. The goal is to provide a guidance in the multi-item-oriented (ordering-type) voting-based attempt at defining a common opinion (ordering), with a "sufficient" degree of agreement (consensus).

4. INPUT FORMS

There are two forms of data input, to be chosen by the judge:

1. ordering of items, and
2. explicit values of d_{ij}^k,

but these two fundamental forms can be presented and mixed in a variety of ways.

Each judge divides the set of items I into nonoverlapping nonempty subsets denoted I_α^{kf}. Index f is composed of two subindices, f´, f", where f´=1 when the contents of I_α^{kf} shall be ordered; f´=2, when preference coefficients shall be given; f´=3 -"unknown precedence" is flagged. Second index specifies choices within these three categories. It is only the contents of each subset that is important at this stage.

Once I_α^{kf} specified, data concering each of them are elicitated.

Within the subsets I_α^{kf} for particular f´, the following further choices can be made, indicated by values of f":

1. f´=1, f"=1: items contained in $I_\alpha^{kf'f''}$ are unequivocally ordered, forming $O_\alpha^k(I_\alpha^k)$,

   f"=2: O_α^k contains an indifferent subsequence (only one), denoted O_α^{kI}, so that for all i,j O_α^{kI} there is d_{ij}^k=0.5, and for all the other i,j I_α^{k12}: d_{ij}^k {0,1},

   f"=3: within O_α^k a subsequence shall be indicated (only one, as before) for which one or more alternative suborderings shall be given, $O_{\alpha\beta}^{kA}$, where β is the serial number of alternative subordering, and for the i,j $O_{\alpha\beta}^{kA}$ there is

$$d_{ij}^k = \frac{1}{B^{k\alpha}} \sum_{\beta=1}^{B^{k\alpha}} d_{ij}^{k\beta}$$

where $B^{k\alpha}$ is the total number of altermative suborderings within the index subset α, while for all the other i,j I_α^{k13}: $d_{ij}^k \in \{0,1\}$.

For the explicit precedence coefficient values there are:

2. f´=2, f"=1: all the d_{ij}^k are specified, one for each pair (i,j), i.e. $\frac{1}{2}n(n-1)$ numbers $\in [0,1]$,

f"=2: only some d_{ij}^k are given, and all those which are not specifid, are deduced from the given ones.

Besides the index combinations given here the system accepts the situation in which during the I_α^{kf} elicitation phase not all $i \in I$ are considered. Thereby, for all the $i \in I - I_a^k$, where

$$I_a^k = \bigcup_\alpha I_\alpha^{kf} \tag{2}$$

"unknown terminal precedence" is assumed, so that, by default,

for all $i,j \in I - I_a^k$ there is $d_{ij}^k=0.5$, and
for all $i \in I_a^k$ and $j \in I - I_a^k$ there is $d_{ij}^k=1$.

It is also possible to set f´=3: "unknown precedence", i.e.:

f´=3 (f"=1): for items contained in I_α^{k3}, $i \in I_\alpha^{k3}$, there is $d_{ij}^k=0.5$, for all $j \in I_a^k$.

As can easily be seen, choices of operations are often quite arbitrary, but they can also be well justified if are right away not self-explanatory. Several questions remain to be answered for f´=2 and shall be dealt with further on.

5. FORMULATIONS OF THE AGGREGATION PROBLEM

Assume that all the d_{ij}^k, i.e. for all $(i,j) \in I \times I$ and $k \in K$, are given. The problem of aggregation of precedences takes on, in general, two forms, depending on whether the aggregate is to have the form of ordering or whether it can allow fuzzy precedences, i.e. whether $d_{ij} \in \{0,1\}$ or $d_{ij} \in [0,1]$.

In the first, "crisp" case the aggregation problem can be formulated along the lines set by Marocótorchino and Michaud (1979), i.e. in the form of the binary LP problem:

$$\max_D\{Q_1(D) = \sum_{i<j} (\hat{d}_{ij}d_{ij} + \hat{d}_{ji}d_{ji}) \tag{3.a}$$

subject to

$$d_{ij} \in \{0,1\} \qquad \forall i,j \in I \tag{3.b}$$

$$d_{ij} + d_{ji} = 1 \qquad \forall i,j \in I \tag{3.c}$$

$$d_{ij} + d_{ji} - d_{il} \leq 1 \qquad \forall i,j \in I \tag{3.d}$$

where $D = \{d_{ij}\}_{n\times n-1}$, and

$$\hat{d}_{ij} = \frac{1}{m} \sum_{k=1}^{m} d_{ij}^k$$

This LP problem can be solved as a continuous one via standard routines, insofar as subsitution of (3.b) by $d_{ij} \in [0,1]$ with application of the standard LP procedures yields solutions satisfying (3.b).

The problem (3) consists in finding of an ordering, defined by d_{ij}, which is the closest one to the common score given by $\hat{d}_{ij}$.

Now, in the second case, i.e. when aggregate precedence coefficients are allowed to be fuzzy, the problem in its most "formal" shape can be presented as follows:

$$\min_D\{Q_2(D) = \sum_{i,j} |d_{ij} - d_{ij}|^2\} \tag{4.a}$$

subject to

$$d_{ij} \in [0,1] \qquad \forall i,j \in I \tag{4.b}$$

$$d_{ij} + d_{ji} = 1 \qquad \forall i,j \in I \tag{4.c}$$

$$d_{ij} + d_{ji} - d_{il} \le 1 \qquad \forall i,j,l \in I \tag{4.d}$$

which means approximation of $\{\hat{d}_{ij}\}$ by antisymmetric and transitive $\{d_{ij}\}$ within the same space.

6. FUZZY DATA CONSISTENCY

There are three questions related to consistency of fuzzy data on precedence coefficients as provided by the judges:

* completeness,
* transitivity,
* determination of the unspecified precedence coefficients (for $f'=2$, $f''=2$).

6.1. Completeness

This question is, for simplicity, solved by assuming

$$d_{ji}^{k} = 1 - d_{ij}^{k} \tag{5}$$

with d_{ij}^{k} previously specified.

6.2. Transitivity

Note first that transitivity requirement may be applied both to input data and to aggregation results, and that it may be relaxed on both these ends. These relaxations have, of course, different meanings. Thus:

Remark 1: If (5) holds for all $k \in K$ and $i,j \in I$, and all $\{d_{ij}^{k}\}$ for $k \in K$ are transitive, i.e.

$$d_{ij}^{k} + d_{ji}^{k} - d_{il}^{k} \le 1 \qquad \forall i,j,l \in I \tag{6}$$

then problem (4) can be solved without the constraints (4.c) and (4.d) to yield $d_{ij} = \hat{d}_{ij}$.

Remark 1 provides the result which diminishes computational effort necessary for aggregation. Since, however, all $\{d_{ij}^{k}\}$, $k \in K$ still have to be transitive, it is more advisable to secure transitivity in (4) while relaxing it for the judge-provided data. Not only is that way of proceeding computationally easier, but it may also be believed that averaging over a greater number of judges would yield $\hat{d}_{ij}$ violating the transitivity constraint to at most a low degree (the similar would happen to completeness, were it necessary). Thus, though the system provides procedures for "transitivization", it is not advised to use them for each D^{k}, since this may make the session too lengthy and cumbersome.

Remark 2: $d_{ij}^{k} + d_{jl}^{k} - d_{il}^{k} \le 2$ for all $d_{ij}^{k}, d_{jl}^{k}, d_{il}^{k} \in [0,1]$.

Thus, it may be useful to introduce the notion of ε-transitivity: a $D^{k} = \{d_{ij}^{k}\}_{i,j}$ shall be referred to as ε-transitive if

$$d_{ij}^{k} + d_{jl}^{k} - d_{il}^{k} \le 2 - \varepsilon \qquad \forall i,j,l \in I \tag{7}$$

where, obviously $\varepsilon \in [0,1]$, with 0-transitivy characterizing all possible D^{k}'s (Remark 2), and 1-transitivity corresponding to (6), i.e. usual transitivity.

Remark 3: If all D^{k}'s, $k \in K$, are complete and ε-transitive then the solution D of the problem (4), obtained in the absence of constraint (4.c) and (4.d) shall also be complete and ε-transitive.

It should be noted that ε-transitivity is quite natural to fuzzy precedence indices. Consider, for instance $d_{ij}^{k}=0.7$ and $d_{jl}^{k}=0.5$, i.e. "i rather precedes j" and "j and l are equally ranked". Then, according to (6), there is $d_{il}^{k} \ge 0.2$, which leaves quite a margin for the feasible values of d_{il}^{k}.

Besides that, however, with continuous logic admitted - what, if only d^k_{ij} and d^k_{jl} are given - "should" be the "proper" value of d^k_{il} ? In the verbalization of before it should be "i rather precedes l", but this leaves, again a wide choice of feasible values of d^k_{il}.

Now, two properties related to ε-transitivity shall be introduced, of importance for algorithmic transitivization.

Remark 4: If $d^k_{ij} \in [\varepsilon, 1-\varepsilon]$ then (7) holds for these i,j and $l \in I-(i \cup j)$.

Remark 5: If $d^k_{ij}+d^k_{jl} \in [\varepsilon, 2-\varepsilon]$ then (7) holds for these i,j and l.

The first of these remarks can be applied, though, only to $\varepsilon \leq 0.5$, i.e. to in fact, quite low ε values.

6.3. Completion of a fuzzy D^k

This subsection is devoted to calculation of implicitly given d^k_{ij}, when f´=2 and f"=2. This option is chosen when specification of all the d^k_{ij} for $i,j \in I$ is not feasible. It should be mentioned, however, that a high degree of arbitraryness is introduced, and that even with this option, if the degree of arbitraryness is not to go beyond acceptability, one must still specify a large share of $\frac{1}{2}n(n-1)$ values of d^k_{ij}.

When d^k_{ij} and d^k_{jl} are given, it seems plausible to calculate d^k_{il} according to

$$d^k_{il} = F(d^k_{ij}, d^k_{jl}) \tag{8}$$

where F is a certain a priori defined function. Even, however, when a function F is acceptable formally and intuitively for index triplets, the question arises whether it is acceptable to superimpose F over itself, for instance

$$d^k_{ii'} = F(F(d^k_{ij}, d^k_{jl}), F(d^k_{li'}, d^k_{i'i}))$$

thereby further increasing arbitraryness of a part of data. The system is prepared for such a case, but certainly this suboption should not be abused. Even in case, however when unlimited superposition of F is allowed, at least n-1 values of d^k_{ij} must be specified.

7. AGGREGATION SOLUTIONS AND AGREEMENT MEASURES

7.1. Aggregation solutions

Solutions to the ordering problem (3) are obtained via methods described elsewhere - Marcotorchino and Michaud (1979), Michaud (1981), and Owsiński and Zadrożny (1986), where algorithms are given.

According to (3) the aggregation problem in its ordering-oriented version is being solved through standard LP. Because, however, of the dimensions of this LP problem, proportional to n^3, application of standard techniques in case of higher n encounters serious difficulties. Thus, the present authors introduced a suboptimization procedure, see Owsiński and Zadrożny (1986), referring to (3.a), but having the simplicity of classical progressive merger procedures, so that they can be used in-session with modest hardware.

The simplified method uses (3.a) in a modified form:

$$Q_1(D,r) = r \sum_{i<j} \hat{d}_{ij} d_{ij} + (1-r) \sum_{i<j} \hat{d}_{ji} d_{ji} \tag{9}$$

where $r \in [0,1]$ is the weighting coefficient, whose value changes along with the algorithm. There is, obviously,

$$\arg\max_D Q_1(D,1) = \{1,2,\dots,n\} = D^{opt}(1) = D_I \tag{10}$$

and if proper d_{ij}´s are equal zero then there are more maximizing D´s than

D_I. On the other hand, there is

$$\arg\max_{D} Q_1(D,0) = \{n,n-1,\dots,2,1\} = D^{opt}(0) = \overline{D}_I \qquad (11)$$

Naturally, $\max_D Q_1(D,\frac{1}{2})$ corresponds to the optimum for (3). By changing r from 1 to 0 one gets a finite series $D^{opt}(r)$ solutions, D_I to $\overline{D}_I$. Each such solution would be valid for a segment Δr^t, where t denotes the step number, with

$$\bigcup_t \Delta r^t = [0,1] \quad \text{and} \quad \Delta r^{t-1} \cap \Delta r^t = r^t, \quad r^0 = 1.$$

The simplified method does not explicitly solve this series of LP problems, which would amount to parametrization of (3), but proceedes in an approximative way: starting with D_I it applies a transformation of orderings taken from a limited class of such tranformations so that at some point $\overline{D}_I$ is finally reached. At every step corresponding $\hat{r}^t$ and $Q_1(D^t,\hat{r}^t)$ are calculated.

As far as (4) is concerned, its solution is reduced to mere transitivization of $\hat{D}$, whenever deemed necessary.

7.2. Agreement measures

Without a knowledge about the scales of Q_1 and Q_2 values actual assessment of the degree of agreement is doubtful.

Thus, with respect to Q_1, when completeness holds and X are subject to usual constraints:

Remark 6:

$$\arg\min_{\hat{D}}\max_{X} Q_1(\hat{D},X) = \{\tfrac{1}{2}\}_{i,j}, \text{ and} \qquad (12)$$

$$\min_{\hat{D}}\max_{X} Q_1(\hat{D},X) = \frac{1}{4}n(n-1)$$

while

$$\max_{\hat{D}}\max_{X} Q_1(\hat{D},X) = \frac{1}{2}n(n-1), \qquad (13)$$

corresponding to any argument $\hat{D}$ representing an ordering. Thus, the following relative measure can be proposed:

$$M_1^1(\hat{D}) = \frac{Q_1^{opt}(\hat{D}) - \frac{1}{4}n(n-1)}{\frac{1}{4}n(n-1)} = \frac{Q_1^{opt}(\hat{D})}{\min_{\hat{D}} Q_1^{opt}(\hat{D})} - 1 \qquad (14)$$

with $M_1^1(\hat{D}) \in [0,1]$ reaching 0 for $D = \{\frac{1}{2}\}_{i,j}$ and 1 for an ordering. Note that $M_1(\hat{D})$ does not reflect that much the agreement among judges as the agreement with respect to ordering of items. Thus $\hat{D} = \{\frac{1}{2}\}_{i,j}$ can be obtained if $d_{ij}^k = \frac{1}{2}$ for all $i,j \in I$ and all $k \in K$. On the other hand, $M_1(\hat{D})$ reaches its maximum when all the judges $k \in K$ give the data corresponding to the same, consensory, ordering.

Another measure may take the form of

$$M_1^2(\hat{D}) = \frac{\max_X Q_1(\hat{D},X) - \min_X Q_1(\hat{D},X)}{\max_X Q_1(\hat{D},X)} \qquad (15)$$

with $M_1^2(\hat{D}) \in [0,1]$, reaching 0 when $\max_X Q_1(\hat{D},X) = \min_X Q_1(\hat{D},X)$, i.e. when $\hat{D} = \{\frac{1}{2}\}_{i,j}$, and 1 when $\hat{D} \in E_0$, E_0 being the space of orderings. The latter results from the fact that,

Remark 7: $$\min_X Q_1(\hat{D},X)\Big|_{\hat{D}\in E_0} = 0 \qquad (16)$$

Both measures are easy to calculate, since the minimum appearing in (15) can be obtained from X maximizing $Q_1(\hat{D},X)$ by reversing the ordering obtained thereby.

With regard to the agreement among judges without any "outer" reference, the proper absolute measure can be based upon the objective function (4.a) yielding:

$$M_2^A(\{D_k\}) = \frac{2}{n(n-1)} \frac{1}{m} \sum_k \sum_{i<j} |d_{ij}^k - \hat{d}_{ij}| \tag{17}$$

and, in relative terms:

$$M_2(\{D_k\}) = \frac{\max\limits_{\{D_k\}} M_2^A(\{D_k\}) - M_2^A(\{D_k\})}{\max\limits_{\{D_k\}} M_2^A(\{D_k\})} \tag{18}$$

so that the measure M_2 ranges from 0 to 1, reaching 0 for the maximum diversity of opinions and 1 when all the precedence coefficients given by all the judges are identical: $d_{ij}^k = \hat{d}_{ij}$ $\forall k \in K,\ i,j \in I$. An approximate measure can also be used, namely

$$\tilde{M}_2(\{D_k\}) = 1 - 2 \cdot M_2(\{D_k\}) \tag{19}$$

resulting from the fact that

<u>Remark 8</u>: for even m: $\max\limits_{\{D_k\}} M_2(\{D_k\}) = \frac{1}{2}$

while for odd m: $\max\limits_{\{D_k\}} M_2(\{D_k\}) = \frac{1}{2} \frac{(m-1)(m+1)}{m^2}$ (20)

Thus, a set of measures is provided for monitoring a session, with no assumptions being made as to potential bicriterion aggregation, since this would hinder easy intuitive assessment of the results obtained.

8. CONCLUSIONS

The paper presents a complete framework for the group decision/decision aiding session software. From the outline presented one may easily conclude that conduct of such a session, though requiring some preparation for the session manager, is not only quite feasible, both hardware- and software-wise, but may lead to valuable results, which could not be obtained with the methods to date.

REFERENCES

Marcotorchino, J.-F. and P. Michaud (1979): Optimisation en Analyse Ordinale des Données. Masson, Paris.

Michaud, P. (1981): Agrégation des préferences. Ph.D. Thesis. University Paris VI.

Owsiński, J.W. and Sł. Zadrożny (1986): Structuring a Regional Problem: Aggregation and Clustering in Orderings. <u>Applied Stochastic Models and Data Analysis</u>, <u>2</u>, 1 and 2, pp. 83-95.

Owsiński, J.W. and Sł. Zadrożny (1986): Clustering for ordinal data: a linear programming formulation. <u>Control & Cybernetics</u>, <u>15</u>, 183-194.

SYSTEMATIC PROCEDURES TO DERIVE BLOCK DIAGRAMS AND STATE EQUATIONS FROM BOND GRAPHS

P.C. Breedveld [1)]

Abstract

It is shown that a certain procedure can be followed to expand a causal bond graph [2] into a block diagram. Another procedure is presented to transform a bond graph into a set of equations which are ready for simulation and next into a set of state equations.

Introduction

A systematic method to derive a bond graph from an IPM

Starting from an ideal physical model (IPM) of an example problem (Fig. 1) a bond graph can be derived in a systematic way [3]. Fig. 1 shows the IPM of an electric motor with a long shaft, a gear box and a flywheel, including an electric source with internal resistance. For reasons of simplicity the example problem is assumed to be linear, but all procedures can be applied to non-linear systems as well. The procedure is presented directly in the form of an algorithm (every action which concerns a mechanical domain is shown between parentheses if it differs from the one in a non-mechanical domain):

1) Identify all physical domains and all elements (Make a list if necessary).
2) Choose a (positive) reference (direction) in the IPM for every domain.
3) Represent the efforts (velocities) of the system by a unique name in the IPM.
4) Represent these efforts (velocities) by 0-junctions (c.q. 1-junctions) **except for the reference efforts** (pos. ref. direct.) and indicate the variable name chosen in step 3.
5) Identify the effort-differences (relative velocities) in the IPM, which are needed to connect all ports of all elements to the junction structure JS.
6) Construct these effort-differences (relative velocities) in the following way:

7) Connect all ports of all elements with the corresponding efforts (velocities) or effort-differences (relative velocities) and indicate the constitutive parameters of the elements using a colon between the mnemonic code and the parameter symbol.
8) Simplify the resulting bond graph with the use of the following graphical rules:

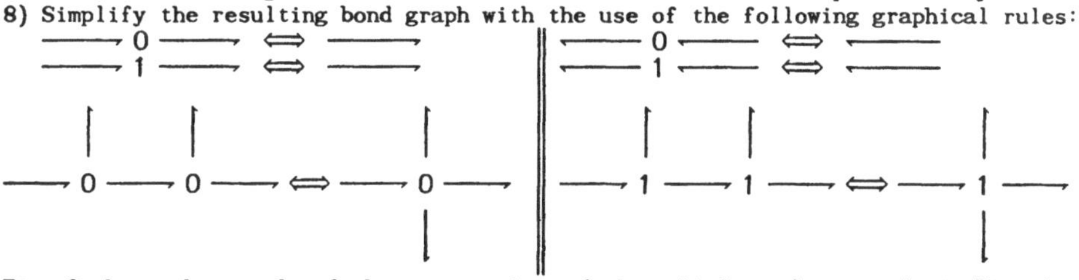

Fig. 2 shows the result of the systematic method applied to the example in Fig. 1.

By definition, every bond in a bond graph represents a bilateral signal-flow [4], of which the opposite computational directions can be indicated in a unique way by a "causal

---

1) University of Twente, Electrical Engineering Department, P.O. Box 217, 7500 AE Enschede, Netherlands, tel.: +31 53 892792, telex: 44200 THTES, telefax: +31 53 354003

Fig. 1: Ideal Physical Model of the example problem

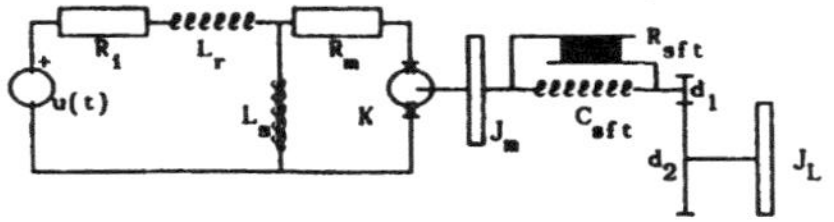

Fig. 2:
Systematic derivation of the bond graph of the example problem.

1) electrical and mechanical (rotation) domain: 1 x Se, 3 x R, 4 x I, 1 x C, 1 x GY (motor), 1 x TF (gear box).

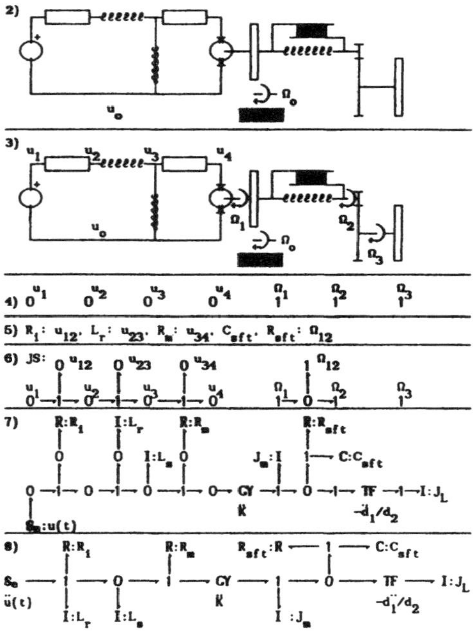

Table 1: CAUSALITY OF THE BASIC ELEMENTS

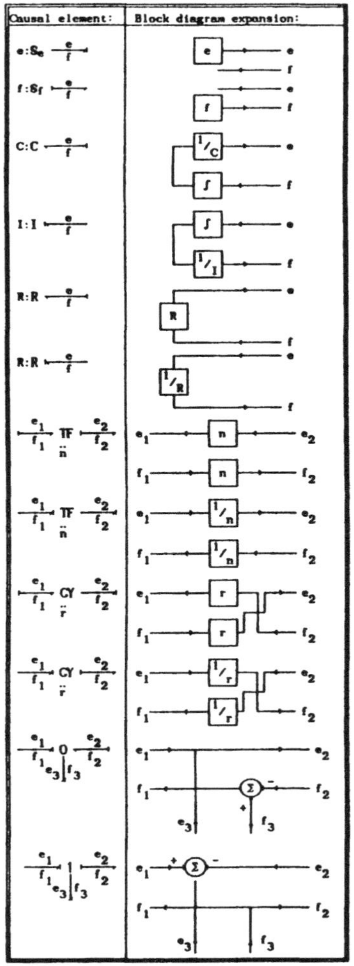

Fig. 3: Causal bond graph of the example problem:

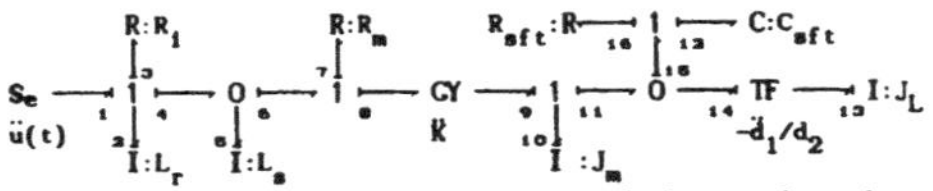

Numbers indicate the order of assignment of the causal strokes.

Fig. 4: Block diagram expansion of the example problem.

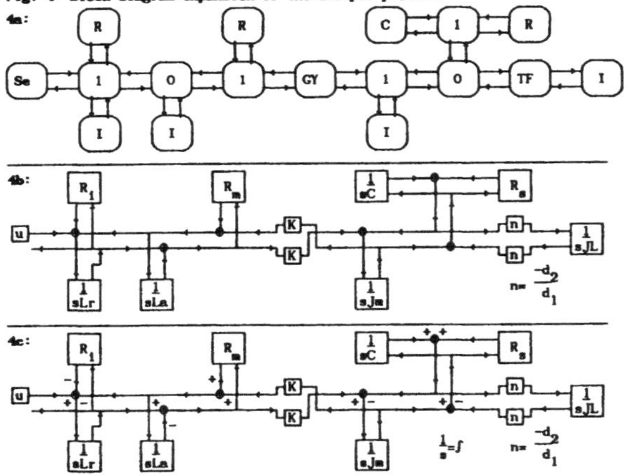

stroke": the causal direction of the effort-signal of a bond is indicated by a small line at the end of the bond and perpendicular to it (e.g. ⊢— or —⊣). Consequently, the open end of a bond indicates the causal direction of the flow-signal. In [2] the different types of causalities are discussed from which an algorithm can be derived:

Algorithm for causality assignment of a bond graph:

1) **fixed causalities** (Se,Sf) and the way they propagate via **causal constraints** (0,1,TF,GY); causal conflicts in this phase mean that the problem is ill-posed;
2) **preferred causalities** (C,I) and the way they propagate via **causal constraints** (0,1,TF,GY); causal conflicts can be solved by deviating from one or more preferred causalities; however, such a situation means that there are dependent state variables;
3) **free causalities** (R) which have not been given a causal assignment in the former steps, mean that there is at least one algebraic loop with accessory numerical complications; a "loop" can be found in a bond graph as a causal path from one element to the other.

The causality assignment of the example problem in Fig. 2 results in Fig. 3.

Expansion of a causal bond graph into a block diagram.

The relation between causal bond graphs and a computable model becomes more clear, if one realizes that every bond consists of opposite signals which are determined by causality. This means that a causal bond graph (e.g. Fig. 3) can be "expanded" into a block diagram by the following procedure:

1) Every bond is represented by two opposite arrows. The **direction of the arrows** is determined by the **causal strokes** (Fig. 4a).
2) All elements are expanded into "computational blocks" or "black boxes", which represent the constitutive relations of the elements as shown in Table IV (Fig. 4b).
3) The junctions consist of a node and a summation, of which the **signs of the ingoing arrows** are determined by the **half-arrows of the bonds** which are connected to the junction: the constitutive relations of the junction have to be satisfied in the proper causal form (Fig. 4c).

Derivation of the equations

Instead of writing a block diagram from a causal bond graph, it is also possible to write the corresponding equations in the proper causal form. This computable set of equations is called the "**mixed set of** (differential or integral and algebraic) **equations**". For the example problem this results in (for the variable identification subscripts based on the bond numbers in Fig. 3 are used):

Se: $e_1 := u(t)$	1: $e_2 := e_1 - e_3 - e_4$	0: $f_6 := f_4 - f_5$
R: $e_3 := R_1 f_3$	$f_1 := f_2$	$e_4 := e_6$
I: $f_2 := L_r^{-1}\int e_2 dt$	$f_3 := f_2$	$e_5 := e_6$
I: $f_5 := L_s^{-1}\int e_5 dt$	$f_4 := f_2$	1: $e_6 := e_7 + e_8$
R: $e_7 := R_m f_7$	GY: $e_8 := K f_7$	$f_7 := f_6$
1: $e_{10} := e_9 - e_{11}$	$e_7 := K f_8$	$f_8 := f_6$
$f_9 := f_{10}$	I: $f_{10} := J_m^{-1}\int e_{10} dt$	0: $f_{15} := f_{11} - f_{14}$
$f_{11} := f_{10}$	1: $e_{15} := e_{12} + e_{16}$	$e_{11} := e_{15}$
C: $e_{12} := C_{sft}^{-1}\int f_{12} dt$	$f_{12} := f_{15}$	$e_{14} := e_{15}$
R: $e_{16} := R_{sft} f_{16}$	$f_{16} := f_{15}$	TF: $e_{13} := -d_2/d_1 \cdot e_{14}$
I: $f_{13} := J_L^{-1}\int e_{13} dt$		$f_{14} := -d_2/d_1 \cdot f_{13}$

Written in the proper input format, these equations can be used directly for computer simulation. In some simulation programs like TUTSIM [1], the input code can be derived directly from a causal bond graph: Identities on the junctions have to be eliminated in the graph and 2-ports (GY,TF) become two blocks. Structure input for TUTSIM of the example problem (N.B. the numbers are now element numbers and ≠ bond numbers):

1,SE,20	6,0,2,-5	11,I,12	16,1,14,15
2,I,4	7,R,6	12,1,9,-16	17,TF,16
3,R,-2	8,1,7,10	13,0,11,-18	18,TF,19
4,1,1,-3,-8	9,GY,6	14,R,13	19,I,17
5,I,8	10,GY,11	15,C,13	20,"block to make u(t)"

The set of state equations in differential form can be derived systematically from the mixed set by differentiation with respect to time of the integral equations, preserving the causal form and by eliminating the algebraic equations in 3 steps (1: identities of junctions; 2: R, TF, GY; 3: summations of junctions):

$$\dot{\underline{x}} = \underline{A}.\underline{x} + \underline{B}.u \qquad \text{with: } \underline{x}^T = [f_2, f_5, f_{10}, e_{12}, f_{13}];\ u = u(t);$$

$$B^T = [L_r^{-1}, 0, 0, 0, 0]$$

$$\underline{A} = \begin{bmatrix} -(R_i+R_m)L_r^{-1} & R_mL_r^{-1} & KL_r^{-1} & 0 & 0 \\ R_mL_s^{-1} & -R_mL_s^{-1} & KL_s^{-1} & 0 & 0 \\ KJ_m^{-1} & -KJ_m^{-1} & -R_{sft}J_m^{-1} & -J_m^{-1} & -R_{sft}J_m^{-1}d_2d_1^{-1} \\ 0 & 0 & C_{sft}^{-1} & 0 & d_2d_1^{-1}C_{sft} \\ 0 & 0 & -R_{sft}d_2(J_Ld_1)^{-1} & -J_L^{-1}d_2d_1^{-1} & -R_{sft}d_2^2(J_Ld_1^2)^{-1} \end{bmatrix}$$

Conclusion

The above procedures have proven themselves in education in the sense that they help to see and to understand the relations between orientation and causality of the bonds in the bond graph at the one hand and the signs of the signals at a summation block in the block diagram and the signal directions at the other hand.

As a side-effect, people who are used to "think in terms of block diagrams", learn to appreciate bond graphs better. Especially it is better understood that a bond graph does not have to be causal to describe a physical system completely, because it is a "physical scheme" [4]. By contrast, a block diagram, which is a computational scheme, has a built-in causality which even may be too restrictive.

References:

[1] Beukeboom, J.J.A.J., Dixhoorn, J.J. van, Meerman, J.W., "Simulation of mixed bond graphs and block diagrams on personal computers using TUTSIM", J. Franklin Inst., Vol. 319, 1985, No. 1/2, pp. 257/268.

[2] Bos, A.M., Breedveld, P.C., "1985 update of the bond graph bibliography", J. Franklin Inst., Vol. 319, 1985, No. 1/2, pp. 269-286.

[3] Breedveld, P.C., "A Systematic Method To Derive Bond Graph Models", Proceedings of the 2nd European Simulation Congress, G.C. Vansteenkiste, E.J.H. Kerckhoffs, I. Dekker, J.C. Zuidervaart (eds.), pp. 38-44, Antwerp, Belgium, September 9-12, 1986.

[4] Paynter, H.M., "Analysis and Design of Engineering Systems", MIT-press, Cambridge (Mass), 1961.

SOFTWARE ENGINEERING ASPECTS OF COMPUTATIONAL SYSTEMS ANALYSIS IN PHYSICS

P. Carl[1)]

1. Introduction

An increasing role is assigned to computational systems analysis in applied physics. From the methodological point of view, the range of related problems may be characterized by two extremal basic tasks:

1. Investigation of phase space topology for well-formulated, physically idealized nonlinear systems in a high-dimensional parameter space.
2. Analysis of the structure, complexity and variability of possible states and routes of evolution of real, incompletely modeled systems.

The theory of operating conditions in thermonuclear magnetic-confinement installations, for instance, is faced with magnetohydrodynamic (MHD) equilibrium and stability problems on the one side and with complex transport processes, comprising particle, momentum, and energy balances, on the other one. MHD studies require extensive sensitivity investigations to trace out the domain of stable solutions - a fundamental task in Controlled Thermonuclear Fusion Research (CTR), which is overshadowed by serious numerical problems. A true challenge to software engineering, however, arises particularly from the second class of problem. Systems of fluid transport equations, interacting with other physical subsystems, form the skeleton of important large-scale tasks in applied physics, covering natural and technical systems and applications that range from forecasting and risk studies up to technology development.

Along with application, regeneration and construction of heterogeneous transport models in CTR and climatology, the coding principles and techniques as well as software tools, which are presented here, were developed and implemented to meet the technological requirements of complex systems analysis. Conceptual framework and available software of the OLYMPUS methodology and programming system [1] proved to offer a sound basis of a methodological and technical abstraction, which is kept close to practice in accompanying code development and application.

2. The challenge of systems analysis: Construction vs. programming

The complexity of real systems requires consideration of manifold dissipative and feedback mechanisms for a qualitatively correct formulation of the physical model, which usually develops only gradually. The modeling stage occupies a major part of work and is re-entered many times during the life cycle of the corresponding computational model. A growing degree of self-consistency and an increasing internal structuring are characte-

---

1) Central Institute for Electron Physics, GDR Academy of Sciences
Hausvogteiplatz 5-7, Berlin-1086, GDR

ristic of such codes. Software engineering is required to prepare both the computational environment and the internal structures which allow for concentration on the physical task and which keep the program clean. Besides that, plasma/solid interaction in CTR and atmosphere/ocean interaction in climatology are examples of model expansion extending the spatial domain, which became inevitable with increasing physical systems understanding. These and other tendencies of modeling call for technological preparation of interactive (including nested and hybrid) operation of existing models. Logically, the ultimate (and desirable) closure consists in a model system which offers free combination of its individual components, i.e. the full scale of tools for problem-orientation. In any case, claiming a systems analytical approach to "real physics" contradicts with the widespread computer- or program-oriented attitude. Oppositely, it demands to dispense with a systems "fetishism" at the other extreme, resulting in over-ambitious generalization and control, which could never meet the evolving knowledge and the resulting requirements of modeling.

As one of the earliest (and most serious) steps towards a problem-oriented attitude in computational physics, the OLYMPUS technology [1] forms an outstanding basis of computational systems analysis. Conventions are set up and software tools are available, which allow for convenient job scheduling, coding, regeneration, and maintenance of computational models. An increasing number of 'olympian' codes is already available in CTR, and the use of OLYMPUS is generally recommended in this field [2]. In moving physical computing away from the specific computer, and in clearly separating and standardizing the distinct subtasks of coding, however, the computational point of view of the two sides of 'computational physics' remains accented. The prescribed main control structure does not relieve the physicist of thinking in the framework of administrative requirements of computer (and FORTRAN) use. Additionally, the main computational cycle of OLYMPUS remained unstructured, whereas the data structures are well-prepared for separation of control, administration, physics, numerics, and I/O. Though this may have been acceptable for standalone operation of large codes, it was regarded insufficient in an approach to unify the solution of medium-scale tasks [3]. The way sketched in [3] proved practicable, it was followed and extended in the context of large-scale tasks in CTR and is now applied to climate modeling [4].

Though both experience and application are obviously not broad enough to arrive at final conclusions here, it is clearly seen that the strong demands posed to software engineering by systems analysis rise the technological level of codes towards minimization of programming efforts. The computational analyst may thus concentrate on construction instead - a work which is largely oriented to the physical task and is nearly independent on administrative structures and even on FORTRAN. Most surprisingly, however, an insight is developing, that the structures which are gradually carved out and the language which begins to form, may potentially connect the practice of systems analysis with possible future knowledge bases and expert systems in applied physics. Ultimately, an interpretative use of input information may additionally offer a guideline towards artificial intelligence in computational systems analysis. In the present stage, the individual user is supported to develop his own private 'knowledge base' in the process of modeling.

3. Coding principles and techniques for computational systems analysis

As the basic principle of architecture, a clear and complete separation of physics and numerics is applied now as before [3]. This task is not trivial and is not solved so far for all the cases under consideration (especially not for higher-dimensional numerical schemes). Maximum use of supporting facilities and utilities, and of generation techniques is aspired to, though also the program and file generator does not yet meet all requirements which arose along with development of the software technology. Among the most consequential technical steps, the extensive use of work storage allows for numerics-independent, storage-saving, overlay-supporting coding, which is indispensable for one of the ultimate tasks in complex systems analysis - the interactive operation of different codes. To dissolve from administative program structures, a level of physics-oriented subroutines is added, and a simple mapping procedure "translates" between both levels - comp. figure 1. In addition to a first standardization of the OLYMPUS main computational module STEPON (comp. [3]), its central part ADVANC is subject now to further deep structuring, with a fractional-step conception of the physics operator (DRIVER), and including physical, numerical, and administrative options of high flexibility. The most serious advancement of the last years, however, arose from generalization of the internal part of an input processor (READIN), which was originally developed to replace the NAMELIST facility [3]. Figure 2 demonstrates, that the access to the complete data structure as far as organized in COMMON, may be used by a much broader scale of internal systems applications. At the lowest level, requirements of parallel processing may easily be met, for instance. Though a practical trial was not undertaken so far, the tools are available on principle (or may be easily generalized to this end) to use parts of the READIN input processor also for model changes via interpretative operation of an input channel. Models and their modifications could be stored, retrieved, displayed etc. as a knowledge base. It is also imaginable here to use the output of a symbolic algebra system as input to a numerical code. At the FORTRAN level, of course, extensive operation in an interpretative mode is not advisable for computational efficiency. It would however be of interest for testing model changes, and especially for further accompanying research on problem structures arising in computational systems analysis. No doubt, this field shapes largely the status of computational physics today.

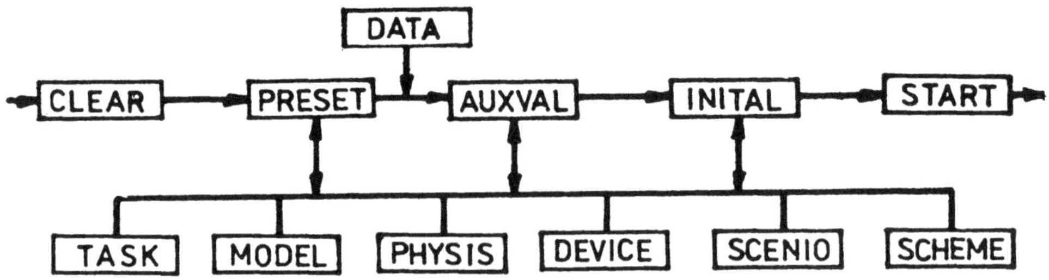

Figure 1: Mapping of administrative to physics-oriented program structures of the initialization stage of OLYMPUS.

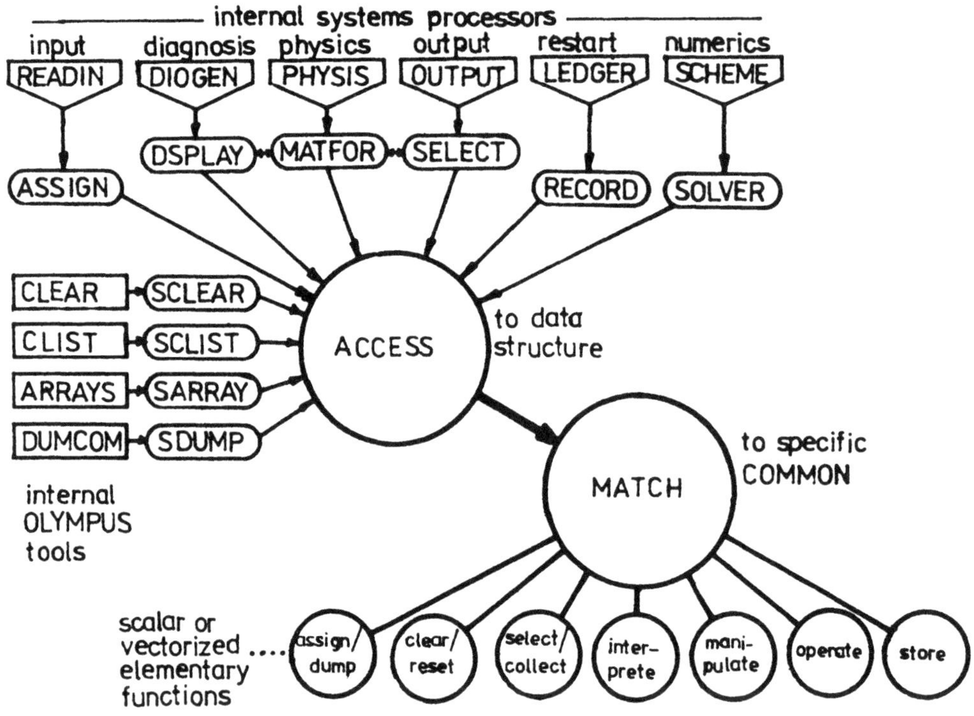

Figure 2: General COMMON data access facility and its broad range of application for internal use in 'olympian' codes.

4. References

[1] ROBERTS, K.V.: "An Introduction to the OLYMPUS System", Computer Phys. Commun. 7(1974)237-243;

[2] KILLEEN, J.: "Report on the IAEA Consultants' Meeting on Computer Codes for Fusion Research", Computer Phys. Commun. 19(1980)273-291;

[3] CARL, P.: "OLYMPUS for a BESM-6 - The Development of Medium-scale Programs within the OLYMPUS Framework: System OLYDES", Computer Phys. Commun. 20(1980)165-179;

[4] CARL, P. and G.L. STENCHIKOV. "Structural Analysis of the Climatic Response to a Nuclear War", this conference.

SET DEVIATIONS MODELLING OF FOREST FIRE SPACE STATE

T.N.Ivanilova [1)]

Abstract. In the paper the probability set methods assigned to set deviations estimation of the forest fire space random states in the function of their mean configuration form are considered. The deviation modelling methods as to union, interseption and as to mean deviation are suggested. The forest fire distribution identification algorithm as to average measure set and average measure deviation are presented.

Key words: random spread process (RSP), average measure set (AMS), finite random set (FRS), deviation as to union ($\cup$ - deviation), deviation as to intersection ($\cap$ - deviation), average measure deviation ($\int$ - deviation), covering probability, local spread probability.

1. Introduction

In the paper intraditional approach to the forest fire spread modelling is used [1,2] . This approach methods are set probable, therefore random spread identification and its configuration development prediction realization becomes possible on the basis of fairly slight experimental data volume [3] . The multiplicity, probability and configuration characteristics of the development fire - are the result of the prediction.
The need of set deviation estimation arises after the attempt to study the forest fire random spread identification process covergence. The $\cup$ - deviations, $\cap$ - deviations, $\int$ - deviations notions are necessary for properties of number statistical estimations study. Becides, $\int$ - deviation modelling permits to evaluate the forest fire real states spread. Let us consider the fire spread over the M territory, being the ($M \subset R^2$) plate square lattice. Assume the M is to be devided into the m areas, which are homogeneous which respect to the ($M = \bigcup_{j=1}^{m} M_j$) combustion. The $\forall M_j$ combustion probability is to be given for $g_j(x)$. The fire development over the M may be explained with the help of the random spread process: $\{K_t, t \in T\}$, K_t - is finite random set. $\forall K_t$ is recurrently presented by the local random finite set ${}^tS_x$:

$$K_{t+1} = \bigcup_{x \in K_t} {}^tS_x \tag{1}$$

The $\forall {}^tS_x$ distribution, at its turn, is fully determined by the local distribution probability $p_{ij} : {}^tS_x = {}^tS_x(p_{ij}, g_j)$. Consequently, the forest fire configuration modelling with the help of RSP becomes possible after the identification step, which consists in p_{ij} seeking. The p_{ij} models such RSP, characteristics of which are approximate to the true state forest fire characteristics.

[1)] Computer Centre, SB AS USSR, Krasnoyarsk, 660036, USSR

2. Set deviations modelling

One of the probability set theory main notion is the average measure set - $\int K_t$, which characterizes the average form of the process under study. As [2] defines AMS is a family of sets satisfying the following relation

$$\int K_t = \{A: \{\pi_{K_t} < h\} \subset A \subset \{\pi_{K_t} \leq h\}\} \tag{2}$$

and shear level h is selected from the conditions

$$\mu\{\pi_{K_t} < h\} \leq E\mu(K_t) \leq \mu\{\pi_{K_t} \leq h\} \tag{3}$$

where $\pi_{K_t} = \pi_{K_t}(x) = P\{x \in K_t\}$ the $x \in M$ point covering probability by K_t - random set, h - covering probability shear level.
$\pi_{K_t}(x)$ estimations are used under statistical modelling with N realizations, and consequently $\hat{\int}_N K_t = \hat{\int}_N K_t(X_0, p)$ - is the AMS estimation which N realizations at the t moment, which begins from the state of X_0 and is defined by the p set.
Now let us consider the $K_t^{\cup n} = K_t^1 \cup K_t^2 \cup \ldots \cup K_t^n$ (K_t^i - let indipendent homogeneously distributed FRS) which is random one as K_t is to be. The estimation of the covering probabilities by the specified set of points from M is defined by the following,

$$\hat{\pi}_{K_t^{\cup n}}(x) = 1 - (1 - \hat{\pi}_{K_t}(x))^n. \tag{4}$$

Then, with the help of (2),(3) making use of $\hat{\pi}_{K_t^{\cup n}}(x)$ from (4), $\hat{\int}_N K_t^{\cup n}$ may be defined. Now we call it deviation as to union.
Covering probabilities estimation $\hat{\pi}_{K_t^{\cap n}}(x)$ for FRS $K_t^{\cap n} = K_t^1 \cap K_t^2 \cap \ldots \cap K_t^n$ (K_t^i - independent homogeneously distributed FRS) is realized with the help of phormulas:

$$\hat{\pi}_{K_t^{\cap n}}(x) = (\hat{\pi}_{K_t}(x))^n, \tag{5}$$

$\hat{\int}_N K_t^{\cap n}$ - is called deviation as to intersection.
Let us introduce in our study a symmetrical difference of the two sets $K_t \Delta \hat{\int}_N K_t = (K_t \cup \hat{\int}_N K_t) \setminus (K_t \cap \hat{\int}_N K_t)$ and while calculate its AMS - $\hat{\int}_N(K_t \Delta \hat{\int}_N K_t)$ using necessary covering probabilities:

$$\hat{\pi}_{K_t \Delta \hat{\int}_N K_t}(x) = \begin{cases} \hat{\pi}_{K_t}(x), & x \notin \hat{\int}_N K_t \\ 1 - \hat{\pi}_{K_t}(x), & x \in \hat{\int}_N K_t \end{cases} \tag{6}$$

$\hat{\int}_N(K_t \Delta \hat{\int}_N K_t)$ set is the average measure deviation and characterizes the random, set spread concerning to its average from. AMS deviation notion may be interpreted as an FRS variance analog.

AMS deviation modelling may be realized for FRS arbitrary form: for incoherent sets, sets with rough bounds, ets.
It is necessary to notice, that RSP which has identifical AMS are differs with different variance for the different FRS spread reason. Set deviations modelling allows to take into consideration this very fact.

3. AMS - deviations modelling according to FRS single realization

Experimental data used for forest fire predictions must be a map of successive real limits and homogeneity map [4]. The first one is usually got with the help of aerospace observations. In practice, however, measurement opportunity is restricted to one or two profiles of the fire spread territory.
For the identification algorithm operation (p.4) it is necessary to know spread estimation of the $\mathfrak{D}$ fire variance real states. That's why estimation task of FRS AMS-deviations is urgent and is realized for violent limited number of its selected values (measured realizations).
Let us consider the $\mathfrak{D}$ real variance estimation way with the help of probability set method. Suppose X_T - is the experimental fire state at the moment T. X_T measurements are to some extent in error, and that's why X_T - is to be the single FRS realization. So calculate AMS deviation of the FRS, it is necessary to know $\hat{\pi}_{X_T}(x)$ covering probability estimation. This estimation may be taken by χ_{X_T} fair indicator method as to family of preset neighbourhoods - i.e. of FRS kernel and base [2].
Then, using the (6) formula we may calculate $\hat{\pi}_{X_T \Delta \int X_T}$ and taking into accound of AMS modelling method as to a single realization [2], we'll calculate AMS deviation estimation $\mathfrak{D} = \hat{\iint} X_T$ as to a single selected value - X_T. $\mathfrak{D}$ estimation (a forest fire real states spread) allows to give a substantiated prediction of the exprected random set - state spread comparatively with its average form.

4. Identification problem definition as to AMS and AMS-deviation

Let two experimental fire configurations X_0, X_T are known, fixed at the sequential time moments $t=0$ and $t=T$. Identification problem (p.1) is in local spread definition. p estimation in identification algorithm for two moments (the first - AMS, the second - AMS-deviation) may be realized with the help of the two-component iterative statistical procedure. At ($n+1$) step the iterative formula has the following form:

$$p_{n+1}^{(\mathcal{N})} = F^{(\mathcal{N})}\left(p_n, t_n, \hat{\int}_{\mathcal{N}} K_{t_n}(X_0, p_n), \hat{\iint}_{\mathcal{N}} K_{t_n}(X_0, p_n), X_T\right) \tag{7}$$

$$t_{n+1}^{(\mathcal{N})} = H^{(\mathcal{N})}\left(p_n, t_n, \hat{\int}_{\mathcal{N}} K_{t_n}(X_0, p_n), \hat{\iint}_{\mathcal{N}} K_{t_n}(X_0, p_n), X_T\right)$$

Where $\iint_{\mathcal{N}} K_{t_n}(X_0, p_n)$ - is estimation of AMS-deviation as to $\mathcal{N}$ -realizations at the RSP t_n moment, which was begin in X_0 and specified by p_n set, t_n - is model time moment.

The first component $F^{(\mathcal{N})}$ - is a local spread identification probabilities procedure as to the first moment - AMS. Probabilities estimation with the help of $F^{(\mathcal{N})}$ is conducted with approximation aim of AMS $\hat{\int}_{\mathcal{N}} K_{t_n}(X_0, p_n)$ and experimantal X_T [4] .

The second component $H^{(\mathcal{N})}$ makes recalculation of modelling time. Calculations as to $H^{(\mathcal{N})}$ are necessary for approximation not only the first moments, but the second ones - AMS deviations and $\mathcal{D}$ experimental state variance. $\mathcal{D} = \hat{\iint} X_T$ - is calculated with the help of probability. Set modelling methods, described in p.3.

Let identification accuracy is to be $\varepsilon(X_0, X_T)$ which varies on X_0 and X_T experimental states. Then the specified accuracy achieved domain will be.

$$\hat{S}^{(\mathcal{N})}(\varepsilon) = \left\{ p : \hat{d}^{(\mathcal{N})}(p, t, X_0, X_T) \leq \varepsilon(X_0, X_T) \right\} , \qquad (8)$$

where $\hat{d}^{(\mathcal{N})}$ - is iterative procedure error, doublecomponent vector of the following form:

$$\hat{d}^{(\mathcal{N})}(p, t, X_0, X_T) = \left[\rho(\hat{\int}_{\mathcal{N}} K_t(X_0, p), X_T), \rho(\hat{\iint}_{\mathcal{N}} K_t(X_0, p), \mathcal{D}) \right] \qquad (9)$$

Making use of introduced notations we formulate identification problem for the two moments with specified accuracy ε and with probability which is not less than α :

Define the $p^{(\mathcal{N})}$ and $t^{(\mathcal{N})}$ values, that $\hat{\mathcal{P}} \left\{ p^{(\mathcal{N})} \in \hat{S}^{(\mathcal{N})}(\varepsilon) \right\} \geq \alpha$.

The forest fire prediction development calculations with the help of the very algorithm will allow to get the better accuracy of the predicted characteristics.

5. References

[1] Vorobiev,O.Yu.,E.N.Valendik: Probability set modelling. Nauka,Novosibirsk 1978.

[2] Vorobiev,O.Yu.; Average measure simulation. Nauka, Moscow 1984.

[3] Vorobiev,O.Yu.,T.N.Ivanilova: Probability set methods of random spread identification. Preprint, Computer centre of SB AS USSR, Krasnoyarsk 1982.

[4] Ivanilova,T.N.: Dialogical system of the forest fire prognosis. Proceedings of the international congress on simulation, Praha 1987.

AN ATTEMPT FOR AN ADAPTED MODEL GENERATION METHODOLOGY : EXAMPLE OF APPLICATION IN BUILDING ENERGY MODELLING

Louis LARET [1/]

INTRODUCTION

Issued only from intelligent beings, the conscient actions are based on prospective analyses and reasoning, and somewhere introduce what can be called modelling. In fact, these "intelligent" behaviours use an inductive processing : a conceptual internal representation is built and processed in order to predict the effects of different possible causes, the causes and their predicted effects have a direct interest for the intelligent being.

These logical goal seeking actions proceed of the general application of problem solving processing. Problem solving is the search of practical goal oriented actions in a given situation.

More intelligence means the consideration of deeper and longer term prospectives : thus a more powerful modelling. It must be underlined that a deeper model means a more accurate evaluation of the predictable and pure random behaviour.

Problem solving can be considered following four different functions :

- the specification which includes the definition of the goals and the situation evaluation, like constraints;
- the synthesis of solution based on the background knowledge ;
- the evaluation of the solution ;
- the interpretation of the solution in comparison with the specifications ;

This decomposition is not so clear in practice and the relations betwwen these functions during a problem solving process are sketched on Figure 1. The true processing is not as sequential as in the scheme, and more often appears as random and opportunist jumps, especially during design applications. But probably, a better understanding of efficient processes will appear when the evaluation phase will be more deeply known, especially for the high level knowledge processing. This phase is directly concerned with modelling and, up to now, automatic processing of models is almost only considered for numerical simulation. This quantitative application of models is a limited resource of the models knowledge resources.

Thus, in this paper, modelling action is not only seen as the classical evaluation mean, in front of a numerical processor. From our point of view, the models are also "behaviour explicitor", this means a way to transfer knowledge from the experts to the user.
In other words, a model will not only give (and sometimes has not to do it), for instance, a good quantitative prediction of the energy consumtion for a given building in a given situation. Moreover, a model will "explain" a lot of other important features, as for the preceeding example, "if outside temperature decreases, the energy consumption will always decrease, faster in this situation than in another reference, ..."
These essential informations are contained in the model, and are often obvious for experts, but should appear explicitly for all users during the model process.

1/ Centre Scientifique et Technique du Bâtiment Etablissement de SOPHIA ANTIPOLIS
B.P. 21 06561 VALBONNE CEDEX FRANCE

This paper only considers mathematical models. The mathematical modelling is a pure design action, where the expert art and creativity is mixed with rigour and method. The validation of many assumptions are often based on heuristics.

The modelling in itself is always oriented to an estimation of a specific behaviour and its adequation can only be analyzed in this context [1]. The modelling always consider the whole world in which only a limited aspect is analized. This restricted observation and its consequences appear by the system boundaries and the boundary conditions as it appears on scheme of Figure 2.

Another main characteristic is the state of the system or the postulate that, for evolutive systems, the past is not the same than the future. Besides, the state implies that the whole history of the system affect its future evolution, and is somewhere contained in the system at any time. The modelling always introduces an approximated representation of the system memory. The consequences of the assumption as for boundary conditions are never completly understood.

Modelling appears as a key point for future developments in all fields. "Better decisions require better evaluations" [1].

Better evaluations means the communication of a more rich information. On this point, a good adequation of the model with the problem to solve is essential. This also implies an interactive and evolutive modelling during the problem solving process. Among others, parsimony appears as a main quality of a model. This is not for numerical problems, although this aspect is not negigeable, mainly for reliability. But the structure of the "simplest" model(s) able to give the required information is the highest level of information exchangeg with the user.

Modelling, as any design process, always starts on the a priori knowledge. The decomposition of the analyzed system into primitives for which models already exist intoduces this reference. These primitives are called the component models.

The decomposition itself and the resulting coupling are not obvious operations and are two main difficulties of the computer aided modelling. But a better appreciation of this aspect is possible only when a component model base is available and analyzed.

This paper focuses on the component model, considered as a well defined cell of knowledge to be processed in a defined range of use. The paper aims on the methodology to produce and analyze this basic cell in order to make its inside knowledge available.

COMPONENT MODELLING

In our study, the component model is an abstract representation of a given behaviour or facet of a well defined object, in relation with an environment. The relation between the component model and its environment comes from the observation and the corresponding experimental frame [1]. The object behaviour is contained in the model structure.

Before to define some concepts, the analysis context has to be made explicit. The component models are the primitives used in the design of a specific global model. The global model is designed for a well defined purpose : to solve a given problem.
So, it will be assumed that for a given problem a solving process or related method can be defined. This method requires a specific model. It means a user oriented model, for which inputs and outputs can be defined.

In fact, many methods use the model in the reverse order, namely computes some inputs for given outputs. Main exemples are the control (synthesis of inputs in order to reach some goals), the parametric identification (best estimates of parameters, provide given input-output sets), ...

Nevertheless, the method always orients the model, very often close to some physical causality. But, the main interest of a primitives base stands in its elements flexibility in the further coupling. Thus, depending on the designed global model, a given component model can be used following different orientations. From a practical point of view, a component model can generally not be reversed completly. It means that the initial formulation introduces strong limitations in the further coupling and processing of a given model, and this point has to be made very clear. Generally, more fundamental or close to phenomenological reprensentation is a model, more oriented is its formulation which express more and more the physical causality.

Formulation

A very pragmatical approach can be proposed, restricted to the mathematical formulation.

A mathematical model is thus a set of equations. So, the model description will be directly based on the mathematical properties. But, considering the further use, the relation between the mathematical formulation and the attached physics (or behavioural property) is essential. This is the main objective of the following.

The more general formulation of an equation set is the implicit form. It means, general relations between variables and parameters. Variables are entities which can have different values on a given set. Parameters have some values, defined on a given set. In principle, the number of variables must be larger than or equal to the number of equations.

But, for component model applications, slightly different definitions of variables and parameters could be given, in order to express the model "meaning".
The relation between the component model and the environment is expressed by the boundary variables. In this definition, boundary variables are quantities which can be directly associated with a measuring system, at the arbitrary boundary between the component and its environment. It implies that all comparable primitives or component models have the same relations or boundary with the environment. The boundary variables will always be shared by the model and a coupling element : another component or a boundary condition, modelling the environment of the global model.

The internal variables are either internal observable variables, assuming in fact a higher level of decomposition, either variables appearing in the mathematical formulation which are not boundary variables or parameters.

The parameters are the other symbols appearing in the formulation which depends on the specific object.

In other words, the component model of an object is the behavioural characterization of a given class of objects. The parameters are the individual characterization discriminating each object of the class. It means that two objects with the same parameters instantion are identical for the considered behaviour, at the model level.

From these definitions, the parameters can be non constant values; but are explicit functions of the model variables and parameters. The distinction between parameters and variables is thus possible.

The choice of the parameters definition is totally submitted to a subjective and opportunist appreciation of the object property characterization at the modelling level. This arbitrary choice influences strongly the component modelling and cannot be avoided. This is a main limitation for the generality of the component models, but it is the only way to maintain a minimum standardization for further assembling and data base modelling. This parameter definition is in fact a first orientation of the model, in function of the further use and it has to be stated explicitly.

A variable (boundary or internal) is a state variable, if it explicitly express the time variation of any term of the model, and a reference value at a given time has to be assumed.

A state variable express a memory modelling but sometimes not directly concerned with the physical meaning of the other variables. For instance, this phenomena appears with time varying parameters, and, particularly , for time varying resistances.

From a general appraoch, resistances are parameters related with a pair of conjugated variables, generalized potential and flow. These variables product is a generalized power dissipated irreversibly (thus without coupling) in an other domain. A time varying resistance means a structure evolution of the modelled element, thus a corresponding state. But, the modelling has considered a perfect discoupling between the state expressed by the resistance time evolution and the states associated with the associated variables.

Spatial and time variables are object independent variables and from the fundamental physics are only related in relativity theories. Also, the fundamental postulate of physics implies that the model behaviour is independent of the arbitrary referential for space and time.

Reference model

In this paper, the reference model is a "physical model" of a given component behaviour. The reference model is considered accurrate enough for the assumed purposes. A "physical model" means a model based on phenomenological principles, in fact a common high level of induction.

There is a main difference between the reference model and the base model defined by Zeigler [3]. The base model is purely a concept. The reference model does exist, and for a given compoent behaviour provide from an experts general consensus.

In fact, it is a practical way to obtain an usable formulation of the fuzzy concept of "accurate enough for defined class of applications". The reference model expresses this knowledge, by this high level induction model.

The reference model is based on phenomenoligical principles, thus the mathematical formulation is directly issued from these principles. The model principles are the most condensated representation of the model and this knowledge processing is essential in the model design.

Presently, the so-called soft sciences (biology, economy and all human sciences) cannot propose a very high level of induction and a single reference model could not be proposed. This is due to the extreme complexity of the system facets compared with other sciences fields [2]. In this case the proposed methodology can still be used, but the difficulty will appear for further assembling.

The emergence of general modelling principles is a key point for a significant progress in these sciences. These rules or principles can be called the "model invariant". This is in fact the only aspect which is untouchable during any model processing.

Empirical model

On an opposite side, the empirical models can be defined as arbitrary mathematical equations whose parameters are fitted by identification techniques on experimental data. Contrarly to the reference model, an empirical model can never be extrapolated.

The accuracy of the empirical models outside their validation range is not known and, of course, there are several empirical models of a component In fact, the empirical models are the direct adequation of mathematical equations to experimental facts. Due to the degree of freedom in the formulation, these models are oriented for specific uses.

Several techniques of parameter identification are now available, but the structure identification remains a not obvious exercise [2].

Adapted models

Between the two extreme bounds of the component model domain, several other models can be defined, mixing experimental fit and some phenomenological rules or heuristic which is a simplified physics, only usable in a restricted range.

This set of rules or expert short circuit reasoning is the main part of the expert knowledge for given uses. These adapted model rules or invariants is a way to formulate explicitly this kind of knowledge. Thus, by their part of physical support, the adapted models can be extrapolated in a well defined range.

MODEL RELATIONSHIPS

The main characteristics of the different model levels previously defined are summarized on figure 3. Now, it appears essential to establish the links between the different comparable models of a given component. As previously assumed, comparable models means that they are concerned with the same behaviour and that the bonds with the environment are the same, which is the most restrictive constraint. In the scope of the application, the limits for model complexity is bounded by, on the one side, the reference model and, on the other hand, the empirical models.

All these comparable models or component representations are, of cause, isomorphisms of a common aspect of a well defined element and a link exist between all the different formulation. Now, it should be essential to express this link in such a way that it can be processed.

The most useful formulation is expressed by mathematical relations. But, as seen before, models are designed and selected following their capabilities to represent a behaviour. Thus, it is essential to associate the behavioural (or physical) meaning to the mathematical link between each pairs of models.

The physical formulation of the link will be made by attributes called invariants. The choice of invariant means, that two models with the same attribute instantiation will have the same behaviour for the aspect concerning this attribute.

Following this definition, the first invariant, for all the models of a component, is the description of the boundary of the component model. The mathematical desription will concern the number of variables and their defintion set (the core of real numbers for instance). But, the physical meaning of this boundary is expressed by this invariant attributes. It means a temperatute, heat flow, information flow and type for instance in control (sensors or actuators), ...

A numerical algorithm principle can be used in order to transform a component mathematical model into a completly oriented input output sequential process. This sequential procedure is nothing else than a new component model. Some of the initial properties are conserved, some others are destroyed. In this case, the mathematical effect is clearly defined, but the behavioural consequence is not always clear.

Nevertheless, an essential improvement would be the clear formulation of this physical link. When this link has not yet be established, it should be pointed clearly, because it could be a source of difficulties during the component model use. The best improvements of a given model would be the establishement of these links with the other concurrent component models, especially the reference model.

In fact, from the behavioural point of view, the reference model is assumed the best for all the class of applications. The other models can only replace or add some behavioural. In all cases, it means different specific behaviour than the accepted reference. Of course, it must be underlined that the reference is never definitive. But at the same time, the reference might only be seldom updated. The reference interest is to bound the to-day expert consensus with an existing fundamental component model.

Rather good exemples of physical meaning equivalence are the so called "numerical viscosity" introduced by the spatial discretisation of a distributed parameter formulation. The fixed time step integration techniques add some kind of"equivalent poles" to the model transfer function.

Physical principles are always translated in mathematical laws. Thus, when a model is generated from another, following a physical rule, the mathematical effect can also be expressed. The reverse is not possible. Many times, the mathematical transform from one model to another can be defined rigourously, but the behavioural effect cannot be obtained.

In fact, this behavioural or physical effect can be obtained from a high level validation whose results are generally expressed in experts natural languages.

In this paper, it is proposed to weave a relational network between the different models for a component. In priority, the relations should be expressed by some bevioural estimators called invariant. The relation will express if the invariant is destroyed, with a quantitative expression if it is possible, or if a new invariant is added. The mathematical effects have also to be indicated, as far as possible, by rules. The partial or complete input output orientation, the linear and non linear characteristics, related to the different variables are esssential informations for the further coupling and processing, and a standard compact formulation should be defined.

For exemples, mathematical techniques for order reduction are based on well defined processing, with a well defined structure of the reduced order model. The equivalent destruction of the initial model properties and their amplitudes cannot be always estimated. Nevertheless, it appears quite essential to know if the reduced system will remain asymptotically stable, minimum of phase, ... At least, the non conservation or ignorance of the corresponding invariant effect should be indicated, especially for the case of the aforementioned properties.

When an expert in model design starts, he build in his mind something like such a network, adding elements, jumping from one model to another. The results will be strongly dependant on its experience of the domain. This paper is a tentative to represent a common network which reasonably reflects the up to date knowledge on a given component behaviour.

This network should be a good platform for expert implementations. First, they can easily place their new models and see at which levels new models should be added. Secondly, new relations between existing models can be established either by theory either by validation in better relation with users experiences.

As they are defined the invariants express a formulation for model characterization. This means a way for automatic selection of models. At another level, the invariants would be the way to express the model specifications. From a rigourous model specification, automatic generation of models can be started.

APPLICATIONS

An attempt of application is now presented very briefly in order to show the interest of this approach. First, some mathematical order reduction techniques are briefly mentioned, following the proposed principles. Then, three exemples of adapted modelling in building applications are presented.

Mathematical order reduction of linear models

The techniques considered are based on linear dynamic continuous models. The principles are based on the approximation of a given input output pair considering some aspects. In this analysis, only the SISO (single input single output) case is considered. The MIMO (multi input multi output) generalisation is not obvious.

The main principles of these techniques allow the direct computation of the parameters of a reduced order model. Very often, the order determination is also possible.

The different techniques can classified into three main groups, following the approach [6] :

- the state space approach considers mainly the singular perturbation and its derivatives. The main principle is based on the split of the system matrix into a "slow" partition and a "fast" partition. The dynamics of the fast partition is neglected or roughly approximated (modified singular perturbation). The slow part and the steady state are maintained and, thus, are the invariants of this process. But, the propagation of high frequencies are considered too fast and not enough damped ;

- the eigen space approach considers the extraction of an eigen value subset from the whole set. Here only the phase minimum property could be destroyed. But, the corresponding physical representation of the reduced model is not possible ;

- the transfer function approach will consider a polynomial fraction in Laplace domain. All these techniques give very good approximations of the transfer function with a very low order. The Pade approximants give asymptotical equivalence for low frequencies at a given order. In this case, only the shape of the transfer function at low frequencies and the steady state are invariants; the asymptotical stability is not insured. The Pade Hurwitz approximants give also an asymptotical equivalence for high frequencies; the phase minimum property and eigen value type are no more preserved.

Figure 4 shows the principles of these methods and some of their characteristics.

Other methods exist, but for all methods, the approximation destroy some property and it is essential in the further coupling to know exactly what is maintained or not. The modification of the pole and zero positions is a very interesting evaluation of the resulting effects.

The main interest of all these methods is their generality : they can be applied automatically provide that the starting model is given in the right formulation. These methods can be now considered as general tools.

Building thermal dynamic model

A reference thermal model of building can be expressed in linear state space differential equations. This model is designed following the classical approximations for these kind of applications. The air flows between thermal zones are considered as input variables.

The different levels of modelling differ by the model order which express the thermal zoning.and the heat conduction approximation. The result is a particular formulation of energy conservation combining air flow conservation [7], [8].

The resulting models have the following behaviour :

- asymptotic stability;
- minimum of phase;
- when the air flows between thermal zones are symmetrical, the system is symmetrical and the eigenvalues are real;

The direct identifications on the reference model for single thermal zone has shown that, for the input output pairs of interest, a second order model gives very accurate results. The second order model can be considered as the base of an empirical modelling [4], [7].

From theoretical considerations on the reference model, a direct computation of the parameters of this second order model can be proposed. The resulting model will conserve the previously mentioned properties of the reference model. Moreover, the physical meaning of the state space parameters allow the model extrapolation for varying air flows, which is a non linear problem [7]. The reduction principle can also be extended for multi zones models [9].

The interest of these reduced models, maintaining a strong link with the physics, have been demonstrated for many applications, especially for design [4], [5].

Building net space heating demand

The empirical models are based on an experimental fit called the building signature. The first step of induction was based on the Degree Day approach. This method is a simple steady state heat balance of a single zone building models on a daily mean basis. The solar effects are neglected and the dynamic is approximatively, mainly by the day averaging.

For long enough period these techniques give an overestimated crude estimates of the building net space heating demand.

However, a more careful analysis with a reference model [8] and the French B regulation [12] have shown that simple models can give very accurate estimation, provide that the solar effects are taken into account.

Thus, different improved models were proposed. All these models are linked by relations which express different levels of physical approximations [5], [10]. At the highest present level of improvement, the dynamic characteristics are introduced with the previous dynamic model [10]. It means that the link is established with the same previous reference model considering a detailed dynamic behaviour as it appears on the scheme of Figure 5.

Boiler model

In building applications fuel oil boiler with hot water network for heat distribution are very common components.

The reference phenomenological models are developped for power plant applications and are extremely complex. This reference could never be used for numerical simulation in building applications, but can be very useful for lower level model implementation. From the empirical level, it is known that a very simple boiler linear characteristic is accurate enough. This characteristic can be obtained directly from the classical boiler tests. Unfortunatly, this characterisic is very sensitive to boiler operation.

On this basis, a simple model was designed considering a simplified combustion process and a simple heat exchanger. The result is a simple formulation, giving the linear characteristic but with a physical meanig for the involved parameters. The parameters can be directly identified on the same classical boiler tests as previously. Then, the model can be used in a very wide range. Some dynamical aspects can also be introduced [11].

CONCLUSIONS

This paper has proposed an approach for model high level knowledge characterization. and future processing The starting point is based on the definition of comparable component models. The model comparability implies a standard boundary with the environment and the modelling of the same behavioural aspect.

For a given component, the different models relations are established on a behavioural or physical basis. The relation characterization is given by attributes which are called invariants. These invariants are a way to qualify as rigourously as possible a high level knowledge.

The level of modelling for a component is bounded by two opposite modelling level. On the one side a reference model issued from expert consensus on a phenomenological "enough accurate" representation for a given wide range of application. On the other side, the empirical models concentrate the expert experimental frames. Between these two different representation level, appears the adapted level modelling.

In conclusion, this approach has attempt for a methodology of a classification of comparable models and characterization with a high level attribute.

This should be have the following advantages :

- a direct and condensated view on present knowledge on a given component modelling and, consequently, a continuous easy implementation ;
- an easy implementation in computer systems for further processing, especially with object oriented representations ;
- the continuous improvement of the relations between models providing from different users observation;
- a starting point for automatic model generation by application of similitude principles ;
- in computer aided modelling, a significant step for the automatic selection and further coupling of models based on the processing of high level knowledge.

The reference model as defined, opens the possibility for a decomposition of the component model into sub-elements. In fact, the phenomenological approach is entirely based on a conceptual split of the phenomena into simpler basic phenomena. By this way, the link between different encased reference models is possible.

Compared with the reference model, adapted models have also some possibilities of decomposition. This property will be expressed by the invariants considering the conserved internal variables pattern.

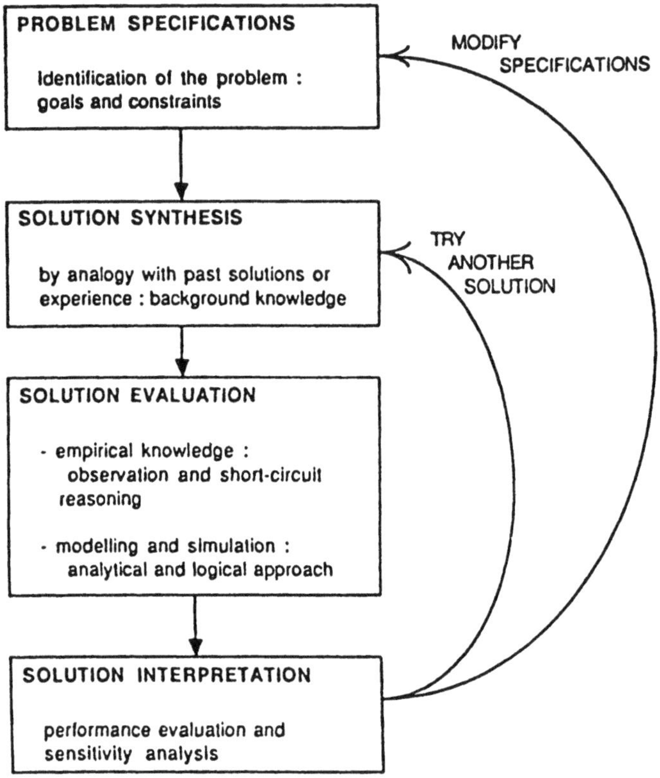

Figure 1 Problem solving scheme

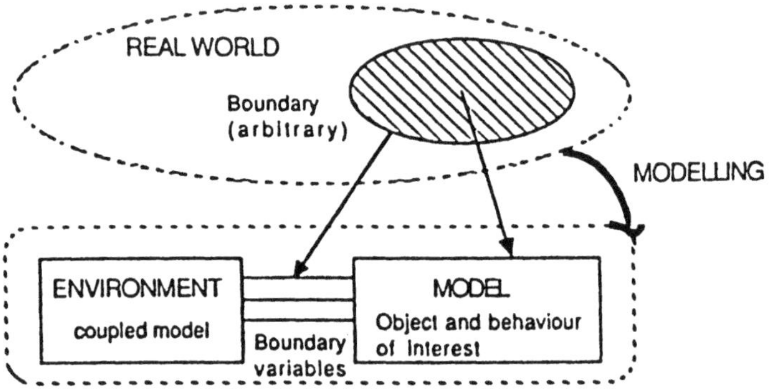

Figure 2 Relation between model and environment.

	REFERENCE MODEL	EMPIRICAL MODEL	ADAPTED MODEL
EQUATIONS	Phenomenological equations of physics	Any arbitrar equation	Based on a "simplified" physics
PARAMETERS	Known by physicals principles (Elementary properties)	Obtained by parametric estimation	Obtained by identification (P2) or by the "simplified" physics (P1) (global properties)
VALIDATION	Validated by the general physics	Validated in the domain of identification	Validated in a defined range
USABILITY	very low	very high	depending on problem complexity

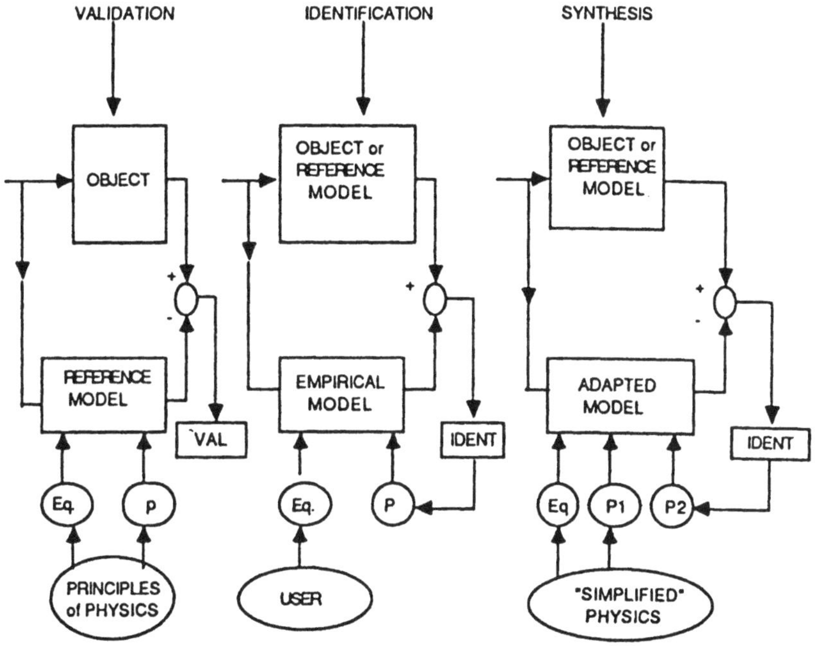

Figure 3 Main characteristics of different modelling levels

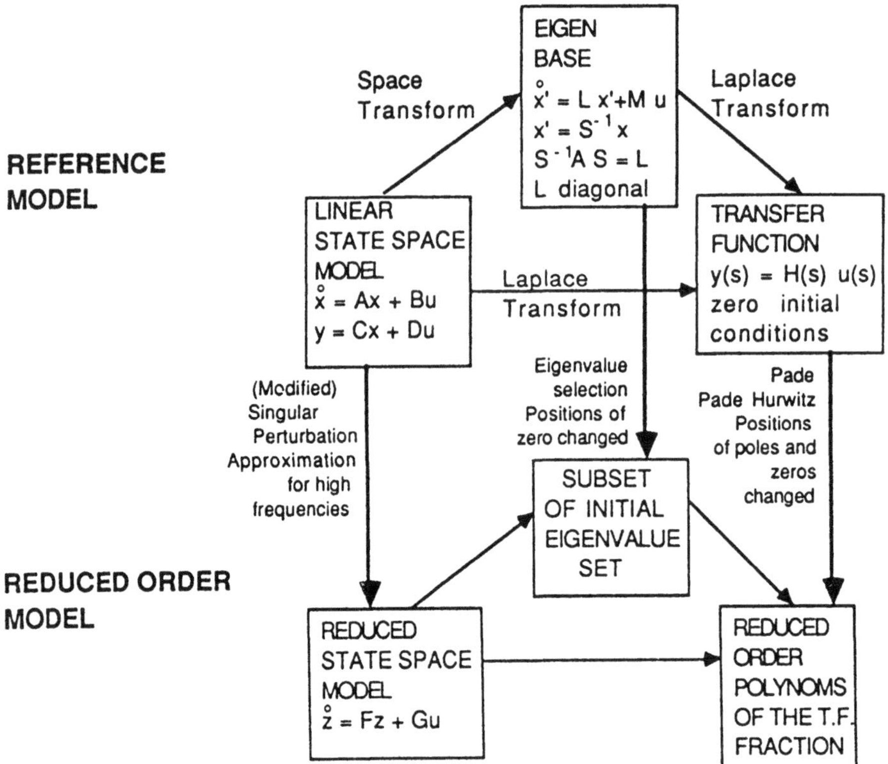

Figure 4 Main principles of mathematical order reduction techniques.

REFERENCES

[1] Methodology in Systems Modelling and Simulation - ZEIGLER(Ed.), ELZAS(Ed.), KLIR(Ed.), OREN(Ed.) Proceedings of the Symposium - Weitzmann Inst. of Science, August 1978

[2] SPRIET, J., VANSTEENKISTE, G. - Computer Aided Modelling and Simulation - Academic Press 1982

[3] ZEIGLER, B. - Theory of Modelling and Simulation - Wiley 1975

[4] LARET, L. - Use of General Models with a Small Number of Parameters Part.1 Theoretical Analysis - International Congress CLIMA 2000 Budapest 1980

[5] Basic Characteristics for Low Cost Houses in order to reduce the Energy Consumption for Building - CEC contract nr.EEA-5-060-B(G) - Final Report, Part 6"New developments in Calculation Methods" 1984

[6] MICHAILESCO ,G. - La Réduction de Modèles en Automatique. Classification des Principales Méthodes - Journées d'Etudes Modélisation et Simulation en Thermique, ENSMA Poitiers (France), September 1984

[7] LARET, L. - Contribution au Développement de Modèles Mathématiques du Comportement Thermique Transitoire de Structures d'Habitation - Doctoral Thesis, University of Liège (Belgium) 1981

[8] DELORME, Y., LARET, L., SOHET, H. - Numerical Predictions of Thermal Behaviour in Dwelling Structures - Universty of Liège (Belgium), Coll. des Publ. del la Fac. des Sc. Appl. nr. 83 1980

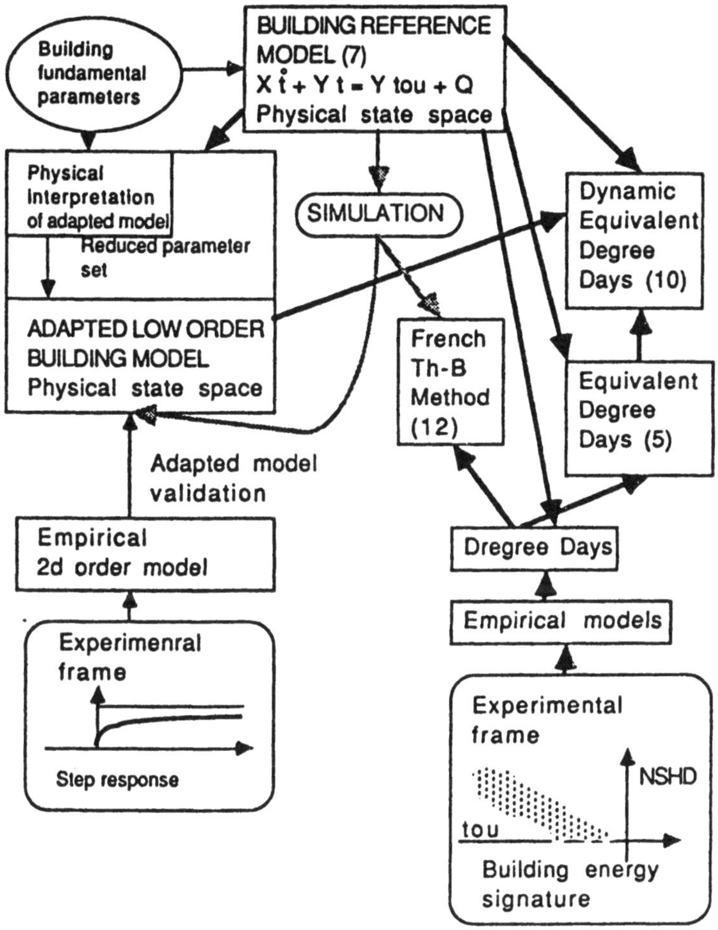

Figure 5 Different modelling levels of building thermal behaviour.

[9] LARET, L. - Component model. Building Specifications - University of Liège (Belgium), International Energy Agency Annex 10, May 1986

[10] LARET, L. - New Developments in Fast Evaluations of Performances for Design Applications Iintegrating Building and External Climate Effects - CEC European Conference on Architecture, Munchen April 1987

[11] LARET, L - Component Model. Boiler Specifications - University of Liège (Belgium), International Energy Agency Annex 10, May 1986

[12] CSTB, Règles Th-B 85, Cahiers du CSTB n°2001, May 1985

A DRAFT OF A COMPONENT MODEL LIBRARY MANAGER, THE "MODELOTHEQUE": ANALYSIS OF AN APPROACH, AND FIRST RESULTS

Anne-Marie DUBOIS 1/

KEY-WORDS : Modelling Methodology, Model Base, Thermal Simulation, Model Documentation

Model users are generally not modelling experts. And modelling experts generally use field-specific language and implicit knowledge. Thus the users often prefer their old usual models, may be somewhat unaccurate, whose well known capabilities and limits allow easy results interpretation.

This contribution is a proposal of a methodology which aims: 1. to produce a general and multifacetted documentation about the model, 2. to process this information for an aid to use the model in a component-based modelling approach.

INTRODUCTION

Building energy thermal analysis is a complex field, because of the great number of data, and deeply coupled phenomenas. Approaches are quite different depending on the classes of problems, and the uses of the users. For instance, HVAC designers don't use the same view on the project as architects or thermal researchers. Another example, for thermal or acoustic behaviour analysis, the decomposition of buiding envelope would be quite different, but some of the characteristics are the same (materials for example), and any decision taken in one domain interact with the other. Moreover, new control mechanisms need more accurate modelling of both equipments and envelope, and their coupling.

In this field, modelling aproaches are usually component-based. Advanced modelling techniques are not used in daily practice. Models are embedded inside the codes, and even in modular codes like TRNSys,[TRNSYS 1979],[CSTBat 1985], it is not easy to know what comes from the model, and what from the solver program. This work is engaged at C.S.T.B., in the framework of GER ALMETH 2/ , towards the development of new tools using advanced computer technologies. It has two components: the first is in the domain of physical analysis, and is aimed at grouping the existing models into a kind of model library, the "Modelotheque". The second concerns the computerized aspect with the goal of making this knowledge accessible to anybody implied in the field (professionals, students, researchers,...) [SORNAY 1985].

What should be the functionalities afforded by advanced computer techniques? Four steps have been identified:

STEP 1

the identification of the problem: objectives, environment, significant results to calculate. This step uses a general modelling knowledge: thermal problem classes, typical solutions if exist, choice of a method (the "demarche"), associated to the class of the given problem. The outputs are: clear goals of the work, main assumptions about precision,phenomena that can be neglected, "crucial facets" [SPRIET, VANSTEENKISTE 82].

1/ Centre Scientifique et Technique du Bâtiment Etablissement de SOPHIA ANTIPOLIS
B.P. 21 06561 VALBONNE CEDEX FRANCE

2/ GER ALMETH gathers for French research teams implied in building energy modelling, AFME, CNRS, CSTB, FNB

STEP 2

- then, but strongly connected to, the building of the global model. We call global model, the representation of the system and of its links with the real world representation, obtained after decomposition, choice of formal elementary models, instantiation of these models in the project data base, and assemblage. In each step, this work uses the problem knowledge established by the first stage: decomposition method, basic objects, assumptions, general framework.

  Here, the outputs are the global model, of course, but also some more precise description of the validity domain of this model, and comprehensive knowledge about the real object: the global model is linked to the project data base, then parameters are connected to the actual already known values.

STEP 3

- different steps may follow: in our first approach, we had in mind execution of simulations run. In this case, the model is send to the solver function: formal manipulations, numerical treatments, according to orders given by the former steps. In fact, some precious informations can already be used, and sometimes they may be sufficient as answers: - well known problems, whose solutions are in a results data base, - assumptions, or shape identification, - deductive rules may be applied and give the expected result. For instance, it may be useful to use qualitative reasoning on the global model, and sometimes it may short-cut heavy calculations [BOBROW 1984]. In anyway, here is a knowledge source.

STEP 4

- Results of experiments on the model may be analysed by specific tools. Some pertinent informations given by previous steps enlighten their meanings, and allow to appreciate them.

THE DOCUMENTATION OF THE MODEL : THE PROFORMA

Most of these functions need knowledge about models : the proforma, documents filled by the authors and given with the model, are basically important for the modelotheque. Inspired with the "DATAPROCESSOR PROFORMA" proposed in 1984 by the ABACUS and LPB Groups to the I.E.A. Working Groups,[CLARKE, LARET 1984], a first draft of the Model Proforma was released by the GER ALMETH after a consensus between the research teams involved in the field (INSA, CSTB, ENSMP, CNRS, FNB, AFME). Already, this structure is used for describing some well-known models, to test its expressive ability.

The goals of this project are:

- a documentation about existing models, a kind of handbook of algorithms.
- help to users for modelling. This will call validity clauses and rules about model use. As fas as possible, automatic modelling will be available.
- ability to give default values, and validated data.

A proforma contains several chapters,

I : a short page, giving essential information for identification of the model (physical object, thermical problem, method,...)

II : formal description, by schemas, and specifications among input, output and parameters

III : validity clauses and rules of uses or practice; in this chapter there wil be also consistence and compatibility rules for solving stage.

IV and V : contain data about validation tests, authors and scientific references.

Together, one model and its associated proforma are an entity of the modelotheque. Access and manipulation primitives are given by the database management. So the modelotheque is extensive and modular. The proforma is a template of thermal modelling knowledge expression, and begin the work to move towards the future with experts systems..[SORNAY, DUBOIS 1986].

DESCRIPTION OF PROFORMA -- (GER ALMETH proposals - may 1986)
<MODEL NAME> Date

CHAP 1: (a short page)
- GENERIC NAME, i.e composed name, (physical object, thermal problem, general hypothesese, method, and indication about (intrinsic, coupled, interfacing)
- ABSTRACT (10 lines max)
- THEORETICAL APPROACH (simplified, detailed,..)
- MODEL TYPE that can be algorithm, numerical, equations
- GENERAL INFORMATION: authors'name, references of the proforma filling (date, translater's name)

CHAP 2: formal description
- NOMENCLATURE: each name will be follow by a comment, units, type (real, integer, character, discrete,..), values domain, default value, and if exists, other constraints
- FUNCTIONAL SCHEME describing the model's core: how the problem is managed, which treatments upon the data, etc...
- BLOCK SCHEME: I/O schematic, distinguishing parameters, input/output, variables (of equations), and external methods used. There we see the plugs of the model box.
- FORMULATIONS than can be equations, algorithms, or logical rules.

CHAP 3: validity clauses
- DETAILED HYPOTHESESE
- VALIDITY BOUNDARIES (on I/O or parameters values, time step, etc..)
- APPLICATIONS LIST using the common user's description language
- EQUIVALENT PROFORMAS
- UPSTREAM MODELS acting as mechanisms for this model
- COMPATIBILITY RULES for the solving stage
- PROBLEMS OR BUGS

CHAP 4:
- VALIDATION TESTS Description or references of validation tests and comments

CHAP 5:
- REFERENCES Explanations and references of the theory.

PROFORMA TESTS

Some models have been documented through this formalism, [ALMETH 1987], some others through a similar one (IEA - [LARET 1986]). In fact, unless people generally agree with the idea, only a few number of proforma are filled in. The filling task seems to be very tedious and awkful, and many of the questions request reflections.

What kind of help would be efficient ? We have two directions:
First, we have to precise the data structure of the model base, so as to give the filler some a priori description of variables, parameters, etc, with thermal domain thesaurus, and menu-driven help. Because of the multifacetted aspects of model, there are differents clasifications according to the various attributes. To show the conceptual trees may be a interesting way to help the user to identify the right place of the model from each point of view. A part of this structure consists in a hierarchy of classes, another part refers to complex objects, or to coupling between objects. The use of object oriented paradigm will facilitate the building of this conceptual structure. We were very interested with the ideas of the paper [ZEIGLER-DE WAEL 1985]. In this case, we have different decomposition depending on the kind of problem: you can see a boiler either as a black box, with very simple representation (heat power vs consumption?), or as a composition of a burner, an exchanger, a chimney etc..., or again by some thermal phenomena: convective, radiative, conductive and mass flows. It is the same for envelope. If you are interested with structural problem, some CAD representation would be adequate. For thermal problem, it depends upon accurateness: K or U-value for walls, B or C coefficients, for global characteristics of envelope, or more detailed analysis. Models are classified not only by aspects, but also by their assumptions.

This conceptual structuration is a part of domain knowledge. It has to be easily expanded: a major requirement is that any new model must be easily stored in the Modelotheque. Both remarks show the exigency and the difficulty of the task.

Second idea is the use of knowledge extraction techniques that are used in the A.I. domain. Experiences show that many answers are not accurate enough. Sometimes, they raise new questions. Otherwise, they point out lack of explanations. So we think it may be useful that together a thermician and a knowledge engineer fill up some proforma. Then the real difficulties will be better identified and with some exercises, a best methodology can be obtained.

THE MODELOTHEQUE PROJECT

The proforma base has two goals: to give access by keywords to the proforma entities, to provide an aid to use the model in modelling problems: this task concerns the step 2.
First functional analysis of Modelotheque [DUBOIS 1986] requires different basic functions:

- on line model consultation, for user documentation;
- selection of models:

  - first function: selection of possibly adequate models by generic name (given object, given thermal aspect, main assumptions), giving a list of a priori usable models ;

  - second function: the choice of the "best model", with detailed analysis and if necessary, user's preference. This function is interactive, asking the user for accuracy of new assumptions and choices. An important aspect is the ordering of model list: it depends upon the criteria for the "best". We should like to say that the best model is the simpler one that give the expected precision on requested results. But this put on the question to "what does mean the simpler?". We can think about complexity of equations, number of variables, etc... Another aspect is the versatility of the model. Such questions concern the simulation goal.

- instantiation of a model: this function may be compared to the compilation of a procedure call, it searchs for the actual arguments, parameters values, etc... It prepares the work for linking the model with the project data base.

- validation: at each elementary step, this function encertains that the building of model obey all assumptions of the general framework. Some assumptions may be considered as weak, and the user may forget them: a warning would be given at the result stage.

- assembling : if the elementary model is accepted, then it has to be connected by its plugs with the actual in-building model. New facts, as assumptions, etc... have to be added to the project data base.

- updating, modifying function, allow to re-use the built model, for a similar problem, or even to test any "what-if" question. It can be compare to backtrack. It needs some inverse-link to know what is implied with local change.

- tracing function, would be very useful to learning about model uses. This function gather all informations about practical uses of the model. It may also retrieve the building of a global model, and this can be useful when the user want to save the global model, as a macro-model to be inserted in the Modelotheque.

A PRELIMINARY DRAFT

After this first stage analysis, it seems necessary to test main ideas and to make some drafts. Objectives are to give a realistic idea of what a proforma base management could be, and to improve the functional specification of Modelotheque.

It was decided to take a problem not too much complicated, but having some features of a real problem: to build a model for a single cell, "IRCOSE cell". In this draft, we would simulate the work of building a global model from a model base and its tools.

This draft is limited to manipulation of models: it simulates only a part of the task, the other beeing supposed to be treated by some powerful software (to identify the problem, to acquire the project description, to solve the equational system, to compute and to analyse the results). Moreover, we suppose that it exists a data structure, object-oriented, and that every external link of models (variables, parameters) are described in terms of the conceptual data definition.

We did it in PROLOG, to benefit of easiness of this declarative language, and internal inference system. Students wrote down some functions: frame-based representation of the objects of the building, here, the envelope of the cell, representation of models, handling functions. [ISI 1987].

What they had to do was:

- to choose a representation of proforma, (remember that a proforma is the representation of a model in the model base)
- to write down in this formalism the elementary models described by thermal experts, - when the description is not clear enough, ask the experts more explanations,
- to simulate an object-oriented representation of the building project,
- to calibrate handling functions, in order to run the whole.

What does a session look like? The user gives the name of his project ("IRCOSE" in this case is described in an existing file), the kind of problem, and the name of the object he wants to model. He may give some assumptions to simplify or to restrict the problem. Then the system proposes him a list of models selected by

their generic name. When the first model is chosen, the system begin to link this model with the project data. The user may decide that some variables will remain variables for the global model.
The algorithm for the building of the global model use the recursive procedure CONNECT, described in the figure 3.

ALGORITHM OF BUILDING OF A GLOBAL MODEL

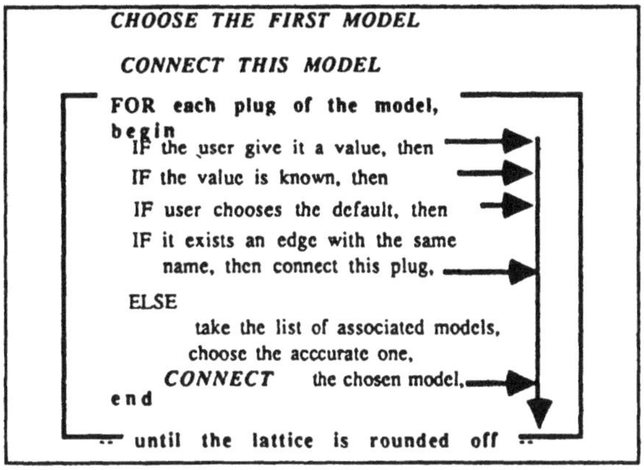

For each choice of model, the user is asked for agreement, and for new assumptions. All steps are interactive.

In this draft, the models are graphically represented as labelled boxes with plugs; generic name, equations, hypothesis and validity clauses are associated to the box. The plugs give the input-output description. They are qualified by their kind: input, output, variables, or parameter, by their name and type in the general data structure. Some plugs are of array type. For example, thermal zone balance model has a plug named "temperature of internal surface of surrounding wall". This plug refers every wall whose one surface encloses the zone. When the model will be instantiate, i.e. connected to the actual zone, then the instantiation function requests the project description for the list of concerned walls. In this draft, they refer also to a list of models that have a plug of this type. For instance, the plug "temperature of internal surface of surrounding wall" referring to an object "wall", by the way wall/surface/temperature, is connected to a model list that may be: (thermal zonal sensible balance.thermal zonal enthalpic balance. wall conductive transferts.surface energy balance.nil).

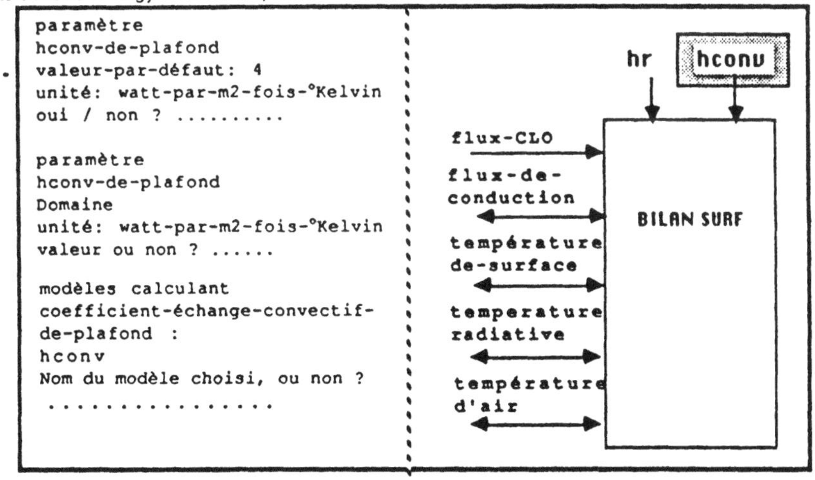

One of the difficulties is the right order of the list of models, that depends on the context. To avoid problems, it has been linked to the model plug and not to a general plug dictionary. The knowledge about "how to connect the plug" is locally stored, and that may cause problem when adding a new model in the model base.

LESSONS FROM THIS DRAFT

These implementation choices, and the algorithm can be discussed. In particular, human modeller does not consider variables one by one, but probably as a whole, and they reason about problems. L. LARET is writing a paper about the model building.

It is worth stating at this point that this draft does not have the purpose to give a new solution, but to show what are the important points of proforma handling. This draft allowed us interesting experiences and raised some questions.

Among them, we will citate some points:

- Although this proforma representation was a restricted one, it has been necessary to get the needed proforma filled in by thermal experts. And computer is very strict! Several working sessions were necessary to get this representation correct and complete enough. As benefits, our model concept, not an easy one to grasp, was enlightened, we gained better understanding between members of the team, and some specifications of the system were improved. These experiences raise again the need for thesaurus of the domain that everybody in the field would agree.

- Another interesting fact is the possible role of such a draft in a knowledge-acquiring methodology. While the expert is using the software, in fact, he is more interested with how the system works, than with the test of the software. Then, he is giving some fundamental comments about the way the model has to be built. The system acts like a mirror. Despite the weaknesses of the draft interfaces, we may obtain some precious informations, invaluable help to grasp the definition of the concept of "demarche" (i.e. modelling processes) that is central to the modelling knowledge.

- Some questions about proforma have been raised. They are under consideration, in the purpose to improve the proforma definition.

- The model building method raises the problem of time, and memory representation, because the built network "seems" to be an instantaneous system of equations. This problem overcomes the limits of this paper. It has been already mentioned in [OREN 1979] and other publications.

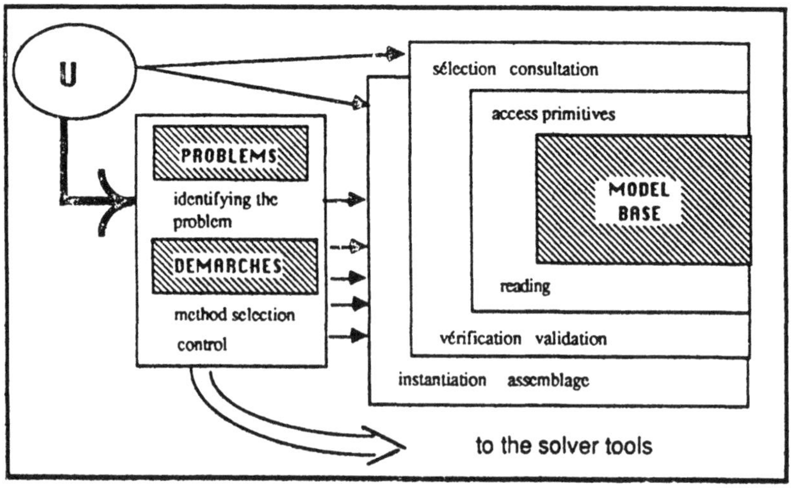

CONCLUSION

The PROFORMA and MODELOTHEQUE ideas seem to be very useful ones. In doing this draft, we raised some interesting and unexpected questions.

Our short-range project is to improve the interfaces of the draft, and to pursue some tests and try to solve some real problems. Among the problems of this part, a difficult aspect is the explicit representation of time dependancies. Of course, this work concems only a part of the job. Besides, we are working on the building of knowledge bases for the prior steps: problem base, "demarche" base., and informations exchanges between them and the "Modelotheque: the link could be the project data base, like a fact base. Thermal modelling is the competence domain of our team: then, for the other aspects, we think the best is to work with domain specialists.

When widening our horizon, we met other domain researchers, and we think we have many thinks to learn from them. Model classification, modelling methodology, and others discussions of Simulation Society, or other simulation, and A.I researchers will be very useful to us.

BIBLIOGRAPHY :

[ALMETH 1987] Rapport d'avancement sur le travail du groupe ALMETH, mars 1985-mars1987.

[ADDANKI SANJAYA, DAVIS 1985] "A representation for complex physical domains". 9ème IJCAI,LOS ANGELES, Août 1985.

[BOBROW 1984] "Qualitative Reasoning about Physical Systems", North-Holland,1984.

[CLARKE, LARET 1984]. "Explanation on the Data Processor Proforma", ABACUS (Glasgow) et LPB (Liège), Working Document, December 1984.

[DUBOIS 1986] "Vers une base de connaissance en modélisation thermique du bâtiment : éléments d'analyse d'une modélothèque", Rapport de recherche CSTB, MGL/609, octobre 1986.

[ISI 1987] ANSIDEI, J-P. "Maquette d'un système d'aide a l'assemblage de modèles thermiques du bâtiment". C.S.T.B., MGL/710, Août 1987.

[LARET 1986] "Synthesis report, 1983-1985 : Component Models Level 1", Agence Internationale de l'Energie, Annexe 10 (System Simulation), Avril 1986.

[LARET 1987] "A future for building energy modelling : the problem of knowledge representation and processing", Workshop on Future Building Energy Modelling, ISPRA, november 1987.

[LARET 1988] "An attempt for adapted model generation methodology : example of application in building energy modelling", in : 3rd Int. Intern. Symp. Systems analysis and simulation, BERLIN, september 1988.

[OREN 1979] "Concepts for Advanced Computer Aided Modelling", Methodology in Systems Modelling and Simulation. North-Holland, 1979.

[SORNAY, LEBRU, BAZALGETTE 1985] CSTBât - Version 1.1 - Code de Simulation de la Physique du Batiment. C.S.T.B. MPE/429, Mai 1985.

[SORNAY 1985] "Recent developpements of building energy simulation tools in Europe", Building Energy Simulation, SEATTLE, August 1985.

[SORNAY, DUBOIS 1986] "Structuration et documentation des modèles en thermique des bâtiments : concept de Modélothèque et de Proforma", CSTB MPE/535, Avril 1986.

[SPRIET, VANSTEENKISTE 1982] Computer-Aided Modelling and Simulation. Academic Press, 1982.

[TRNSYS 1979] "A Transient System Simulation Program3, Solar Energy Laboratory, Univ. of Wisconsin. MADISON USA, 1979.

[ZEIGLER, DE WAEL 1985] "Towards a knowledge-based implementation of multifacetted modelling methodology", Artificial Intelligence applied to Simulation, GHENT, February1985.

MODELS FOR LAKE EUTROPHICATION

Tibor Kutas[1/]

1. Introduction

Eutrophication is the answer of an ecosystem on increased nutrient loading. In the case of lake ecosystems it can threaten the ability of the lake for recreation or for using its water as drinking water. For meliorating the water quality the first step is the understanding the system to be modelled, the second step is forecasting its future in different nutrient loading conditions, revealing the response of the lake on certain measures. The first goal needs detailed models, large amount of state variables and processes, etc. while for fulfilling the second goal far simpler models are needed. At present the majority of models are deterministic, here one possible way of constructing stochastic model and its adventage is discussed.

In the process of understanding an ecosystem behaviour, to quantify the processes an important tool is the mathematical model describing the mass flows in space and among somehow defined compartments. At first step this has to be a deterministic model - for its reasons see later - simulating daily changes in the ecosystem. In most cases these are systems of ordinary or partial differential or difference equations. They can reveal parts in the ecosystems which are important in present and forecasting th future but the information is not enough for modelling their quantitative behaviour. This can be a lack of structural information /e.g. which phosphorus forms can be found in the sediment of a lake/ or dynamical information /e.g. which phosphorus forms are available for algae in a lake/ or quantitative one /e.g. how deep the active phosphorus layer in the sediment/. Cooperating with ecologists a model can give advices or directions for field and laboratorial experiments. Sometimes the model can estimate such processes which are unmeasurable either in situ or in laboratory /e.g. the internal - coming from the sediment - nutrient load of a shallow lake/.

The disadvantage of these models is that they need too much parameters and some of them - depending upon the model and data - can't be estimated they can be only calibrated. These models can

[1/]Computer and Automation Institute of HAS, Budapest, XI. Kende u.13-17. Hungary simulate daily and seasonal changes in the ecosystem so maximal values of certain biomasses, concentrations or processes and structural information can be obtained from them. Depending upon the external loadings these models can forecast the ecosystem for some years too. If the process of eutrophication is in its developing period - which is the most interesting and the same time the most sensible one - these models can simulate the system for a restricted period of time as the structural changes are quite frequent and unpredictable and some parameters having been supposed constant are variable for longer period.

It's wellknown, that a complex biological - ecological system has an intrinsic stochastic behaviour, so from mathematical point of view a stochastic model is preferable. As stochastic models need more data and information as a first step the deterministic model is recommended but as second step soon after the first one the stochastic model is unavoidable.

For studying the long term behaviour of an ecosystem /e.g. in the case of forecasting the effect of a certain measure on the water quality for some decades/ one day as time step is too short, too much processes have to be involved into the model with large amount of parameters to be estimated. Models with time step one year are far simpler and in dynamic case they are able to follow the water quality for several years. Their stochastic counterpart can model the fluctuations, they can increase the efficiency of these simple models.

The case of Lake Balaton

The process of eutrophication reached Lake Balaton at the beginning of seventies. As thé lake is large it has been divided into four basins, they were modelled separatly, but the models of basins are the same either in structure or in parameters. They are concatenated by hydrological throughflow. Here the models of one basin is discussed.

For understanding the ecosystem series of deterministic models with time step one day had been constructed. In this case series meant that the model developed with ecological knowledge, a failure in model resulted in thinking after the problem and in changing parameter set or in changing model structure or process formulation. The final version of Balaton Eutrophication Model /BEM/ consists of 10 state variables: winter-spring, summer, autumn phytoplankton, blue-green algae, bacterioplankton, dead organic matter, dissolved inorganic phosphorus and nitrogen, organic matter and exchangeable phosphorus in sediment. The choice of state variables reflects some specific features of the lake or of the ecosystem. As the lake is shallow the sediment plays an important role in mass transport and it stores the incoming phosphorus. An other consequence of shallowness is that the water temperature varies from 0 up to 28°C, so several groups of algae had been selected based upon their optimal temperature. Zooplankton and fish don't play important role in this ecosystem, so they were neglected from the model. Using BEM we could simulate the seasonal changes in all the four basins which have different trophic states from mezotrophic to highly eutrophic one. Detailed description of the BEM and its biological background can be read in [2].

The second step in the process of modelling and the first step in finding new mathematical tools for ecosystem modelling was the developing the stochastic counterpart of BEM. The stochastic model is a Markovian pure jump process, the state variables and the processes to be modelled are the same that in deterministic one. To introduce this new model was simple, as it doesn't need further parameters and the assumptions made are very natural. These were the following: the probability of a state transition - which is the stochastic equivalent of a subprocess - is proportional to its deterministic rate; the state transitions are independent of each other; the probability of no change in a certain time interval can be obtained by subtracting the probability of sub-processes from unity. Simulation results were very surprising. The average of stochastic runs coincided the deterministic one more or less, but in the case of blue-green algae the maximum value was ten times higher than in deterministic run. Namely the yearly maximal biomass was 0.1 in deterministic model, which means that the water quality is acceptable; the maximum of maximal value over simulation runs were 1.0 in stochastic realizations, in this case the water quality is not acceptable. If the external nutrient load was reduced in simulation runs the average of stochastic simulations and the deterministic values decreased but the maximum of stochastic simulation remained or even increased. This uncertainty could be detected in two consecutive years where the meteorological and load conditions were similar but in the first one an algal bloom was observed and in the next one the algal biomass wasn't acceptable. Exact definition and detailed simulation results can be read in [3].

After understanding the major part of quantitative behaviour of the ecosystem and after government's decision to take measures for improving the water quality the importance of forecasting the long term behaviour came to the light. A new simple dynamic model has been developed which has only two state variables: biologically available phosphorus in the water and in the sediment. The first describes the water quality while the second one shows the stored nutrient in the lake. The modelling time step is one year, so this model gives two values for every simulated year, so it's easy to overlook its results. The only loading function of this model the incoming phosphorus, so several loading scenarios can be simulated even by a microcomputer. Detailed description and simulation results can be read in [1].

The last step in our modelling process was the stochastic counterpart of this simple model. The construction from the deterministic model was the same as in the case of BEM. The results are not so surprising than that in the daily model but this model can offer probability distributions for water quality of every simulated year, so improving the usability of this simple model.

Literature

[1] Herodek, S., Kutas, T., Heder, A.: A simple dynamic eutrophication model. MTA SZTAKI Working Paper MS/26, Budapest 1987.

[2] Kutas, T., Herodek, S.: Effects of load reductions on the water quality of a large shallow lake. Ecological Modelling, 39 /1987/ 85-99.

[3] Kutas, T., Tóth, J.: A stochastic model of phytoplankton dynamics in Lake Balaton. J. of Statistical Computation and Simulation, 21 /1985/ 241-264.

WAYS TO REDUCE TIME CONSUMPTION OF STATISTICAL MODELLING OF INTEGRATED CIRCUITS BY THE MONTE CARLO METHOD

A.I. Baškys, D.J. Zanevitčius, V.J. Gobis[1)]

Of the most widespread methods to determine parametric reliability the Monte Carlo method is the best to consider the specific character of integrated circuits (IC). Such common problems of statistical modelling as the inclusion of non-linearity of a circuit model, large number of random input variables and their statistical dependence for the Monte Carlo methods amount essentially to the problems of computation time and memory, i.e. time consumption.

The main way to reduce the time consumption of the method is the reduction of the time consumption of one test which involves the generation of input variable random vector and the calculation of input parameters of the circuit according to its mathematical model.

One of the ways to reduce the output parameter calculation time is to employ the models of the circuits constructed on the basis of the functions readily calculated by a computer. The paper suggests to use non-linear transistor models in which not an exponential function (Shockley model) but a factional-rational one is employed to describe the volt-ampere characteristics of p-n junction

$$I = I_0 U/(\Psi - U) \tag{1}$$

or polynomical

$$I = \sum_{i=0}^{m} A_i U_i \ , \tag{2}$$

where I_0 , Ψ , A are the parameters of the p-n junction, I is the current through the junction, U is the voltage applied to the junction. And if at high injection levels at which integrated transistors mainly operate the Shockley model loses its physical meaning, the model (1) is not devoided of physical significance.

Another way is to present the circuit model in a more convenient for a computer from than that of nowadays IC simulation programs, e.g. in terms of a separate subprogram. For lack of topological matrices the computation memory and time are saved in the process. The algorithm is suggested on the basis of which the arrangement of integrated circuit subprograms-models can be automated by analytical calculation

1) Academy of Sciences of the Lithuanian SSR, 232600 Vilnius, K.Poželos 52, USSR.

systems. The algorithm is meant for the case when the transistor model is constructed on the basis of the functions (1) and (2) [2] . If is based depends on (fig. 1) the splitting of the circuits into the similar type parts referred to as circuit elements (CE). There is a universal model of a circuit element presented in terms of mathematical expression. The model of each specific circuit element is obtained by substituting appropriate symbolic expressions into the universal model. By this means the models of all circuit elements are obtained, the totality of which is the circuit model. To realize the algorithm the system of analytical calculations "VIHRAN" is used allowing to obtain the circuit model, i.e. equation system in terms of the subprogram in the programming language FORTRAN.

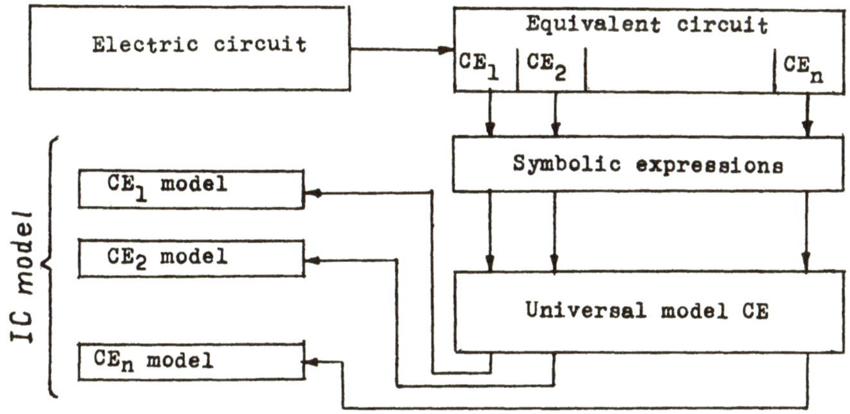

Fig. 1

With statistical modelling of IC it is essential that a host of random vector X realizations of circuit element parameters should be generated. The existence of the probable relations between the majority of vector X components requires to consider the correlative matrix R of the vector. We approximate the distribution of vector X with the multidimensional Gaussian distribution and cousider the fact that a lot of elements similar in geometric size are used in IC. We combine such element parameters of the same type into groups and presuppose the identity of parameter distribution in each group. Vector X broken down into groups takes the form

$$X = ((x_{11}, x_{12}, \ldots, x_{1n_1}), (x_{21}, x_{22}, \ldots, x_{2n_2}), \ldots, (x_{m1}, x_{m2}, \ldots x_{mn_m})).$$

Set of numbers $(n_1, n_2, \ldots, n_m)$ where $\sum_{i=1}^{m} n_i = N$ determines the breakdown of matrix R to blocks consisting of similar correlation coefficients ρ_{ij}, $i,j = 1, 2, \ldots, m$. It turns out that matrix with dimensions

$$Q = \begin{pmatrix} \rho_{11} + \frac{1-\rho_{11}}{n_1} & \rho_{12} & \ldots & \rho_{1m} \\ \rho_{21} & \rho_{22} + \frac{1-\rho_{22}}{n_2} & \ldots & \rho_{2m} \\ \ldots & \ldots & \ldots & \ldots \\ \rho_{m1} & \rho_{m2} & \ldots & \rho_{nm} + \frac{1-\rho_{nm}}{n_m} \end{pmatrix}$$

contains all the information about the block correlation matrix and can be employed to break down the latter into multipliers.

$$R = C \cdot C' \tag{3}$$

Where C' denotes the transposed matrix C. Such a break down is necessary to generate Gaussian correlated random values. [1] presents various variants of breakdown (3) on the basis of the analogous decomposition of matrix Q, which give matrix C with a large number of zeroth elements. This allows considerable saving of the time for generating circuit elements parameters and necessary computer memory for this.

References

1. Gobis V.J., Bashkys A.I., and Sakalauskas L.L.: The methods of generation correlated random values for statistical IC modelling. Avtomatizacija projektirovanija v elektronike. Kiev: Texnika, 1987, Nr. 35, p. 44-49.
2. Zanevitchius D.J., Bashkys A.I., and Sakalauskas L.L.: System modelling of integrated circuits. Proc. of the 2nd symp. IFAC. Systems analysis and simulation. Berlin, 1985, vol. II, p. 297-299.

MATHEMATICAL MODELS OF NON-LINEAR ELEMENTS IN THE THEORY OF SYSTEMS

D. Zanevičius

Theory of systems is known to investigate objects of various physical nature and to base itself on mathematical simulation and a physical experiment. Methodologically the theory of systems is constructed in terms of the abstract theory of the complex systems in which there is a host of interrelated elements. Mathematical models of elements determine in many ways the system model.

In the theory of systems the concept "mathematical model" of an element in distinction to "mathematical model" in mathematics, must conform with the principle of physical nature [1] . In this respect mathematical physics asks for still more rigid requirements here, the basis of model construction is formed by fundamental physics lows [2] .

Thus, from the gravitation law we shall obtain that the work necessary to remove the body from the centre of the Earth (R is the radius of the Earth) for a distance τ can be determined by $W = W_0(1-\frac{R}{\tau})$, where W_0 = const. Assuming the surface of the Earth as a reference system $H = \tau - R$ we obtain

$$\frac{H}{R} = \frac{\frac{W}{W_0}}{1-\frac{W}{W_0}} . \tag{1}$$

Turning to kinetic energy which depends on the initial velocity V_0 , from Binet equation if follows [3] that the motion trajectory will be in terms of a circle (the first cosmic velocity) when $V_0 = V_1 = \sqrt{gR}$, which is equivalent to $\frac{W}{W_0} = 0{,}5$ in (1); the motion trajectory will be as a parabola (the second cosmic velocity) when $V_0 = V_2 = \sqrt{2gR}$ which is equivalent to $\frac{W}{W_0} = 1$ in (1).

When the height (z) and distance (x) of flight are small in comparison with the radius of the Earth, gravitation of the Earth can be regarded as constant. Then the motion trajectory [3] will be obtained from the second Newton's law

$$z = z_0\left(1-\frac{x}{x_0}\right)\frac{x}{x_0} , \tag{2}$$

where z_0 = const., x_0 = const.

What is the place the rational functions (1) and (2) occupy

Academy of Sciences of the Lithuanian SSR, 232600 Vilnius, K.Poželos 52, USSR among a host of functions?

Already in the works of Euler the integral

$$x = \int_0^y \frac{d\varepsilon}{\sqrt{a_0 + 4a_1\varepsilon + 6a_2\varepsilon^2 + 4a_3\varepsilon^3 + a_4\varepsilon^4}} \qquad (3)$$

has been investigated.

"Inversion" of the integral (3) determines the elliptical sine $Y = sn\,x$. Of in (3) $a_0 = I$, $a_2 = \frac{1}{6}$, $a_1 = a_3 = a_4 = 0$ then he hyperbolic sine $y = sh\,x$ is obtained. When $a_0 = 1$, $a_2 = -\frac{1}{6}$, $a_1 = a_3 = a_4 = 0$.in (3), $y = \sin x$. However, if in (3) $a_0 = a_1 = a_2 = a_3 = a_4 = 1$ the parabolic sine $y = spx$ is obtained, where

$$spx = \frac{x}{1-x}, \quad cpx = \frac{1}{(1-x)^2}, \quad tpx = \frac{spx}{cpx}, \quad sapx = \frac{1}{1-x}, \quad x \leq 1. \qquad (4)$$

The parabolic sine can be determined in terms of the non-linear differential equation solution.

$$\frac{d^2\eta}{d\varepsilon^2} - 2(1+\eta)^3 = 0$$

or as a series of geometric progression. Function sp x has a geometric interpretation.

Transcond is the non-linear two-pole network the current-voltage characteristic of which is determined by

$$i = J_0\, sp \frac{u}{\psi_0},$$

where J_0, ψ_0 are the transcond parameters.

Transcond conductivity is $y = \frac{J_0}{\psi_0} sap \frac{u}{\psi_0}$. Non-linear two-pole network, the resistivities of which are

$$R = \frac{\psi_0}{J_0} sap \frac{u}{\psi_0}$$

is referred to as the antitranscond. Current-voltage characteristic of antitranscond is

$$i = J_0\, tp \frac{u}{\psi_0} \qquad (5)$$

If is significant that the relative power of the transcond $p' = \frac{iu}{J_0 \psi_0}$ is equal to one where $u' = \frac{u}{\psi_0} = 0{,}618...$ and

$i' = \frac{i}{I_0}$ = 1,618... i.e. when the relative current and voltage correspond to the numbers of Fibonacci.

If the principal characteristics of elements are determined by one of the fluctuations (4), such elements will be refered to as parabolic. The parabolic elements form rather a big class of elements in the theory of systems.

In the theory of electric field 4 the Coulomb's law (as well as the gravitation law in the gravitational field) founds the power to be inversely related to the square of the distance, therefore the work (voltage difference) which is being performed by the movement of a charge in a spheric condenser will be also determined by expression (1).

As it is known one of the ways to approximate the magnetization curve is $B = Arsp\frac{u}{a}$. The important element in the theory of systems is the return coupling. If transfer coefficient of the system without return coupling is k, and the transfer coefficient of the return coupling circuit is β, the transfer coefficient of the system with return coupling (positive force) is known to be

$$K_0 = \frac{1}{\beta} k\beta$$

Parabolic characteristics are exhibited by a number of other physical and chemical systems. Thus, the operation of a heat engine at adiabatic expansion (from volume V_0 to volume V, when the adiabat index α = 2) is determined by $W = W_0 Arsp\frac{V}{V_0}$

Paper [5] shows that both pn junction model and bipolar transistor model can be presented in terms of spx, and its equivalent circuit is presented to be on transconds.

I-V characteristic of a MOS transistor is known to be presented in terms of tg x, and its equivalent circuit is shown to be on antitransconds.

The attention should be given to some peculiarities of parabolic models. First, parabolic models unlike the transcendent ones are formed by the solution of non-linear differential equations. Second, algebraic nature of the models allows to use the programs of computer algebra, on the other hand the reduction of the computation time of system modelling allows to pass from the determined to statistical analysis [6] and to optimizing system parameters:

References

[1] Дружинин В.В., Конторов Д.С. Системотехника. Изд.Радио и связь. М., 1985.

[2] Вигнер Е. Непостижимая эффективность математики в естественных науках. Сб. Этюды о симметрии. Изд. Мир. М., 1971.

[3] Космодемьянский А.А. Курс теоретической механики, ч.I, Изд. Высшая школа. М., 1965.

[4] Калашников С.Г. Электричество. Изд. Наука., М., 1977.

[5] Занявичюс Д.Ю. Системный анализ интегральных схем. Математическое и машинное моделирование в микроэлектронике (ММММ-86). Научный совет АН СССР по комплексной проблеме "Математическое моделирование", Академия наук Литовской ССР Институт физики полупроводников. Вильнюс, 1987, с.16-22.

[6] Baschkys A., Zanevitchius D. Statistic modeling of integrated circuits Prcceedings of the 2nd IFAC Symposium Stochastic Control. Pergamon Press Oxford, 1986, p. 461-466.

The Medial Section as a Proportion for the Thermodynamical Equilibrium of Selforganizing Systems

A. P. Stachov[1]

A fundamental role in living systems is beeing played by the conception of thermodynamical equilibrium. In the paper /1/ the state of thermodynamical equilibrium is defined as such state of a system in which the value of the thermodynamical probability (and the entropy respectively) of the system remains unchanged under random fluctuations of the elements.

The following notations are used:

n = number of elements of the system (e.g. atóms or molecules)

m = number of classes into which the elements fall (e.g. energy niveaus for atoms or molecules)

$n_1, n_2, \ldots n_m$ = number of elements in the corresponding class.

As shown in reference /1/ the following relations exist between the number of elements in neighbouring classes:

$$\frac{n_m}{n_{m-1}} = \frac{n_{m-1}}{n_{m-2}} = \qquad = \frac{n_3}{n_2} = \frac{n_2}{n_1}$$

It can be easily be seen /2/, that these relations correspond to the well known medial section:

$$\alpha = \frac{1+\sqrt{5}}{2},$$

i.e.

$$\frac{n_m}{n_{m-1}} = \frac{n_{m-1}}{n_{m-2}} = \ldots = \frac{n_2}{n_1} = \alpha = \frac{1+\sqrt{5}}{2}$$

In this paper we derivate a formula for the entropy of a system which is in the state of thermodynamical equilibrium using yhe medial section /3/:

$$S_p = A \cdot n \qquad (1)$$

Here A is a physical constant for thermodynamical equilibrium given by

[1]Politechnical Institute
286021 Vinnitza, Chmelnitzkoye Chaussee 133, USSR

$$A = k \cdot \ln\left[(2+\sqrt{5})\,\alpha^{\frac{1}{\alpha}}\right] \qquad (2)$$

with k = Boltzmann-constant

and α = medial section

From $S_p \leq S_{max} = k \cdot n \cdot \ln(m)$ follows for the number of classes m:

$$m \leq (2+\sqrt{5})\,\alpha^{\frac{1}{\alpha}} \qquad (3)$$

The physical interpretation of that inequality allows to define the minimal number of classes $m_{min} = 6$ for which the system is able to pass over into the state of thermodynamical equilibrium.

The main conclusion drawn in this paper is that the value of entropy of a system in the state of thermodynamical equilibrium grows lineary with rising n and does not depend on the number of classes m of under the condition that $m \geq 6$.

Supposing that:

a) complex systems in living and nonliving nature show the effect of selforganization which inhibits the growth of the entropy of the system and

b) selforganization of a system is connected with the transition into the state of thermodynamical equilibrium

than the results described in this paper can be used for the following concretisation of the 2nd law of thermodynamics:

An arbitrary isolated system for which the number of classes is not less than 6 tends to reach the state of thermodynamical equilibrium in which the ratio between the numbers of elements in neighbouring classes is equal to the medial section.

Reaching the state of thermodynamical equilibrium the systems entropy remains a constand value given by the formulae (1) and (2) which does not depend on changes of the number of classes as long as the inequality $m - 6$ holds.

References:

1. Glaser R. Einführung in die Biophysik, Gustav Fischer Verlag, Jena, 1971, S. 301.
2. Grzedzielski J. Energetyczno-geometrycny kod prirody, Warschau , 1986, S. 136.
3. Stachov A.P. Kody solotoy proporzii, Moskwa, Radio i swyas, 1984, c. 153.

RECENT ADVANCES IN DECISION TECHNOLOGIES FOR MANAGEMENT

Madan G. Singh[1)]

ABSTRACT

In this paper we provide an overview of the Decision Making Problems which face companies in a complex competitive environment. We then explore the technologies of expert systems, decision support systems and a hybrid between the two as potential tools for helping managers to tackle these decision making issues. Finally, a number of specific decision technology elements developed by the author are described and one of these is illustrated for helping with the solution of an important problem in Retail Banking i.e. determining interest rate mixes for savings and lending products in a competitive environment.

INTRODUCTION

The Business world today is very complex due to the:

- o Globalisation of Markets
- o Explosion of Technology
- o Short product life cycles
- o Aggressiveness of the competition

There is a clear recognition of the importance of information to survive and prosper in a complex business environment by taking "smarter" Decisions. This raises a number of questions:

- o How can Decision Makers take "smarter" decisions? The information explosion provides "too much" information, much of it inaccurate or irrelevant!
- o What technologies exist today or are coming to the market soon to help Senior Managers to take better decisions.
- o What kinds of tools are appropriate for which kinds of decisions?

This paper attempts to answer some of the above questions by examining the nature of the decisions which face senior decision makers in Industry and Commerce, the information requirements for these decisions, the technology of decision making which could help to improve the abilities of Senior Decision makers and the increasing use of these technologies by go ahead companies.

In the next section, we outline the decisions which require knowledge support, then the tools which are emerging to help the Decision Maker and, finally, we give an illustrative example of the use of these aids in the field of Retail Banking.

Professor of Information Engineering, U.M.I.S.T., Manchester, U.K.

MANAGERIAL DECISION MAKING PROBLEMS

Vast numbers of interrelated decisions are made by organisations. It is possible to group these in various ways e.g. through a hierarchy of impact, time horizon of interest, data needs, degree of structure, etc.

Anthony (1965) describes four levels of Decision Making.

- o Strategic Planning Decisions: These are related to choosing the highest level policies and objectives and associated resource allocations.
- o Management Control Decisions: These are decisions made for the purpose of assuring effectiveness in the acquisition and use of resources.
- o Operational Control Decisions: These are decisions made for the purpose of assuring effectiveness in the performance of operations and
- o Operational performance Decisions which are the day to day decisions made while performing operations.

Simon (1960) has described decisions as structured or unstructured depending on whether the Decision Making Process can be explicitly described prior to the time when it is necessary to make a decision. Generally, as Sage points out (Sage 1987) operational performance decisions are more likely than strategic planning decisions to be prestructured. Thus

- o Expert Systems can usually be expected to be more appropriate for operational performance and operational control decisions than they are for strategic planning and management control decisions.
- o DSS will be more appropriate for Strategic Planning and Management Control than for operational control and operational performance.

Note, however that expertise is a relative term which depends upon familiarity with the task and the operational environment into which it is imbedded. Since decision environments do change and since novices become experts through learning and feedback, clearly, there will exist many areas where the proper form of knowledge Base Support is a Hybrid of a Decision Support System and an Expert System.

Hybrid Knowledge Based Systems represent a new generation of Business Decision Technologies for tactical and, to some extent, Strategic Decision Making. They bring together ideas from Systems Engineering , Operations Research, Management Science and Cognitive psychology.

To understand the precise nature of the decisions for which one is seeking knowledge support, let us consider a typical manufacturing hierarchy as shown in Fig. 1. Hybrid Knowledge based Systems would have a major role to play at the Tactical Decision Making level in order to help with issues like:

- o Product Pricing
- o Sales Force Allocation
- o Marketing Budget Allocation

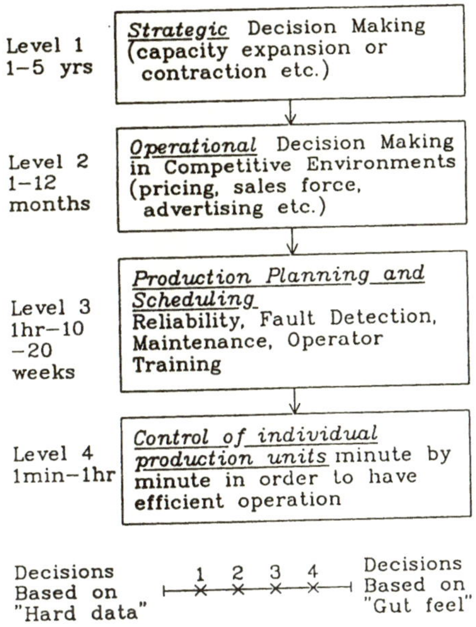

Fig.1 Decision making hierarchy in manufacturing

Fig. 2 shows a similar hierarchy in the Retail Banking Sector. Here, Hybrid Systems would have a major role to play at the operational or tactical decision making levels in order to help with issues like:

- o Interest Rate Mixes for Savings and Lending products
- o Pricing of Services
- o Marketing Budget Allocation.

Similar hierarchies can be developed for any other industry.

BUSINESS DECISION TECHNOLOGIES FOR THE HIGHER DECISION MAKING TASKS

Expert Systems and Decision Support Systems

So far, we have highlighted the various kinds of decisions which Senior Managers make and pointed out that there is a spectrum of Decision aids to help the manager which go from Expert Systems to Decision Support Systems as we go up the managerial decision making hierarchy. We note that

(a) In the case of Expert Systems:
- o The user is generally a novice
- o Such systems are primarily of value in diagnostic types of applications

(b) In the case of Decision Support Systems, the user is assumed to be a Senior Manager or a group of Senior Managers who might be assisted by a facilitator.

Fig. 3 shows the structure of Hybrid Knowledge Based Systems which attempt to marry together Expert Systems and Decision Support Systems.

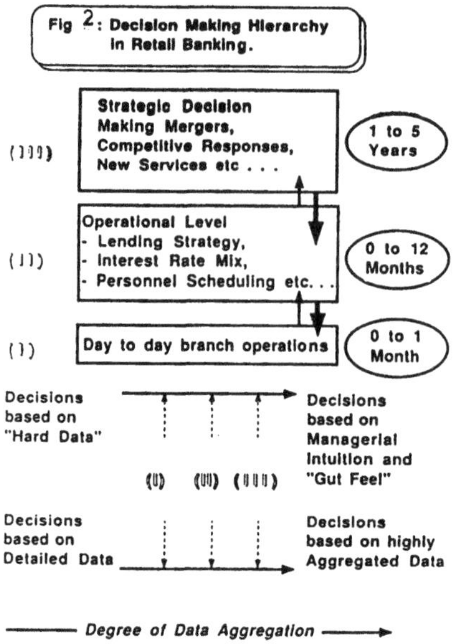

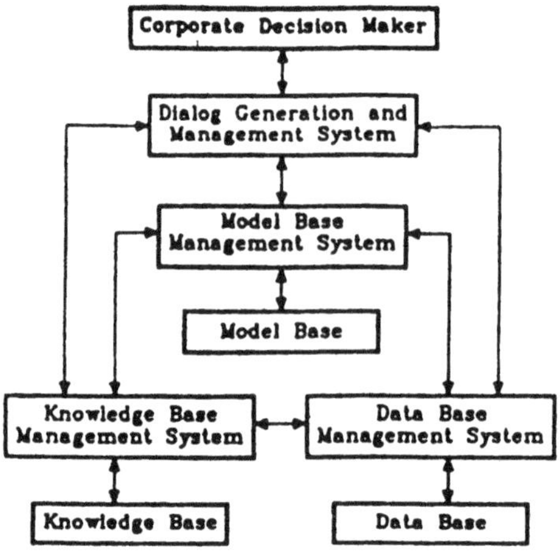

Fig. 3 Structure of Hybrid Knowledge Based Systems (HKBS)

<u>Hybrid Knowledge Based Systems</u>

A number of Hybrid Knowledge Based Systems have been developed in UMIST by the author to help companies to tackle key specific problems in different industries e.g.

- <u>Price-Strat</u>: A system for helping companies to price their products in competitive markets [Singh (1988)][7]
- <u>BS-opt</u>: A system for Banks and Building Societies to determine interest rate mixes in the face of competitors. [Singh and Cook (1986)[3], Singh and Cook (1987)[4], Singh and Cook (1988)[6]
- <u>Resource-opt</u>: A system for
  - Retail Space Allocation
  - Sales Force Allocation Singh, Cook and Corstjens (1988)[5]
- <u>Mark-Mix</u>: A system for marketing budget allocation and sizing

and for Strategic Decision Making: <u>Mark-opt</u> (Singh, Bhondi and Corstjens (1985)[8] and <u>Pharmplan</u> (Corstjens, Bhondi and Singh (1986)]

In each of these systems, a three phase process is used as shown in Figs. 4,5 and 6. In phase 1, the system extracts expertise from Senior Managers about the market and competitors through their responding to a series of "what-if" scenarios. These are combined with historical data and market research data to make up an initial <u>Market Knowledge Base</u> (Fig. 4).

In phase 2, an inference mechanism interrogates the knowledge base to determine the profit maximising decisions and their sensitivity. Sales and profit forecasts are also provided (Fig. 5).

In phase 3 (Fig. 6), new data which emerges after the application of the

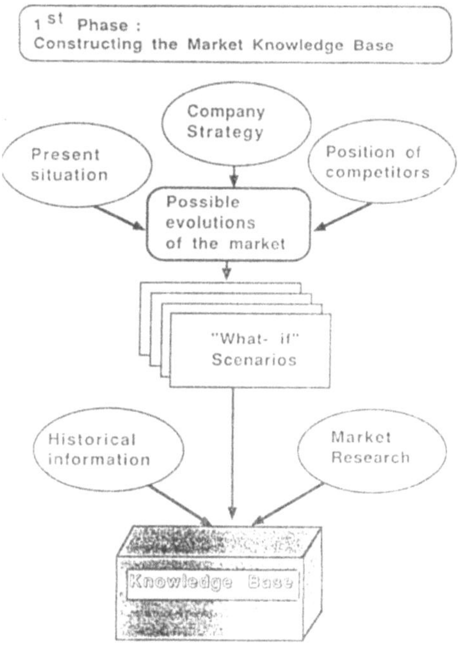
1 st Phase :
Constructing the Market Knowledge Base
Company Strategy
Present situation
Position of competitors
Possible evolutions of the market
"What- if" Scenarios
Historical information
Market Research
Knowledge Base

Fig. 4

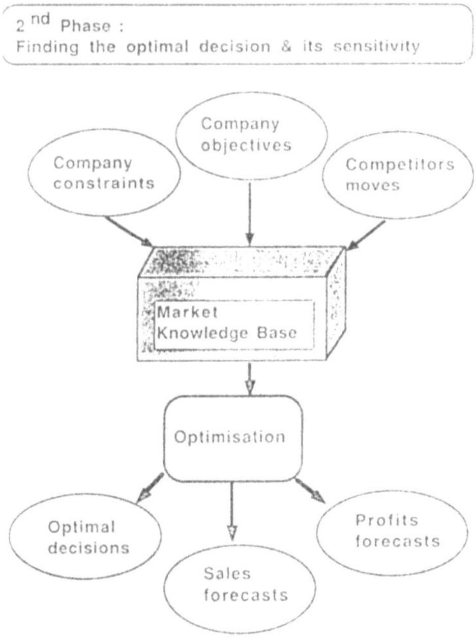
2 nd Phase :
Finding the optimal decision & its sensitivity
Company objectives
Company constraints
Competitors moves
Market Knowledge Base
Optimisation
Optimal decisions
Sales forecasts
Profits forecasts

Fig. 5

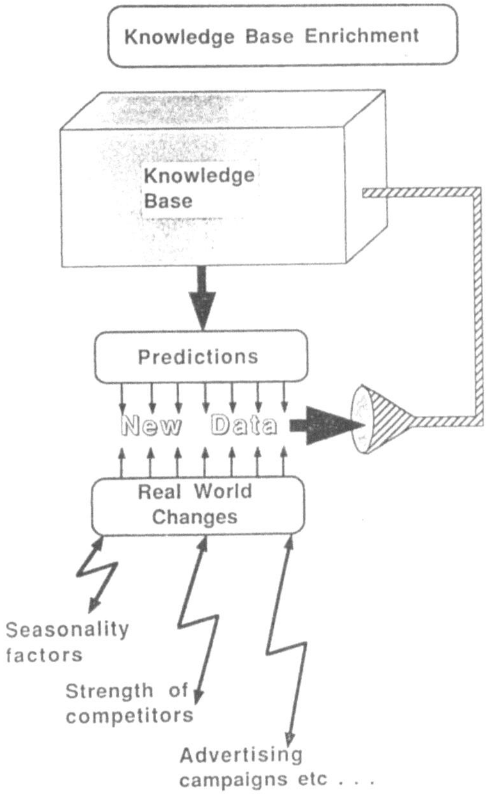

Fig. 6

profit maximising decisions, is used as a reality check and the system uses this to learn about the environment and to enrich its knowledge base. Thus, decision making is improved over time.

EXAMPLE OF HKBS FOR BANKS: BS-OPT SYSTEM

The problem here is to determine a Bank's most profitable interest rates for Saving and Lending products so as to maximise profitability in a competitive environment whilst meeting strategic objectives.

We note that at the AGGREGATE LEVEL (Industry or Economy) Pricing Research has been in the forte of Economists. For the firm, such analysis is useful in determining STRATEGY.

In the Management Science literature, theoretical research on Price Setting for a firm is Very Sparse.

Current State of the Art: Seat of the pants approach!

For all pricing problems we need to know the firm's Long Term STRATEGY, PRICE-POSITIONING, COMPETITOR OFFERING (now and in the future) and to estimate DEMAND and COST STRUCTURES.

Difficulties in tackling the interest rate mix problem

There are a number of difficulties which arise when attempting to tackle the interest rate mix problem e.g.

- o These are dynamic problems
- o The historical date set is very meagre and often of doubtful accuracy
- o We need to know both past and future competitor decisions: will their future decisions be like those in the past?
- o We need to integrate our long term strategy with short term decision making on interest rates such that our long term strategic aims are fulfilled and, at the same time, our short term profit is maximised. Of course, this needs to be done within a non-cooperative competitive context.

Traditional Aproaches

The traditional way of tackling this problem is to use econometric analysis to build a dynamic model based on historical data. Because of the meagreness of this data set, and the difficulties of predicting future competitor decisions, there is no literature on the successful solution of this problem.

The BS-opt Hybrid Knowledge Based Systems Approach has succeeded in getting around most of the difficulties to produce a user friendly system to help in this decision making which runs on a desk top computer.

New Ideas used in BS-opt

The BS-opt System gets around most of the problems by using new ideas from Control theory, Artificial Intelligence and Management Science. In particular, the Dynamic problem is decomposed using the notion of Singular perturbations into a slow and a fast part. This allows one to consider a short term problem where competitor interest rates can be fixed and a long term problem where the slowly changing factors can be tracked.

The meagre historical data set problem is resolved through ideas taken from artificial intelligence where structured interviews are used to elicit "what-if" knowledge from a group of managers. This knowledge is represented in such a way that optimisation can be carried out in order to produce profit maximising interest rate mixes.

The B.S. Opt System for Banks

Fig. 7 shows the structure of the system. Essentially, a Knowledge Base is initially constructed (in a few days) into which the expertise of several managers has been fed in through their responding to a series of "what-if" interest rate scenarios as shown in Fig. 4. The Knowledge Base is enriched with market research information and any historical information which may be available. The Bank also feeds in its long term strategic aims and these are converted into the appropriate short term goals for the current period. Starting balances are also fed in. The system then calculates the best mix of rates for the Bank's products so as to increase the weighted average

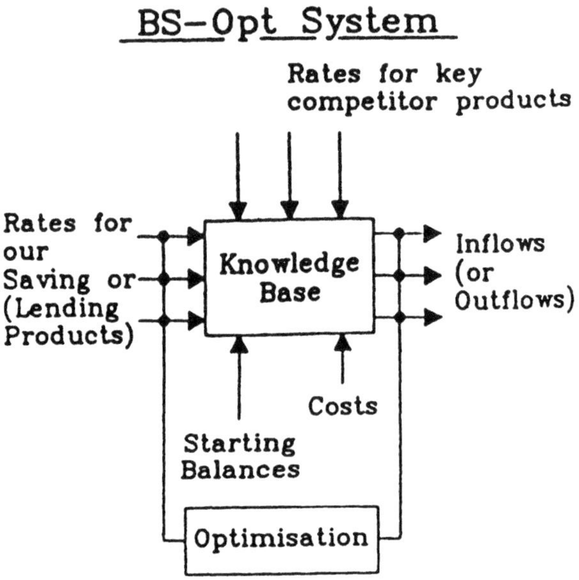

Fig. 7

rate for lending products and decrease it for savings products given a set of key competitor rates. If the latter change , the Bank's best new rate structure can be recalculated within a matter of minutes.

As an example, the system was tried for an institution with three savings products (named higher, five star, seven day) where the current rates were 9.75, 9.25, 8.85. At these rates, expected inflows/outflows from these accounts were respectively £46million, £57million and minus £34million. The longer term strategic considerations were converted into strategic interest rate ranges and those with the starting balances were used to initialise the knowledge base. The Knowledge Base itself was constructed (within a few hours) and the Price-Strat System was set the task of attracting a similar end balance to the current one (£2068million) at a cheaper weighted average rate.

Within a matter of minutes, the system calculated the optimal rate structure. In the current, the weighted aveage rate for having a closing balance at the end of the month of £2068million was 9.39%. The optimal rate structure of 9.63, 9.38, 9.60 yielded the same closing balance at a weighted average rate of 9.33 i.e. a saving of 6 basis points or £1.2million on an annualised basis.

CONCLUSIONS

In this paper, we have given an overview of the approaches which are being developed to help Senior Decision Makers in industry and commerce to take better decisions. After outlining the Decision Problems facing Managers and the technological tools available to help them, we highlighted one area of special interest i.e. tactical decision making within a competitive environment where the proper form of decision support was a hybrid of Decision Support Systems and Expert Systems. The methodology behind a number of these Hybrid Knowledge Based Systems developed by the author was outlined and the approach has been illustrated on a key problem i.e. determining interest rate mixes in Retail Banking.

Decision Technologies, like HKBS, Expert Systems, Decision Support Systems, etc. are being increasingly used by go ahead companies to improve their profitability in a competitive environment.

REFERENCES

1. Anthony, "Planning and Control Systems: a framework for analysis" Harvard University Press, Cambridge, Mass, 1965.

2. A.P. Sage "Decision Support Systems" in Singh M.G. (editor) Encyclopedia of Systems and Control, Volumes 1-8, Pergamon Books Ltd. 1987.

3. M.G. Singh and R. Cook "Price-Strat: a decision support system for determining Bank and Building Society Interest Rate Mixes", International Journal of Bank marketing, Vol 4,4,1986.

4. M.G. Singh and R. Cook "Improving the Profitability of Bank Lending in a Competitive Environment" International Journal of Bank Marketing, Vol. 5.3, 1987.

5. M.G. Singh, R. Cook and M. Corstjens "A Hybrid Knowledge Based System for Retail Space and Other Allocation Problems" Interfaces, May 1988.

6. M.G. Singh and R. Cook "Decision Support Systems for Assets and Libaility Management in Retail Banking", Quarterly Journal of the Girocentrale Bank, Austria, 1988.

7. M.G. Singh "Price-Strat: A Hybrid Knowledge based System for Competitive Pricing", Proceedings IX Euro TIMS, Paris, 1988.

8. M.G. Singh, J. Bhondi and M. Corstjens "Mark-opt: A Negotiating Tool for Manufacturers and Retailers", IEEE Trans SMC, Vol. 15, 14, 1988, pp. 483-495.

Band 1

Inverse and Improperly Posed Problems in Differential Equations

Proceedings of the Conference on Mathematical and Numerical Methods, held in Halle/Saale (GDR) from May 29 to June 2, 1979

Edited by Gottfried Anger
296 Seiten, 1979 (vergriffen)

Band 2

Fundamentals of Computation Theory FCT '79

Proceedings of the Conference on Algebraic, Arithmetic and Categorial Methods in Computation Theory, held in Berlin/Wendisch-Rietz (GDR) September 17–21, 1979

Edited by Lothar Budach
576 Seiten, 1979

Band 3

Decidability and Generalized Quantifiers

By Andreas Baudisch, Detlef Seese, Hans-Peter Tuschik, and Martin Weese
XII, 235 Seiten, 1980

Band 4

Zur Elastizitäts- und Thermoelastizitätstheorie

Potentialtheoretische Methoden in Anwendung auf verschiedene Modelle

Von Lothar Jentsch und Johannes Maul
157 Seiten, 1980

Band 5

Systems Analysis and Simulation 1980

Proceedings of the International Symposium held at Berlin (GDR) September 1–5, 1980

Edited by Achim Sydow
496 Seiten, 1980 (vergriffen)

Band 6

Nichtlineare diskrete Optimierung

von Eberhard Girlich und Michail M. Kowaljow
218 Seiten, 1981

Band 7

Completeness Problems in the Structural Theory of Automata

By Jürgen Dassow
148 Seiten, 1981

Band 8

Darstellung von Lösungen linearer elliptischer Differentialgleichungen

Von Günther Wildenhain
92 Seiten, 1981

Band 9

Introduction to the Spectral Theory of Operators in Spaces with an Indefinite Metric

By I. S. Iohvidov, M. G. Krein, and H. Langer
120 Seiten, 1982 (vergriffen)

Band 10

Nonlinear Time-discrete Systems

A General Approach by Nonlinear Superposition

By Michael Gössel
115 Seiten, 1982 (vergriffen)

Band 11

Preprimal Algebras

By Klaus Denecke
162 Seiten, 1982

Band 12

Ergodic Theory and Related Topics

Proceedings of the Conference held in Vitte/Hiddensee (GDR) October 19–23, 1981

Edited by Horst Michel
232 Seiten, 1982

Band 13

Singularly Perturbed Differential Equations

By H. Goering, A. Felgenhauer, G. Lube, H.-G. Roos, and L. Tobiska
176 Seiten, 1983

Band 14

Leben wir in einer Volterra-Welt?

Ein ökologischer Zugang zur angewandten Systemanalyse

Von Manfred Peschel und Werner Mende
233 Seiten, 1983

Band 15

Markov Decision Problems with Countable State Spaces

Optimality Criteria – Algorithms – Clustering

By Hans Michael Dietz and Volker Nollau
174 Seiten, 1983

Band 16

Stochastische Verfahren zur Suche nach dem Minimum einer Funktion

Von Ryszard Zieliński und Peter Neumann
133 Seiten, 1983

Band 17

Invertibility and Asymptotics of Toeplitz Matrices

By Albrecht Böttcher and Bernd Silbermann
200 Seiten, 1983

Band 18

Structural Induction on Partial Algebras

Introduction to Theory and Application of Partial Algebras – **Part II**

By Horst Reichel
205 pages, 1984

Band 19

Algebraic Methods for Toeplitz-like Matrices and Operators

By Georg Heinig and Karla Rost
212 pages, 1984

Band 20

Frege Conference 1984

Proceedings of the International Conference held at Schwerin (GDR), September 10–14, 1984

Edited by Gerd Wechsung

408 pages, 1984

Band 21

Numerical Analysis of Selected Semilinear Differential Equations

Edited by Christian Großmann, Thomas Riedrich, and Hartmut Schönheinz

203 pages, 1984

Band 22

Singular Ordinary Differential Operators and Pseudodifferential Equations

By Johannes Elschner

200 pages, 1985

Band 23

Lotka-Volterra-Approach to Cooperation and Competition in Dynamic Systems

Proceedings of the 5th Meeting of UNESCO's Working Group on Systems Theory held on the Wartburg, Eisenach (GDR) at March 5–9, 1984

Edited by Werner Ebeling and Manfred Peschel

271 pages, 1985

Band 24

Convergence Structures 1984

Proceedings of the Conference on Convergence held in Bechyně (Czechoslovakia), September 24–28, 1984

Edited by Josef Novák, Werner Gähler, Horst Herrlich, and Jan Mikusiński

254 pages, 1985

Band 25

Parcella '84

Proceedings of the II. International Workshop on Parallel Processing by Cellular Automata and Arrays held in Berlin (GDR), September 25–27, 1984

Edited by Wolfgang Händler, Tamás Legendi, and Gottfried Wolf

239 pages, 1985 (vergriffen)

Band 26

Multiobjective and Stochastic Optimization Based on Parametric Optimization

By Jürgen Guddat, Francisco Guerra Vasquez, Klaus Tammer, and Klaus Wendler

175 pages, 1985

Band 27/28

Systems Analysis and Simulation 1985
I: Theory and Foundations
II: Applications

Proceedings of the International Symposium held in Berlin (GDR), August 26–31, 1985

Edited by Achim Sydow, Manfred Thoma, and Robert Vichnevetsky

per volume 495 pages, 1985

Band 29

Parcella '86

Proceedings of the III. International Workshop on Parallel Processing by Cellular Automata and Arrays held in Berlin (GDR), September 9–11, 1986

Edited by Tamás Legendi, Dennis Parkinson, Roland Vollmar, and Gottfried Wolf

279 pages, 1986

Band 30

Fuzzy Sets Applications, Methodological Approaches, and Results

Proceedings of the International Workshop on Fuzzy Sets Applications cosponsored by IIASA and convened in cooperation with IAEA held on the Wartburg, Eisenach (GDR), at March 3–8, 1985

Edited by Steffen Bocklisch, Sergei Orlovski, Manfred Peschel, and Yasushi Nishiwaki

233 pages, 1986

Band 31

Mathematical Methods of Specification and Synthesis of Software Systems '85

Proceedings of the International Spring School, held in Wendisch-Rietz (GDR), April 22–26, 1985

Edited by Wolfgang Bibel and Klaus P. Jantke

245 pages, 1986 (vergriffen)

Band 32

A Model Theoretic Oriented Approach to Partial Algebras

Introduction to Theory and Application of Partial Algebras – Part I

By Peter Burmeister

319 pages, 1986

Band 33

Finite Difference Method on Irregular Networks

A Generalized Approach to Elliptic Problems of Second Order

By Bernd Heinrich

206 pages, 1986

Band 34

Modern Algorithms for Large Sparse Eigenvalue Problems

By Arnd Meyer

125 pages, 1987

Band 35

Parametric Optimization and Related Topics

Edited by Jürgen Guddat, Hubertus Th. Jongen, Bernd Kummer, and František Nožička

407 pages, 1987

Band 36

Discretization in Differential Equations and Enclosures

Edited by Ernst Adams, Rainer Ansorge, Christian Großmann, and Hans-Görg Roos

250 pages, 1987

Band 37

Factorization of Measurable Matrix Functions

By Georgi Semenovich Litvinchuk and
Ilya Matveyevich Spitkovskiĭ

372 pages, 1987

Band 38

Parallel Algorithms and Architectures

Proceedings of the International Workshop
on Parallel Algorithms and Architectures
held in Suhl (GDR), May 25–30, 1987

Edited by Andreas Albrecht, Hermann Jung,
and Kurt Mehlhorn

205 pages, 1987

Band 39

Parametric Integer Optimization

By Bernd Bank and Reinhard Mandel

138 pages, 1988

Band 40

Computer Analysis of Images and Patterns 1987

Proceedings of the II. International Conference
CAIP '87 on Automatic Image Processing,
held in Wismar (GDR), September 2–4, 1987

Edited by Leonid P. Jaroslavskiĭ,
Azriel Rosenfeld, and Wolfgang Wilhelmi

282 pages, 1987

Band 41

Multivariable Control

A Graph-theoretic Approach

By Kurt Reinschke

274 pages, 1988

Band 42

Equilibrium Distributions of Branching Processes

By Alfred Liemant, Klaus Matthes, and Anton
Wakolbinger

240 pages, 1988

Band 43

Andreotti-Grauert Theory by Integral Formulas

By Gennadi Markovich Henkin and
Jürgen Leiterer

270 pages, 1988

Band 44

Continual Means and Boundary Value Problems in Function Spaces

By Efim Mikhailovich Polishchuk

160 pages, 1988

Band 45

Advances in Mathematical Optimization

Invited papers dedicated to Prof. Dr. Dr. h. c.
F. Nožička on occasion of his 70th birthday

Edited by Jürgen Guddat et al.

232 pages, 1988

Band 46/47

Systems Analysis and Simulation 1988

I: Theory and Foundations
II: Applications

Proceedings of the International Symposium
held in Berlin (GDR),
September 12–16, 1988

Edited by Achim Sydow, Spyros G. Tzafestas,
and Robert Vichnevetsky

428 pages per volume, 1988

Systems analysis and simulation become evident in an increasing degree as an effective method and tool for mastering large complex systems. Among the scientific foundations methods for qualitative analysis, for decision support systems, modelling, control, and simulation of large-scale systems, respectively, knowledge processing for systems analysis and simulation by means of parallel processes are mostly developed. Object-oriented simulation languages and -systems are useful conditions for an effective problem solving of model user and designer.

This volume presents newest applications of modelling, simulation, and decision making concerning the following fields:

- Environmental problems
- Forestry
- Water systems
- Industrial automation
- Robotics
- Industrial and technical systems
- Agricultural systems
- Economic systems
- Energy systems
- Medical and biological systems

Volume I contains mathematical methods and methodological foundations.

Systemanalyse und Simulation erweisen sich in zunehmendem Maße als wirkungsvolle Methode und Instrument zum Verständnis der Wirkungsmechanismen und zur Beherrschung großer komplexer Systeme. Als wissenschaftliche Grundlagen werden die Methoden zur qualitativen Analyse, zu rechnergestützten Entscheidungshilfesystemen, zur Optimierung, Steuerung und Modellierung komplexer Systeme, zur Anwendung von Methoden der Künstlichen Intelligenz in der Systemanalyse sowie zur Simulation mittels paralleler Prozesse intensiv entwickelt.

Objektorientierte Simulationssprachen und Simulationssysteme bieten dem Modellentwickler und dem Modellnutzer wesentliche Voraussetzungen für eine effektive Problemlösung.

Im vorliegenden Band werden neueste Anwendungen der Modellierung, Simulation und Entscheidungsfindung für die nachfolgenden Fachgebiete vorgestellt:

- Umweltschutzprobleme
- Forstsysteme
- Wassersysteme
- Industrielle Automation
- Robotik
- Industrielle und technische Systeme
- Landwirtschaftliche Systeme
- Ökonomische Systeme
- Energiesysteme
- Medizinische und biologische Systeme

Der erste Band enthält mathematische Methoden und methodologische Grundlagen.

ISBN 3-05-500 655-0
ISSN 0138-3019